Artificial Life Models in Software

Second Edition

Maciej Komosinski • Andrew Adamatzky
Editors

Artificial Life Models in Software

Second Edition

Editors
Maciej Komosinski
Poznan University of Technology
Poznan
Poland

Andrew Adamatzky
University of the West of England
Bristol
UK

ISBN 978-1-84882-284-9 e-ISBN 978-1-84882-285-6
DOI 10.1007/978-1-84882-285-6
Springer Dordrecht Heidelberg London New York

British Library Cataloguing in Publication Data
A catalogue record for this book is available from the British Library

Library of Congress Control Number: 2009928773

Cover design: KünkelLopka GmbH; *Cover artwork*: Szymon Ulatowski

Printed on acid-free paper

Springer is part of Springer Science+Business Media (www.springer.com)

Preface

Artificial Life is an interdisciplinary field of science comprised of experts from computer science, biology, physics, chemistry, and mathematics, as well as philosophers and artists. Artificial Life focuses on studying the phenomena of life at all levels of complexity and organization – molecular, cellular, organismic, and population. These studies not only employ conventional computers (using both software and hardware) but also take place in wetware, using techniques from the biochemical laboratory. Artificial Life research is not limited to life-forms that currently exist on the Earth. Rather, it attempts to study the general principles of life that are common to all instances of life, those both already recognized and yet unknown.

Contemporary models of life that are implemented in the hardware medium are covered in the sister book *Artificial Life Models in Hardware*, which is also published by Springer and edited by Andrew Adamatzky and Maciej Komosinski. The book you are reading now is dedicated to software, the most popular and widely employed medium in the artificial life research. Software environments offer nearly unlimited opportunities for experiments that are relatively inexpensive, and easily arranged and modified. Such experiments can be performed with very precise descriptions of experimental conditions and a high degree of reproducibility. Advances in computational performance and efficiency allow for the collection of large amounts of data at a density and resolution that are unavailable in the wetware biological research environment.

This monograph[1] provides an introduction and guide to modern software tools for modeling and simulating life-like phenomena. Researchers, academicians, and students use this software to verify models and concepts related to evolutionary dynamics, self-organization, the origins of life, the development of multicellular organisms, natural and artificial morphogenesis, autonomous

[1] The present book is an advancement of earlier publications:
A. Adamatzky and M. Komosinski, editors. *Artificial Life Models in Software*, first edition. Springer, 2005.
A. Adamatzky, editor. *Kybernetes*, 32 (1/2, special issue on artificial life software), 2003.

and evolutionary robotics, evolvable hardware, emergent collective behaviors, swarm intelligence, the evolution of communication, and the evolution of social behaviors. The artificial life software packages described here are essential components in undergraduate and postgraduate courses in artificial life, complex adaptive systems, multi-agent systems, evolutionary biology, collective robotics, and nature-inspired computing.

Software systems covered in this book are actively developed and supported by world experts in artificial life and software design. In most cases, these programs are created with both professional and amateur users in mind, and the simulators are employed in published research. Each chapter describes a single, usually free-to-use software platform, but references to other similar software packages and related scientific works are included as well. The origins of software packages, milestones in their development, and the most important or interesting experiments are also reported.

Every chapter is self-contained and can be read independently. While the compendium of chapters is divided into four sections, the properties of the individual software models overlap so there are no distinct boundaries between parts of the book. Furthermore, while each package has its own specific properties, the models presented in each chapter are quite general and can be extended relatively easily to develop and utilize additional features depending upon their desired use.

The first part of the book – Virtual Environments – presents original settings for agent-based simulation. The chapters introduce *Avida*: a digital laboratory for studying populations of evolving programs; *Repast*: an advanced agent-based simulation toolkit for studying development of natural and artificial social structures; *Sodarace*: an online-based interface, learning support, and environment for construction and competition between two-dimensional mobile structures; *breve*: a package that facilitates building three-dimensional simulations of multi-agent systems; and *Framsticks*: a platform for modeling 3D creatures that allows for simulation, evolution, and advanced analysis of artificial life-forms and prototypes of bio-inspired robots.

Two- and three-dimensional grid models are dealt with in the second part – Lattice Worlds. The following software platforms are discussed in this section – *StarLogo*: an educational programming language for simulation of life-like phenomena, from population dynamics to emergent behavior of complex systems; *NetLogo*: a general purpose agent-based environment for modeling complex systems, described here from a specific biomedical perspective; *Discrete Dynamics Lab*: an interactive tool for designing and investigating dynamics of discrete dynamical networks, including simulation of cellular automata and decision networks, generating attraction basins, and automatic discovery of mobile patterns; and *EINSTein*: a multi-agent simulator of land combat, modeling individual traits of combatants.

The third part, Artificial Chemistries, comprises models capable of representing diverse systems of interacting and reacting molecules while abstracting natural molecular processes. This powerful approach is used to study phenomena in a variety of fields, including social and ecological modeling, and evolutionary and chemical computing. Two chapters on artificial chemistries discuss modeling of complex molecules in a mixed reactor and the behavior of simple molecules in spatially extended non-stirred reactors. Both chapters provide a review of available software tools.

The fourth and concluding section – Artificial Life Arts – includes three chapters focusing on aesthetic issues of artificial life. The first chapter in this section shows how to breed images and sounds using *SBEAT* (for graphics) and *SBART* (for music) software packages; the computer programs allow the selection and breeding of genotypes that represent graphical and musical pieces. The next chapter brings out *Eden*: an interactive artwork including hardware implementation of cellular automata and agents, allowing observers to interact with the installation and to influence the development of the artificial ecosystem. The last chapter of the book searches for a phenomenological understanding of what makes artificial life appealing to scientists, artists, and laymen and presents an explanation of why people become attracted to certain forms of creative computer art.

The Appendix contains a table that summarizes software packages described in this book. For each environment, the table includes a short description, information about availability on various platforms and operating systems, software requirements, license type, and the Internet web site address.

The book covers hot topics related to computer science, evolutionary and computational biology, simulation, robotics, cognitive science, cybernetics, artificial intelligence, optimization, multi-agent societies, virtual worlds, computer graphics and animation, neuroscience, and philosophy. We believe that academics, researchers, graduate students, and amateurs interested in these fields will find this monograph a valuable guide to Artificial Life and excellent supplementary reading.

Maciej Komosinski
May 2009 *Andrew Adamatzky*

Contents

Part I Virtual Environments

Part II Lattice Worlds

Part III Artificial Chemistries

Part IV Artificial Life Arts

List of Contributors

Andrew Adamatzky
Faculty of Computing, Engineering and Mathematical Sciences, University of the West of England, Bristol BS16 1QY UK, e-mail: andrew.adamatzky@uwe.ac.uk

Gary An
Department of Surgery, Northwestern University Feinberg School of Medicine, Chicago, IL, USA, e-mail: docgca@gmail.com

David M. Bryson
Department of Computer Science and Engineering, Michigan State University, East Lansing, MI 48824 USA, e-mail: brysonda@egr.msu.edu

Edward J. Burton
Soda Creative Ltd., London E2 8HD UK, e-mail: ed@soda.co.uk

Alan Dorin
Center for Electronic Media Art, Faculty of Information Technology, Monash University, Clayton, Australia, 3800, e-mail: alan.dorin@infotech.monash.edu.au

Harold Fellermann
Institute for Physics and Chemistry, University of Southern Denmark, Campusvej 55, 5230 Odense M, Denmark, and Complex Systems Lab (ICREA-UPF), Barcelona Biomedical Research Park (PRBB-GRIB), Dr Aiguader 88, 08003, Barcelona, Spain, e-mail: harold@ifk.sdu.dk

Andrew Ilachinski
Center for Naval Analyses, Alexandria, VA, USA, e-mail: ilachina@cna.org

Wendy Huang
Massachusetts Institute of Technology, Cambridge, MA, USA, e-mail: wh1@mit.edu

Christoph Kaleta
Bio Systems Analysis Group, Jena Centre for Bioinformatics (JCB), and School of Mathematics and Computer Science, Friedrich-Schiller-University, D-07743 Jena, Germany, e-mail: ckaleta@minet.uni-jena.de

Jon Klein
School of Cognitive Science, Hampshire College, 893 West Street, Amherst, MA, USA, e-mail: jk@spiderland.org

Eric Klopfer
Massachusetts Institute of Technology, Cambridge, MA, USA, e-mail: klopfer@mit.edu

Maciej Komosinski
Institute of Computing Science, Poznan University of Technology, Piotrowo 2, 60-965 Poznan, Poland, e-mail: maciej.komosinski@cs.put.poznan.pl

Charles M. Macal
Argonne National Laboratory, 9700 S. Cass Avenue, Argonne, IL, USA, e-mail: macal@anl.gov

Jon McCormack
Centre for Electronic Media Art, Clayton School of Information Technology, Monash University, Victoria, 3800 Australia, e-mail: Jon.McCormack@infotech.monash.edu.au

Peter W. McOwan
Computer Science, Queen Mary, University of London, London, E1 4NS UK, e-mail: pmco@dcs.qmul.ac.uk

Michael J. North
Argonne National Laboratory, 9700 S. Cass Avenue, Argonne, IL, USA, e-mail: north@anl.gov

Charles Ofria
Department of Computer Science and Engineering, Michigan State University, East Lansing, MI 48840 USA, e-mail: ofria@msu.edu

Hal Scheintaub
The Governor's Academy, 1 Elm Street, Byfield, MA 01922, USA, e-mail: hscheintaub@govsacademy.org

Lee Spector
School of Cognitive Science, Hampshire College, 893 West Street, Amherst, MA, USA, e-mail: lspector@hampshire.edu

Szymon Ulatowski
Toyspring, Kolbuszowa, Poland, www.toyspring.com, e-mail: sz@toyspring.com

Tatsuo Unemi
Department of Information Systems Science, Soka University, 1-236 Tangi-machi, Hachiōji, Tokyo, 192-8577 Japan, e-mail: unemi@iss.soka.ac.jp

Daniel Wendel
Massachusetts Institute of Technology, Cambridge, MA, USA, e-mail: djwendel@mit.edu

Uri Wilensky
Departments of Learning Sciences and Computer Science, Center for Connected Learning and Computer-Based Modeling, Northwestern Institute on Complex Systems, Northwestern University, Chicago, IL, USA, e-mail: uri@northwestern.edu

Claus O. Wilke
Center for Computational Biology and Bioinformatics, University of Texas at Austin, Austin, TX 78712 USA, e-mail: cwilke@mail.utexas.edu

Andrew Wuensche
Discrete Dynamics Lab, London, UK, www.ddlab.org, e-mail: andy@ddlab.org

Part I
Virtual Environments

Chapter 1
Avida: A Software Platform for Research in Computational Evolutionary Biology

Charles Ofria, David M. Bryson and Claus O. Wilke

Avida[1] is a software platform for experiments with self-replicating and evolving computer programs. It provides detailed control over experimental settings and protocols, a large array of measurement tools, and sophisticated methods to analyze and post-process experimental data. This chapter explains the general principles on which Avida is built, its main components and their interactions, and gives an overview of some prior research.

1.1 Introduction to Avida

When studying biological evolution, we have to overcome a large obstacle: Evolution is extremely slow. Traditionally, evolutionary biology has therefore been a field dominated by observation and theory, even though some regard the domestication of plants and animals as early, unwitting evolution experiments. Realistically, we can carry out controlled evolution experiments only with organisms that have very short generation times, so that populations can undergo hundreds of generations within a time frame of months or years. With the advances in microbiology, such experiments in evolution have become feasible with bacteria and viruses [18, 49]. However, even with microorganisms, evolution experiments still take a lot of time to complete and are often cumbersome. In particular, some data can be difficult or impossible to obtain, and it is often impractical to carry out enough replicas for high statistical accuracy.

According to Daniel Dennett, "...evolution will occur whenever and wherever three conditions are met: replication, variation (mutation), and differential fitness (competition)" [13]. It seems to be an obvious idea to set up these conditions in a computer and to study evolution *in silico* rather than

[1] Parts of the material in this chapter previously appeared in other forms [38, 37].

in vitro. In a computer, it is easy to measure any quantity of interest with arbitrary precision, and the time it takes to propagate organisms for several hundred generations is only limited by the processing power available. In fact, population geneticists have long been carrying out computer simulations of evolving loci, in order to test or augment their mathematical theories (see [21, 22, 28, 35, 40] for some examples). However, the assumptions put into these simulations typically mirror exactly the assumptions of the analytical calculations. Therefore, the simulations can be used only to test whether the analytic calculations are error-free or whether stochastic effects cause a system to deviate from its deterministic description, but they cannot test the model assumptions on a more basic level.

An approach to studying evolution that lies somewhere in between evolution experiments with biochemical organisms and standard Monte Carlo simulations is the study of self-replicating and evolving computer programs (digital organisms). These digital organisms can be quite complex and interact in a multitude of different ways with their environment or each other, so that their study is not a simulation of a particular evolutionary theory but becomes an experimental study in its own right. In recent years, research with digital organisms has grown substantially ([5, 9, 17, 19, 24, 29, 53, 55, 56, 57], see [3, 51] for reviews) and is being increasingly accepted by evolutionary biologists [39]. (However, as Barton and Zuidema [6] note, general acceptance will ultimately hinge on whether artificial life researchers embrace or ignore the large body of population-genetics literature.) Avida is arguably the most advanced software platform to study digital organisms to date and is certainly the one that has had the biggest impact in the biological literature so far. Having reached version 2.8, it now supports detailed control over experimental settings, a sophisticated system to design and execute experimental protocols, a multitude of possibilities for organisms to interact with their environment (including depletable resources and conversion from one resource into another), and a module to post-process data from evolution experiments (including tools to find the line of descent from the original ancestor to any final organism, to carry out knock-out studies with organisms, to calculate the fitness landscape around a genotype and to align and compare organisms' genomes).

1.1.1 History of Digital Life

The most well-known intersection of evolutionary biology with computer science is the genetic algorithm or its many variants (genetic programming, evolutionary strategies, and so on). All these variants boil down to the same basic recipe: (1) create random potential solutions, (2) evaluate each solution assigning it a fitness value to represent its quality, (3) select a subset of solutions using fitness as a key criterion, (4) vary these solutions by making random changes or recombining portions of them, then (5) repeat from step 2 until you find a solution that is sufficiently good.

This technique turns out to be an excellent method for solving problems, but it ignores many aspects of natural living systems. Most notably, natural organisms must replicate themselves, as there is no external force to do so; therefore, their ability to pass their genetic information on to the next generation is the final arbiter of their fitness. Furthermore, organisms in a natural system have the ability to interact with their environment and with each other in ways that are excluded from most algorithmic applications of evolution.

Work on more naturally evolving computational systems began in 1990, when Steen Rasmussen was inspired by the computer game "Core War" [14]. In this game, programs are written in a simplified assembly language and made to compete in the simulated core memory of a computer. The winning program is the one that manages to shut down all processes associated with its competitors. Rasmussen observed that the most successful of these programs were the ones that replicated themselves, so that if one copy were destroyed, others would still persist. In the original "Core War" game, the diversity of organisms could not increase, and hence no evolution was possible. Rasmussen designed a system similar to "Core War" in which the command that copied instructions was flawed and would sometimes write a random instruction instead on the one intended [43]. This flawed copy command introduced *mutations* into the system, and thus the potential for evolution. Rasmussen dubbed his new program "Core World," created a simple self-replicating ancestor, and let it run.

Unfortunately, this first experiment was only of limited success. While the programs seemed to evolve initially, they soon started to copy code into each other, to the point where no proper self-replicators survived – the system collapsed into a non-living state. Nevertheless, the dynamics of this system turned out to be intriguing, displaying the partial replication of fragments of code and repeated occurrences of simple patterns.

The first successful experiment with evolving populations of self-replicating computer programs was performed the following year. Thomas Ray designed a program of his own with significant, biologically inspired modifications. The result was the Tierra system [44]. In Tierra, digital organisms must allocate memory before they have permission to write to it, which prevents stray copy commands from killing other organisms. Death only occurs when memory fills up, at which point the oldest programs are removed to make room for new ones to be born.

The first Tierra experiment was initialized with an ancestral program that was 80 lines long. It filled up the available memory with copies of itself, many of which had mutations that caused a loss of functionality. Yet other mutations were neutral and did not affect the organism's ability to replicate – and a few were even beneficial. In this initial experiment, the only selective pressure on the population was for the organisms to increase their rate of replication. Indeed, Ray witnessed that the organisms were slowly shrinking the length of their genomes, since a shorter genome meant that there was less genetic material to copy, and thus it could be copied more rapidly.

This result was interesting enough on its own. However, other forms of adaptation, some quite surprising, occurred as well. For example, some organisms were able to shrink further by removing critical portions of their genome and then to use those same portions from more complete competitors, in a technique that Ray noted was a form of parasitism. Arms races transpired where hosts evolved methods of eluding the parasites, and they, in turn, evolved to get around these new defenses. Some would-be hosts, known as hyper-parasites, even evolved mechanisms for tricking the parasites into aiding them in the copying of their own genomes. Evolution continued in all sorts of interesting manner, making Tierra seem like a choice system for experimental evolution work.

In 1992, Chris Adami began research on evolutionary adaptation with Ray's Tierra system. His intent was to have these digital organisms evolve solutions to specific mathematical problems, without forcing them use a predefined approach. His core idea was the following: If he wanted a population of organisms to evolve, for example, the ability to add two numbers together – he would monitor organisms' input and output numbers. If an output ever was the sum of two inputs, the successful organisms would receive extra CPU cycles as a bonus. As long as the number of extra cycles was greater than the time it took the organism to perform the computation, the leftover cycles could be applied toward the replication process, providing a competitive advantage to the organism. Sure enough, Adami was able to get the organisms to evolve some simple tasks, but he faced many limitations in trying to use Tierra to study the evolutionary process.

In the summer of 1993, Charles Ofria and C. Titus Brown joined Adami to develop a new digital life software platform, the Avida system. Avida was designed to have detailed and versatile configuration capabilities, along with high-precision measurements to record all aspects of a population. Furthermore, whereas organisms are executed sequentially in Tierra, the Avida system simulates a parallel computer, allowing all organisms to be executed effectively simultaneously. Since its inception, Avida has had many new features added to it, including a sophisticated environment with localized resources, an events system to schedule actions to occur over the course of an experiment, multiple types of CPUs to form the bodies of the digital organisms, and a sophisticated analysis mode to post-process data from an Avida experiment. Avida is under active development at Michigan State University, led by Charles Ofria and David Bryson.

1.2 The Scientific Motivation for Avida

Intuitively, it seems that natural systems should be used to best understand how evolution produces the variation in observed in nature, but this can be prohibitively difficult for many questions and does not provide enough detail. Using digital organisms in a system such as Avida can be justified on five grounds:

(1) *Artificial life-forms provide an opportunity to seek generalizations about self-replicating systems* beyond the organic forms that biologists have studied to date, all of which share a common ancestor and essentially the same chemistry of DNA, RNA, and proteins. As John Maynard Smith [27] made the case: "So far, we have been able to study only one evolving system and we cannot wait for interstellar flight to provide us with a second. If we want to discover generalizations about evolving systems, we will have to look at artificial ones." Of course, digital systems should always be studied in parallel with natural ones, but any differences we find between their evolutionary dynamics open up what is perhaps an even more interesting set of questions.

(2) *Digital organisms enable us to address questions that are impossible to study with organic life-forms.* For example, in one of our current experiments we are investigating the importance of deleterious mutations in adaptive evolution by explicitly reverting all detrimental mutations. Such invasive micromanaging of a population is not possible in a natural system, especially without disturbing other aspects of the evolution. In a digital evolving system, every bit of memory can be viewed without disrupting the system, and changes can be made at the precise points desired.

(3) *Other questions can be addressed on a scale that is unattainable with natural organisms.* In an earlier experiment with digital organisms [26], we examined billions of genotypes to quantify the effects of mutations as well as the form and extent of their interactions. By contrast, an experiment with *E. coli* was necessarily confined to one level of genomic complexity. Digital organisms also have a speed advantage: Population with 10,000 organisms can have 20,000 generations processed per day on a modern desktop computer. A similar experiment with bacteria took over a decade [25].

(4) *Digital organisms possess the ability to truly evolve, unlike mere numerical simulations.* Evolution is open-ended and the design of the evolved solutions is unpredictable. These properties arise because selection in digital organisms (as in real ones) occurs at the level of the whole organism's phenotype; it depends on the rates at which organisms perform tasks that enable them to metabolize resources to convert them to energy and on the efficiency with which they use that energy for reproduction. Genome sizes are sufficiently large that evolving populations cannot test every possible genotype, so replicate populations always find different local optima. A genome typical consists of 50 to 1000 sequential instructions. With commonly 26 possible instructions at each position, there are many more potential genome states than there are atoms in the universe.

(5) *Digital organisms can be used to design solutions to computational problems* where it is difficult to write explicit programs that produce the desired behavior [20, 23]. Current evolutionary algorithm approaches are based on a simplistic view of evolution, leaving out many of the factors that are believed to make it such a powerful force. Thus, there are new opportunities for biological concepts to have a large impact outside of biology, just as principles of physics and mathematics are often used throughout other fields, including biology.

1.3 The Avida Software

The Avida software[2] is composed of two main components: The first is the *Avida core*, which maintains a population of digital organisms (each with their own genomes, virtual hardware, etc.), an environment that maintains the reactions and resources with which the organisms interact, a scheduler to allocate CPU cycles to the organisms, and various data collection objects. The second component is a collection of *analysis and statistics* tools, including a test environment to study organisms outside of the population, data manipulation tools to rebuild phylogenies and examine lines of descent, mutation and local fitness landscape analysis tools, and many others, all bound together in a simple scripting language. In addition to these two primary components, two forms of interactive *user interface* (UI) to Avida are currently available: a text-based console interface (`avida-viewer`) and an education focused graphical UI, *Avida-ED* [41]. These interfaces allow the researcher to visually interact with the rest of the Avida software during an experiment.

In this chapter, we will discuss the two primary modules of Avida that are relevant for experiments with digital organisms: the Avida core and the analysis and statistics tools.

1.3.1 Avida Organisms

In Avida, each digital organism is a self-contained computing automaton that has the ability to construct new automata. The organism is responsible for building the genome (computer program) that will control its offspring automaton and handing that genome to the Avida world. Avida will then construct virtual hardware for the genome to be run on and determine how this new organism should be placed into the population. In a typical Avida experiment, a successful organism attempts to make an identical copy of

---

[2] Avida packages are available at [1]. For additional information, see [2].

its own genome, and Avida randomly places that copy into the population, typically by replacing another member of the population.

In principle, the only assumption made about these self-replicating automata in the core Avida software is that their initial state can be described by a string of symbols (their genome) and that it is possible through processing these symbols to autonomously produce offspring organisms. However, in practice, our work has focused on automata with a simple von Neumann architecture that operate on an assembly-like language inspired by the Tierra system. Future research projects will likely have us implement additional organism instantiations to allow us to explore additional biological questions.

In the following sub-sections, we describe the default hardware of our virtual computers, and explain the principles of the language on which these machines work.

1.3.1.1 Virtual Hardware

The structure of a virtual machine in Avida is depicted in Fig. 1.1. The core of the machine is the central processing unit (CPU), which processes each instruction in the genome and modifies the states of its components appropriately. Mathematical operations, comparisons, and so on can be done on three registers, `AX`, `BX`, and `CX`. These registers each store and manipulate data in the form of a single, 32-bit number. The registers behave identically, but different instructions may act on different registers by default (see below). The CPU also has the ability to store data in two stacks. Only one of the two stacks is active at a time, but it is possible to switch the active stack, so that both stacks are accessible.

The program memory is initialized with the genome of the organism. Execution begins with the first instruction in memory and proceeds sequentially: Instructions are executed one after the other, unless an instruction (such as a jump) explicitly interrupts sequential execution. Technically, the memory space is organized in a circular fashion, such that after the CPU executes the last instruction in memory, it will loop back and continue execution with the first instruction again. However, at the same time, the memory has a well-defined starting point, important for the creation and activation of offspring organisms.

Somewhat out of the ordinary in comparison to standard von Neumann architectures are the four CPU components labeled *heads*. Heads are essentially pointers to locations in the memory. They remove the need of absolute addressing of memory positions, which makes the evolution of programs more robust to size changes that would otherwise alter these absolute positions [36]. Among the four heads, only one, the instruction head (IP), has a counterpart in standard computer architectures. The instruction head corresponds to the instruction pointer in standard architectures and identifies the instruction currently being executed by the CPU. It moves one instruction forward whenever the execution of the previous instruction has been completed, unless that instruction specifically moved the instruction head elsewhere.

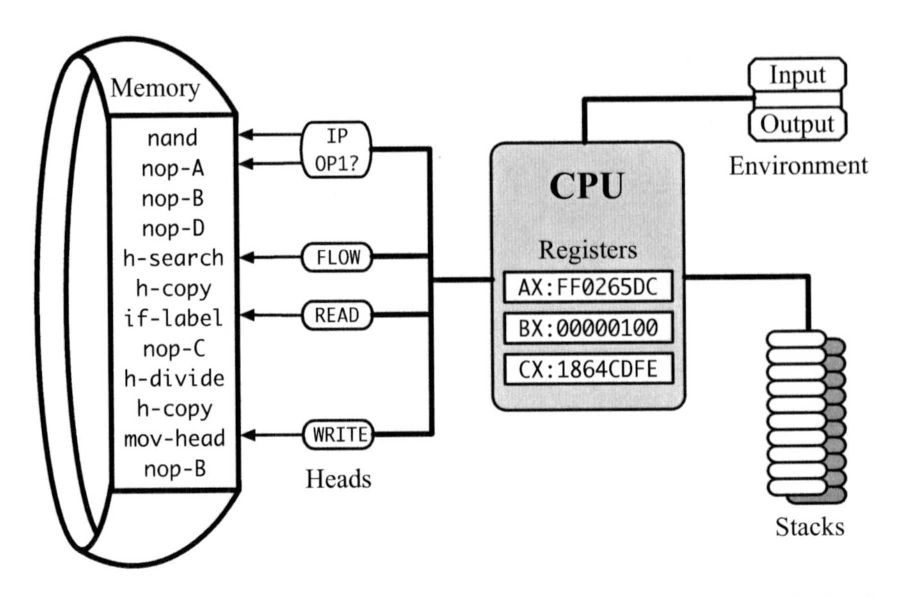

Fig. 1.1 The standard virtual machine hardware in Avida: CPU, registers, stacks, heads, memory (genome), and environment I/O functionality.

The other three heads (the READ head, the WRITE head, and the FLOW head) are unique to the Avida virtual hardware. The READ and WRITE heads are used in the self-replication process. In order to generate a copy of its genome, an organism must have a means of reading instructions from memory and writing them back to a different location. The READ head indicates the position in memory from which instructions are currently being read, and the WRITE head likewise indicates the position to which instructions are currently being written. The positions of all four heads can be manipulated with special commands. In that way, a program can position the read and write heads appropriately in order to self-replicate.

The FLOW head is used for controlling jumps and loops. Several commands will reposition the flow control head, and other commands will move specific heads to the same position in memory as the flow control head.

Finally, the virtual machines have an input buffer and an output buffer, which they use to interact with their environment. The way in which this communication works is that the machines can read in one or several numbers from the input buffer, perform computations on these numbers with the help of the internal registers AX, BX, CX, and the stacks, and then write the results to the output buffer. This interaction with the environment plays a crucial role in the evolution of Avida organisms, and will be explained in detail in Sect. 1.3.2.4.

1.3.1.2 Genetic Language

It is important to understand that there is not a single language that controls the virtual hardware of an Avida organism. Instead, we have a collection of different languages. The virtual hardware in its current form can execute hundreds of different instructions, but only a small fraction of them are used in a typical experiment. The instructions are organized into subsets of the full range of instructions. We call these subsets *instruction sets*. Each instruction set forms a logical unit and can be considered a complete genetic programming language.

Each instruction has a well-defined function in any context; that is, there are no syntactically incorrect programs. Instructions do not have arguments per se, but the behavior of certain instructions can be modified by succeeding instructions in memory. A genome is therefore nothing more than a sequence of symbols in an alphabet composed of the instruction set, similar to how DNA is a sequence made up of 4 nucleotides or proteins are sequences with a standard alphabet of 20 amino acids.

Here, we will give an overview of the default instruction set, which contains 26 instructions. This set is explained in more detail in the Avida documentation, for those who wish to work with it.

Template Matching and Heads: One important ingredient of most Avida languages is the concept of *template matching*. Template matching is a method of indirectly addressing a position in memory. This method is similar to the use of labels in many programming languages: Labels tag a position in the program, so that jumps and function calls always go to the correct place, even when other portions of the source code are edited. The same reasoning applies to Avida genomes, because mutations may cause insertions or deletions of instructions that shift the position of code and would otherwise jeopardize the positions referred to. Since there are no arguments to instructions, positions in memory are determined by series of subsequent instructions. We refer to a series of instructions that indicates a position in the genome as a *template*.

Template-based addressing works as follows. When an instruction is executed that must reference another position in memory, subsequent *nop* instructions (described below) are read in as the template. The CPU then searches linearly through the genome for the first occurrence of the complement to this template and uses the end of the complement as the position needed by the instruction. Both the direction of the search (forward or backward from the current instruction) and the behavior of the instruction if no complement is found are defined specifically for each instruction.

Avida templates are constructed out of no-operation (nop) instructions; that is, instructions that do not alter the state of either CPU or memory when they are directly executed. There are three template-forming nops, `nop-A`, `nop-B`, and `nop-C`. They are circularly complementary; that is, the complement of `nop-A` is `nop-B`, the complement of `nop-B` is `nop-C`, and the complement of `nop-C` is `nop-A`. A template is composed of consecutive nops only. A template will end with the first non-nop instruction.

Non-linear execution of code ("jumps") has to be implemented through clever manipulation of the different heads. This happens in two stages. First, the instruction `h-search` is used to position the FLOW head at the desired position in memory. Then, the IP is moved to that position with the command `mov-head`. Fig. 1.2 shows an example of this.

	⋮	Some code.
10	`h-search`	Prepare the jump by placing the FLOW head at the end of the complement template in forward direction.
11	`nop-A`	This is the template. Let's call it α.
12	`nop-B`	
13	`mov-head`	The actual jump. Move the FLOW head to the position of the IP.
14	`pop`	Some other code that is skipped.
	⋮	
18	`nop-B`	The complement template $\bar{\alpha}$.
19	`nop-C`	
	⋮	The program continues ...

Fig. 1.2 Example code demonstrating flow control with heads-based instruction set.

Although this example looks somewhat awkward on first glance, evolution of control structures such as loops are actually facilitated by this mechanism. In order to loop over some piece of code, it is only necessary to position the FLOW head correctly once and to have the command `mov-head` at the end of the block of code that should be looped over. Since there are several ways in which the FLOW head can be positioned correctly, of which the above example is only a single one, there are many ways in which loops can be generated.

nop's as Modifiers: The instructions in the Avida programming language do not have arguments in the usual sense. However, as we have seen above for the case of template matching, the effect of certain instructions can be modified if they are immediately followed by nop instructions. A similar concept exists for operations that access registers. The `inc` instruction, for example, increments a register by one. If `inc` is not followed by any nop, then by default it acts on the `BX` register. However, if a nop is present immediately after the `inc`, then the register on which `inc` acts is specified by the type of the nop. For example, inc nop-A increments the AX register, and inc nop-C increments the CX register. Of course, inc nop-B increments the BX register and, hence, works identical to a single inc command. Similar nop modifications exist for a range of instructions, such as those that perform arithmetic like inc or dec, stack operations such as push or pop, and comparisons such as if-n-equ. The details can be found in [38] or in the Avida documentation. For some instructions that work on two registers – in particular, comparisons – the concept of the complement nop is important because the two registers are specified in this way. Similar to nops in the template matching, registers are cyclically complementary to each other (i.e., BX is the complement to AX, CX to BX, and AX to CX). The instruction if-n-equ, for example, acts on a register and its complement register. By default, if-n-equ compares whether the contents of the BX and CX registers are identical. However, if if-n-equ is followed by a nop-A, then it will compare AX and BX. Fig. 1.3 shows a piece of example code that demonstrates the principles of nop modification and complement registers.

```
01  pop       We assume the stack is empty. In that case,
              the pop returns 0, which is stored in BX.
02  pop       Write 0 into the register AX as well.
03  nop-A
04  inc       Increment BX.
05  inc       Increment AX.
06  nop-A
07  inc       Increment AX a second time.
08  nop-A
09  swap      The swap command exchanges the contents
              of a register with the one of its complement
              register. Followed by a nop-C, it exchanges
              the contents of AX and CX. Now, BX= 1, CX= 2,
              and AX is undefined.
10  nop-C
11  add       Now add BX and CX and store the result
              in AX.
12  nop-A     The program continues with BX= 1, CX= 2,
              and AX= 3.
     ⋮
```

Fig. 1.3 Example code demonstrating the principle of nop modification.

nop modification is also necessary for the manipulation of heads. The instruction mov-head, for example, by default moves the IP to the position of the FLOW head. However, if it is followed by either a nop-B or a nop-C, it moves the READ head or the WRITE head, respectively. A nop-A following a mov-head leaves the default behavior unaltered.

Memory Allocation and Division: When a new Avida organism is created, the CPU's memory is exactly the size as its genome; that is, there is no additional space that the organism could use to store its offspring-to-be as it makes a copy of its program. Therefore, the first thing an organism has to do at the start of self-replication is to allocate new memory. In the default instruction set, memory allocation is done with the command `h-alloc`. This command extends the memory by the maximal size that an offspring is allowed to have. As we will discuss later, there are some restrictions on how large or small an offspring is allowed to be in comparison to the parent organism, and the restriction on the maximum size of an offspring determines the amount of memory that `h-alloc` adds. The allocation always happens at a well-defined position in the memory. Although the memory is considered to be circular in the sense that the CPU will continue with the first instruction of the program once it has executed the last one, the virtual machine nevertheless keeps track of which instruction is the beginning of the program and which is the end. By default, `h-alloc` (as well as all alternative memory allocation instructions, such as the old `allocate`) inserts the new memory between the end and the beginning of the program. After the insertion, the new end is at the end of the inserted memory. The newly inserted memory is either initialized to a default instruction, typically `nop-A`, or to random code, depending on the choice of the experimenter.

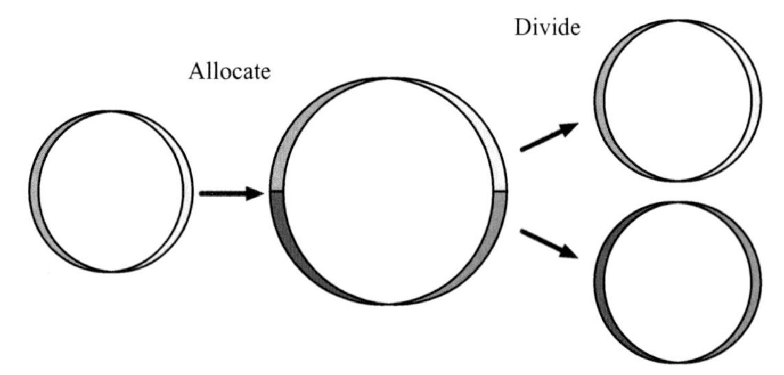

Fig. 1.4 The `h-alloc` command extends the memory, so that the program of the offspring can be stored. Later, upon successful execution of `h-divide`, the program is split into two parts, one of which becomes the genome of the offspring.

Once an organism has allocated memory, it can start to copy its program code into the newly available memory block. This copying is done with the help of the control structures we have already described, in conjunction with the instruction `h-copy`. This instruction copies the instruction at the position of the READ head to the position of the WRITE head and advances both heads. Therefore, for successful self-replication, an organism mainly has to assure that, initially, the READ head is at the beginning of the memory and the WRITE head is at the beginning of the newly allocated memory, and then it has to call `h-copy` for the correct number of times.

After the self-replication has been completed, an organism issues the `h-divide` command, which splits off the instructions between the READ head and the WRITE headand uses them as the genome of a new organism. The new organism is handed to the Avida world, which takes care of placing it into a suitable environment and so on. If there are instructions left between the WRITE head and the end of the memory, these instructions are discarded, so that only the part of the memory from the beginning to the position of the READ head remains after the divide.

In most natural asexual organisms, the process of division results in organisms literally splitting in half, effectively creating two offspring. As such, the default behavior of Avida is to reset the state of the parent's CPU after the divide, turning it back into the state it was in when it was first born. In other words, all registers and stacks are cleared, and all heads are positioned at the beginning of the memory. The full allocation and division cycle is illustrated in Fig. 1.4.

Not all `h-divide` commands that an organism issues lead necessarily to the creation of an offspring organism. There are a number of conditions that have to be satisfied, otherwise the command will fail. Failure of a command means essentially that the command is ignored, and a counter keeping track of the number of failed commands in an organism is increased. It is possible to configure Avida to punish organisms with failed commands. The following conditions are in place: An `h-divide` fails if either the parent or the offspring would have less than 10 instructions, the parent has not allocated memory, less than half of the parent was executed, less than half of the offspring's memory was copied into, or the offspring would be too small or too large (as defined by the experimenter).

1.3.1.3 Mutations

So far, we have described all of the elements that are necessary for self-replication. However, self-replication alone is not sufficient for evolution. There must be a source of variation in the population, which comes from random *mutations*.

The principal form of mutations in typical Avida experiments are so-called copy mutations, which arise through erroneously copied instructions. Such miscopies are a built-in property of the instruction `h-copy`. With a certain probability, chosen by the experimenter, the command `h-copy` does not properly copy the instruction at the location of the READ head to the location of the WRITE head, but instead writes a random instruction to the position of the WRITE head. It is important to note that the instruction written will always be a legal one, in the sense that the CPU can execute it. However, the instruction may not be meaningful in the context in which it is placed in the genome, which, in the worst case, can render the offspring organism non-functional.

Another commonly used source of mutations are insertion and deletion mutations. These mutations are applied on `h-divide`. After an organism has successfully divided off an offspring, an instruction in the daughter organism's memory may, by chance, be deleted or a random instruction may be inserted. The probabilities with which these events occur are again determined by the experimenter. Insertion and deletion mutations are useful in experiments in which frequent changes in genome size are desired. Two types of insertion/deletion mutations are available in the configuration files; they differ in that one is a genome-level rate and the other is a per-site rate.

Next, there are point (or cosmic ray) mutations. These mutations affect not only organisms as they are being created (like the other types described above), but all living organisms. Point mutations are random changes in the memory of the virtual machines. One of the consequences of point mutations is that a program may change while it is being executed. In particular, the longer a program runs, the more susceptible it becomes to point mutations. This is in contrast to copy or insertion and deletion mutations, whose impact depends only on the length of the program, not on the execution time.

Finally, it is important to note that organisms in Avida can also have *implicit* mutations. Implicit mutations are modifications in a offsping's program that are not directly caused by any of the external mutation mechanisms described above, but rather by an incorrect copy algorithm of the parent organism. For example, the copy algorithm might skip some instructions of the parent program or copy a section of the program twice (effectively a gene duplication event). Another example is an incorrectly placed READ head or WRITE head on divide. Implicit mutations are the only ones that cannot easily be controlled by the experimenter. They can, however, be turned off completely by using the `FAIL_IMPLICIT` option in the configuration files, which gets rid of any offspring that will always contain a deterministic difference from its parent, as opposed to one that is associated with an explicit mutation.

1.3.1.4 Phenotype

Each organism in an Avida population has a phenotype associated with it. Phenotypes of Avida organisms are defined in the same way as they are defined for organisms in the natural world: The phenotype of an organism comprises all observable characteristics of that organism. As an organism in Avida goes through its life cycle, it will self-replicate and, at the same time, interact with the environment. The primary mode of environmental interaction is by inputting numbers from the environment, performing computations on those numbers, and outputting the results. The organisms receive a benefit for performing specific computations associated with resources as determined by the experimenter (see Sect. 1.3.2.4).

In addition to tracking computations, the phenotype also monitors several other aspects of the organisms behavior, such as the organism's gestation length (the number of instructions the organism executes to produce an offspring, often also called *gestation time*), its age (the total number of CPU cycles since it was born), if it has been affected by any mutations, how it interacts with other organisms, and its overall fitness. These data are used both to determine how many CPU cycles should be allocated to the organism and for various statistical purposes.

1.3.1.5 Genotypes

In Avida, organisms are classified into several taxonomic levels. The lowest taxonomic level is called *genotype*. All organisms that have exactly the same initial genomes are considered to have the same genotype. Certain statistical data are collected only at the genotype level. We pay special attention to the most abundant genotype in the population – the *dominant* genotype – as a method of determining what the most successful organisms in the population are capable of. If a new genotype is truly more fit than than the dominant one, organisms with this higher fitness will rapidly take over the population.

We classify a genotype as *threshold* if there are three or more organisms that have ever existed of that genotype (the value 3 is not hard-coded, but configurable by the experimenter). Often, deleterious mutants appear in the population. These mutants are effectively dead and disappear again in short order. Since these mutants are not able to successfully self-replicate (or at least not well), there is a low probability of them reaching an abundance of three. As such, for any statistics we want to collect about the *living* portion of the population, we focus on those organisms whose genotype has the threshold characteristic.

1.3.2 The Avida World

In general, the Avida world has a fixed number N of positions or *cells*. Each cell can be occupied by exactly one organism, such that the maximum population size at any given time is N. Each of these organisms is being run on a virtual CPU, and some of them may be running faster than others. Avida has a scheduler that divides up time from the real CPU such that these virtual CPUs execute in a simulated parallel fashion.

While an Avida organism runs, it may interact with the environment or other organisms. When it finally reproduces, it hands its offspring organism to the Avida world, which places the newborn organism into either an empty or an occupied cell, according to rules described below. If the offspring organism is placed into an already occupied cell, the organism currently occupying that cell is killed and removed, irrespective of whether it has already reproduced or not.

1.3.2.1 Scheduling

In the simplest of Avida experiments, all virtual CPUs run at the same speed. This method of time-sharing is simulated by executing one instruction on each of the N virtual CPUs in order, then starting over to execute a second instruction on each one, and so on. An *update* in Avida is defined as the point where the average organism has executed k instructions (where $k = 30$ by default). In this simple case, for one update, we carry out k rounds of execution.

In more complex environments, however, the situation is not so trivial. Different organisms will have their virtual CPUs executing at different speeds (the details of which are described below) and the scheduler must portion out cycles appropriately to simulate that all CPUs are running in parallel. Each organism has associated with it a value that determines its *metabolic rate* (sometimes referred to as *merit*). The metabolic rate indicates how fast the virtual CPU should run. Metabolic rate is a unitless quantity and is only meaningful when compared to the metabolic rates of other organisms. Thus, if the metabolic rate organism A is twice that of organism B, then A should, on average, execute twice as many instructions in any given time frame as B.

Avida handles this with two different schedulers (referred to as the `SLICING_METHOD` in the configuration files). The first one is a perfectly integrated scheduler, which comes as close as possible to portioning out CPU cycles proportional to each organisms' metabolic rate. Obviously, only whole time steps can be used; therefore, perfect proportionality is not possible in general for small time frames. For time frames long enough such that the granularity of individual time steps can be neglected, the difference between the number of cycles given to an organism and the number of cycles the organism should receive at its current metabolic rate is negligible.

The second scheduler is probabilistic. At each point in time, the next organism to be executed is chosen at random, but with the probability of an individual being chosen proportional to its metabolic rate. Thus, *on average*, this scheduler is perfect, but there are no guarantees.

The perfectly integrated scheduler can be faster under various experimental configurations, but occasionally it can cause odd effects, because it is possible for the organisms to become synchronized, particularly at low mutation rates where a single genotype can represent a large portion of the population. The probabilistic scheduler avoids this effect and, in practice, is comparable in performance with recent versions of Avida. The default configuration used the probabilistic scheduler.

1.3.2.2 World Topologies and Birth Methods

The N cells of the Avida world can be assembled into different topologies that affect how offspring organisms are placed and how organisms interact (as described below). Currently, there are three basic world topologies: a 2D bounded grid with Moore neighborhood (each cell has eight neighbors), a 2D toroidal grid with Moore neighborhood, and a fully connected, clique topology. In the latter, the fully connected topology, each cell is a neighbor to every other cell. New topologies can easily be implemented by listing the neighbors associated with each cell. A special type of meta-topology, called *demes*, is described below.

When a new organism is about to be born, it will replace either the parent cell or another cell from either its topological neighborhood or any cell within the population (sometimes called *well stirred* or *mass action*). The specifics of this placement strategy are set up by the experimenter. The two most commonly used methods are *replace random*, which chooses randomly from the potential cells, or *replace oldest*, which picks the oldest organism from the potential organisms to replace (with a preference for empty cells if any exist).

Mass action placement strategies are used in analogy to experiments with microbes in well-stirred flasks or chemostats. These setups allow for exponential growth of new genotypes with a competitive advantage, so that transitions in the state of the population can happen rapidly. Two-dimensional topological neighborhoods, on the other hand, are more akin to a Petri dish, and the spatial separation between different organisms puts limits on growth rates and allows for a slightly more diverse population [8].

In choosing which organism in a neighborhood to replace, a random placement matches up well with the behavior of a chemostat, where a random portion of the population is continuously drawn out to keep population size constant. Experiments have shown [4], however, that evolution occurs more rapidly when the oldest organism in a neighborhood is the first to be killed off. In such cases, all organisms are given approximately the same chance to prove their worth, whereas in random replacement, about half the organisms are killed before they have the opportunity to produce a single offspring. Interestingly, when replace oldest is used in 2D neighborhoods, 40% of the time it is the parent that is killed off. This observation makes sense, because the parent is obviously old enough to have produced at least one offspring.

Note that in the default setup of Avida, replacement by another organism is not the only way for an organism to die. It is also possible for an organism to be killed after it has executed a specified number of instructions, which can either be a constant or proportional to the organism's genome length, the default. Without this setting, it is possible in some cases for a population to lose all ability to self-replicate, but it persist since organisms have no means by which to be purged.

1.3.2.3 Demes

Demes, a relatively new feature of Avida, subdivide the main population into sub-populations of equal size and structure. Each deme is isolated, although the population scheduler is shared among all demes. Typical experiments using demes provide a mechanism for deme-level replication. Such mechanisms will either test for the completion of a group activity or replicate demes based on a measured value (the latter being akin to mechanisms used in a genetic algorithm). There are several possible modes of deme replication. The default replication method creates a genome-level copy of each organism in the parent deme, placing the offspring into the target deme. The experimenter can configure Avida to perform a variety of alternative replication actions, including *germline* replication, where each deme has base genotype that is used to seed new copies with a single organism.

1.3.2.4 Environment and Resources

All organisms in Avida are provided with the ability to absorb a default resource that gives them their base metabolic rate. An Avida environment can, however, contain other resources that the organisms can absorb to modify their metabolic rate. The organisms absorb a resource by carrying out the corresponding computation or *task*.

An Avida environment is described by a set of resources and a set of reactions that can be triggered to interact with those resources. A reaction is defined by a computation that the organism must perform to trigger it, a resource that is consumed by it, a metabolic rate effect on the organism (which can be proportional to the amount of resource absorbed or available), and a byproduct resource if one should be produced. Reactions can also have restrictions associated with them that limit when a trigger will be successful. For example, another reaction can be required to have been triggered first, or a limit can be placed on the number of times an organism can trigger a certain reaction.

A resource is described by an initial quantity (which can be infinite if a resource should not be depletable), an inflow rate (the amount of that resource that should come into the population per update), and an outflow rate (the fraction of the resource that should be removed each update.) If resources are made to be depletable, then the more organisms trigger a reaction, the less of that resource is available for each of them. This setup allows multiple, diverse sub-populations to stably coexist in an Avida world [10].

The default Avida environment rewards nine boolean logic operations, each associated with a non-depletable resource, but organisms can receive only one reward per computation. Other pre-built environments that come with Avida include one with 77 different logic operations rewarded, one similar to the default 9-resource environment, but with the resources set up to be depletable, with fixed inflow and outflow rates, and one with 9 computations rewarded, and where only the resources associated with the simplest computations have an inflow into the system, and those for more complex operations are produced as byproducts, in sequence, from the reactions using up resources associated to simpler computations.

An important aspect of Avida is that the environment does not care *how* a computation is performed, only that the output of the organism being tested is correct given the inputs it took in. As a consequence, the organisms find a wide variety of ways of computing their outputs, some of which can be surprising to a human observer, seeming to be almost inspired.

Even though organisms can carry out tasks and collect associated resources at any time in their gestation cycle, these reactions typically do not immediately affect the speed at which their virtual CPU runs. The CPU speed (metabolic rate) is set only once at the beginning of the gestation cycle and then held constant until the organism divides. At that point, both the organism and its offspring have their metabolic rates adjusted, reflecting the resources the organism collected during the gestation cycle it just completed. In a sense, the organisms collect resources for their offspring, rather than for themselves. The reason why we do not change an organism's metabolic rate during its gestation cycle is to level the playing field between old and young organisms. If organisms were always born with a low initial CPU speed, then they may never execute enough instructions to carry out tasks in the first place. At the same time, mutants specialized in carrying out tasks but not dividing could concentrate all CPU time on them, thus effectively shutting down replication in the population. It can be shown that the average fitness of a population in equilibrium is independent of whether organisms get the bonuses directly or collect them for their offspring [50].

1.3.2.5 Organism Interactions

As explained above, populations in Avida have a firm cap on their size, which makes space for the fundamental resource for which the organisms must compete. In the simplest Avida experiments, the only interaction between organisms is that an organism is killed when another gives birth, in order to make room for the offspring. In slightly more complex experiments, the organisms collect resources that increase their metabolic rate and hence earn a larger share of the CPU cycles for performing tasks. Since there are only a fixed number of CPU cycles given out each update, the competition for them becomes a second level of indirect interactions among the organisms. As the environment becomes more complex still, multiple resources take the place of fixed metabolic rate bonuses for performing tasks, and the organisms must now compete over each of these resources independently. In the end, however, all these interactions boil down to the indirect competition for space: More resources imply a higher metabolic rate, which, in turn, grants the organisms a larger share of the CPU cycles, allowing them to replicate more rapidly and claim more space for their genotype.

In most Avida experiments, indirect competition for space is the only level of interaction we allow; organisms are not allowed to directly write to or read from each other's genomes, so that Tierra-style parasites cannot form (although the configuration files do allow the experimenter to enable them). The more typical way of allowing parasites in Avida is to enable the `inject` command in the Avida instruction set. This command works similar to divide, except that instead of replacing an organism in a target cell, the would-be offspring is inserted into the memory of the organism occupying the target cell; the specific position in memory to which it is placed is determined by the template that follows the `inject`.

In Tierra, parasites can replicate more rapidly than non-parasites, but an individual parasite poses no direct harm to the host whose code it uses. These organisms could, therefore, be thought of more directly as cheaters in the classic biological sense, as they effectively take advantage of the population as a whole. In Avida, a parasite exists directly inside of its host and makes use of the CPU cycles that would otherwise belong to the host, thereby slowing down the host's replication rate. Depending on the type of parasite, it can either take all of the host's CPU cycles (thereby killing the host) and use them for replicating and spreading the infection, or else spread more slowly by using only a portion of the hosts CPU cycles (sickening it) but reducing the probability of driving the hosts, and hence itself, into extinction.

Two additional forms of interaction, resource *sensors* and direct *communication*, can be enabled by the experimenter. Resources sensors allow organisms to detect the presence of resources in the environment, a capability that could be used to exchange chemical signals. Direct communication can allow organisms to send numbers to each other and possibly distribute computations among themselves to solve environmental challenges more rapidly. Avida supports a variety of communication forms, including directional messaging to adjacent organisms, organism constructed communication networks, and population-wide broadcast messaging.

1.3.3 Test Environments

Often when examining populations in Avida, the user will need to know the fitness or some other characteristic of an organism that has not yet gone through a full gestation cycle during the course of the run. For this reason, we have constructed a *test environment* for the organisms to be run in, without affecting the rest of the population. This test environment will run the organism for at least one gestation and can either be used during a run or as part of post-processing.

When an organism is loaded into a test environment, its instructions are executed until it produces a viable offspring or a time-out is reached. Unfortunately, it is not possible to guarantee identification of non-replicative organisms (this is known as the Halting Problem in computer science), so at some point we must give up on any program we are testing and assume it to be dead. If age-based death is turned on in the actual population, this becomes a good limit for how long a CPU in the test environment should be run.

The fact that we want to determine if an organism is viable can also cause some problems in a test environment. For example, we might determine that an organism does produce an offspring but that this offspring is not identical to itself. In this case we take the next step of continuing to run the offspring in the test environment and, if necessary, its offspring until we either find a self-replicator or a sustainable cycle. By default, we will only test three levels of offspring before we assume the original organism to be non-viable and move on. Such cases happen very rarely, and not at all if you turn off implicit mutations from the configuration file.

Two final problems with the test environments include that they do not properly reflect the levels of limited resources (this can be difficult to know, particularly if we are post-processing) and that they do not handle any special interactions with other organisms since only one is being tested at a time. Both of these issues are currently being examined and we plan to have a much improved test environment in the future. Test environments do, however, work remarkably well in most circumstances.

In addition to reconstructing statistics about organisms as they existed in the population, it is also possible to determine how an organism would have fared in an alternate environment, or even to construct entirely new genomes to determine how they perform. This last approach includes techniques such as performing all single-point mutations on a genome and testing each result to determine what its local fitness landscape looks like or artificially crossing over pairs of organisms to determine their viability. Test environments are most commonly used in the post-processing of Avida data, as described in the next section.

1.3.4 Performing Avida Experiments

Currently, there are two main methods of running Avida – either with one of the user interfaces described above or via the command line executable (which is faster and full featured but requires the user to pre-script the complete experimental protocol). Researchers will often use one of the user interfaces to get an intuitive feel of how an experiment works, but then they will shift to the command line executable when they are ready to perform more extensive data collection.

The complete configuration of an Avida experiment consists of five different initialization files. The most important of these is the main *configuration* file, called `avida.cfg` by default and typically referred to as simply the 'config' file. The config file contains a list of variables that control all of the basic settings of a run, including the population size, the mutation rates, and the names of all of the other initialization files necessary. Next, we have the *instruction set*, which describes the specific genetic language used in the experiment. Third is the *ancestral organism* with which the population should be seeded. Fourth, we have the *environment* file that describes which resources are available to the organisms and defines reactions by the tasks that trigger them, their value, the resource that they use, and any byproducts that they produce. The final configuration file is *events*, which is used to describe specific actions that should occur at designated time points during the experiment, including most data collection and any direct disruptions to the population. Each of these files is described in more detail in the Avida documentation.

Once Avida has been properly installed and the configuration files set up, it can be started in command line mode by simply running the `avida` executable from within the directory that contains the configuration files. Some basic information will scroll by on the screen (specifically, current update being processed, number of generations, average fitness, and current population size). When the experiment has completed, the process will terminate automatically, leaving a set of output files that described the completed experiment. These output files are, by default, placed in a subdirectory called `data`. Each output file begins with a comment header describing the contents of file.

1.3.5 Analyze Mode

Avida has an analysis-only mode (short: *analyze mode*), which allows for powerful post-processing of data. Avida is brought into the analyze mode by the command-line parameter "-a". In analyze model, Avida processes the analyze file specified in the configuration file ("analyze.cfg" by default). The analyze file contains a program written in a simple scripting language. The structure of the program involves loading in genotypes in one or more *batches* and then either manipulating single batches or doing comparisons between batches.

In the following subsections, we present a couple of example programs that will illustrate the basics of the analyze scripting language. A full list of commands available in analysis mode is given in the Avida documentation.

1.3.5.1 Testing a Genome Sequence

The following program will load in a genome sequence, run it in a test environment, and output the results of the tests in a couple of formats:

```
VERBOSE
LOAD_SEQUENCE rmzavcgmciqqptqpqctletncoggqxutycuastva
RECALCULATE
DETAIL detail_test.dat fitness length viable sequence
TRACE
PRINT
```

The program starts off with the VERBOSE command, which causes Avida to print to the screen all details of what is going on during the execution of the analyze script; the command is useful for debugging purposes. The program then uses the LOAD_SEQUENCE command to define a specific genome sequence in compressed format. (The compressed format is used by Avida in a number of output files. The mapping from instructions to letters is determined by the instruction set file and may change if the instruction set file is altered.)

The RECALCULATE command places the genome sequence into the test environment and determines the organism's fitness, metabolic rate, gestation time, and so on. The DETAIL command that follows prints this information into the file "detail_test.dat". (This filename is specified as the first argument of DETAIL.) The TRACE and PRINT commands will then print individual files with data on this genome, the first tracing the genome's execution line by line and the second summarizing several test results and printing the genome line by line. Since no directory was specified for these commands, the resulting output files are created in "archive/", a subdirectory of the "data" directory. If a genotype has a name when it is loaded, then that name will be kept. Otherwise, it will be assigned a name starting with "org-S1", then "org-S2", and so on. The TRACE and PRINT commands add their own suffixes (".trace" and ".gen") to the genome's name to determine the filenames they will use.

1.3.5.2 Finding Lineages

The portion of an Avida run that we will often be most interested in is the lineage from a genotype (typically the final dominant genotype) back to the original ancestor. There are tools in the analyze mode to obtain this information, provided that the necessary population and historical data have been written out with the events SavePopulation and SaveHistoricPopulation. The following program demonstrates how to make use of these data files.

```
FORRANGE i 100 199
  SET d /Users/brysonda/research/instset/is_ex_$i
  PURGE_BATCH
  LOAD $d/detail-100000.pop
  LOAD $d/historic-100000.pop
  FIND_LINEAGE num_cpus
  RECALCULATE
  DETAIL lineage-$i.html depth parent_dist html.sequence
END
```

The FORRANGE command runs the contents of the loop once for each possible value in the range, setting the variable i to each of these values in turn. Thus the first time through the loop, 'i' will be equal to the value 100, then 101, 102, and so on, all the way up to 199. In this particular case, we have 100 runs (numbered 100 through 199) with which we want to work.

The first thing we do once inside the loop is to set the value of variable 'd' to be the name of the directory with which we are going to be working. Since this directory name is long, we do not want to have to type it every time we need it. If we set it to the variable 'd', then all we need to do is type "$d" in the future.[3] Note that in this case we are setting a variable to a string instead of a number; that is fine, and Avida will figure out how to handle the contents of the variable properly. The directory we are working with changes each time the loop is executed, since the variable 'i' is part of the directory name.

We then use the command PURGE_BATCH to get rid of all genotypes from the last execution of the loop (lest we are accumulating more and more genotypes in the current batch) and refill the batch by using LOAD to read in all genotypes saved in the file "detail-100000.pop" within our chosen directory. A detail population (".pop") file contains all of the genotypes that were currently alive in the population at the time the detail file was printed, whereas a historic file contains all of the genotypes that are ancestors of those that are still alive. The combination of these two files gives us the lineages of the entire population back to the original ancestor. Since we are only interested in a single lineage, we next run the FIND_LINEAGE command to pick out a single genotype and discard everything else except for its lineage. In this case, we pick the genotype with the highest abundance (i.e., the highest number of organisms, or virtual CPUs, associated with it) at the time of output.

As before, the RECALCULATE command gets us any additional information we may need about the genotypes, and then we print that information to a file using the DETAIL command. The filenames that we are using this time have the format "lineage-$i.html" that is, they are all being written to the "data" directory, with filenames that incorporate the run number. Also, because the filename ends in the suffix ".html", Avida prints the file in html format, rather than in plain text. Note that the specific values that we choose to print take

[3] Analyze mode variable names are currently limited to a single letter.

advantage of the fact that we have a lineage (and hence have measured things like the genetic distance to the parent) and are in html mode (and thus can print the sequence using colors to specify where exactly mutations occurred).

These examples are only meant to present the reader with an idea of the types of analyses available in this built-in scripting language. Many more are possible, but a more exhaustive discussion of these possibilities is beyond the scope of this chapter.

1.4 A Summary of Avida Research

Avida has been used in several dozen peer-reviewed scientific publications, including *Nature* [26, 24, 53] and *Science* [7]. We describe a few of our more interesting efforts ahead.

1.4.1 The Evolution of Complex Features

When Darwin first proposed his theory of evolution by natural selection, he realized that it had a problem explaining the origins of vertebrate eye [11]. Darwin noted that "In considering transitions of organs, it is so important to bear in mind the probability of conversion from one function to another." That is, populations do not evolve complex new features *de novo*, but instead modify existing, less complex features for use as building blocks of the new feature. Darwin further hypothesized that "Different kinds of modification would [...] serve for the same general purpose," noting that just because any one particular complex solution may be unlikely, there may be many other possible solutions, and we only witness the single one lying on the path evolution took. As long as the aggregate probability of all solutions is high enough, the individual probabilities of the possible solutions are almost irrelevant.

Substantial evidence now exists that supports Darwin's general model for the evolution of complexity (e.g., [12, 20, 31, 32, 54]), but it is still difficult to provide a complete account of the origin of any complex feature due to the extinction of the intermediate forms, imperfection of the fossil record, and incomplete knowledge of the genetic and developmental mechanisms that produce such features. Digital evolution allowed us to surmount these difficulties and track all genotypic and phenotypic changes during the evolution of a complex trait with enough replication to obtain statistically powerful results [24]. We isolated the computation EQU (logical equals) as a complex trait and showed that at least 19 coordinated instructions are needed to perform this task. We then performed an experiment that consisted of 100 independent populations of digital organisms being evolved for approximately 17,000 generations. We evolved 50 of these populations in a control environment where EQU was the only task rewarded; we evolved the other 50 in a more complex environment where an assortment of 8 simpler tasks were rewarded as well, to test the importance of intermediates in the evolution of a complex feature.

Results: In 23 of the 50 experiments in the complex environment, the EQU task was evolved, whereas *none* of the 50 control populations evolved EQU, illustrating the critical importance of features of intermediate complexity ($P \approx 4.3 \times 10^{-9}$, Fisher's exact test). Furthermore, all 23 implementations of the complex trait were unique, with many quite distinct from each other in their approach, indicating that, indeed, this trait had numerous solutions. This observation is not surprising, since even the shortest of the implementations found were extraordinarily unlikely (approximately 1 in 10^{27}). We further analyzed these results by tracing back the line of decent for each population to find the critical mutation that first produced the complex trait. In each case, these random mutations transformed a genotype unable to perform EQU into one that could, and even though these mutations typically affected only 1 to 2 positions in the genome, a median of 28 instructions were required to perform this complex task – a change in any of these instruction would cause the task to be lost, thus it was complex from the moment of its creation. It is noteworthy to mention that in 20 of the 23 cases the critical mutations would have been detrimental if EQU were not rewarded, and in 3 cases the prior mutation was actively detrimental (causing the replication rate for the organisms to drop by as much as half), yet it turned out to be critical for the evolution of EQU; when we reverted these seemingly detrimental mutations, EQU was lost.

1.4.2 Survival of the Flattest

When organisms have to evolve under high mutation pressure, their evolutionary dynamics is substantially different from that of organisms evolving under low mutation pressure, and some of the high-mutation-rate effects can appear paradoxical at first glance. Most of population genetics theory has been developed under the assumption that mutation rates are fairly low, which is justified for the majority of DNA-based organisms. However, RNA viruses, the large class of viruses that cause diseases such as the common cold, influenza, HIV, SARS, or Ebola, tend to suffer high mutation rates, up to 10^{-4} substitutions per nucleotide and generation [16]. The theory describing the evolutionary dynamics at high mutation rates is called quasispecies theory [15].

The main prediction for the evolutionary process at high mutation rates is that the selection acts on a cloud of mutants, rather than on individual sequences. We tested this hypothesis in Avida [53]. First, we let strains of digital organisms evolve to both a high-mutation-rate and a low-mutation-rate environment. The rationale behind this initial adaptation was that strains that evolved at a low mutation rate should adapt to ordinary individual-based selection, whereas strains that evolved at a high mutation rate should adapt to selection on mutant clouds, which means that these organisms should maximize the overall replication rate of their mutant clouds, rather than their individual replication rates. This adaptation to maximize overall replication rate under high mutation pressure takes place when organisms trade individual fitness for mutational robustness, so that their individual replication rate is reduced, but, in return, the probability that mutations cause further reduction in replication rate is also reduced [52]. Specifically, we took 40 strains of already evolved digital organisms and let each evolve for an additional 1000 generations in both a low-mutation-rate and a high-mutation-rate environment. As result, we ended up with 40 pairs of strains. The two strains of each pair were genetically and phenotypically similar, apart from the fact that one was adapted to a low mutation rate and one to a high mutation rate. As expected, we found that in the majority of cases the strains evolved at a a high mutation rate had a lower replication speed than the ones evolved at a low mutation rate.

Next, we let the two types of strains compete with each other, in a setup where both strains would suffer from the same mutation rate, which was either low, intermediate, or high. Not surprisingly, at a low mutation rate the strains adapted to that mutation rate consistently outcompeted the ones adapted to a high mutation rate, since, after all, the former ones had the higher replication rate (we excluded those cases in which the strain that evolved at a low mutation rate had a lower or almost equal fitness to the strain evolved at a high mutation rate). However, without fail, the strain adapted to a high mutation rate could win the competition if the mutation rate during the competition was sufficiently high [53]. This result may sound surprising at first, but it has a very simple explanation. At a high mutation rate (one mutation per genome per generation or higher), the majority of an organism's offspring differ genetically from their parent. Therefore, if the parent is genetically very brittle, so that most of these mutants have a low replication rate or are even lethal, then the overall replication rate of all of the organism's offspring will be fairly moderate, even though the organism itself may produce offspring at a rapid pace. If a different organism produces offspring at a slower pace, but is more robust toward mutations, so that the majority of this organism's offspring have a replication rate similar to that of the parent, then the overall replication rate of this organism's offspring will be larger than the one of the first organism. Hence, this organism will win the competition, even though it is the slower replicator. We termed this effect the "survival of the flattest," because at a sufficiently high mutation rate, a strain that is located at a low but flat fitness peak can outcompete one that is located on a high but steep fitness peak.

1.4.3 Evolution of Digital Ecosystems

The experiments discussed above have all used single-niche Avida populations, but evolutionary design is more interesting (and more powerful) when we consider ecosystems. The selective pressures that cause the formation and diversity of ecosystems are still poorly understood [46, 48]. In part, the lack of progress is due to the difficulty of performing precise, replicated, and controlled experiments on whole ecosystems [30]. To study simple ecosystems in a laboratory microcosm (reviewed in [49]), biologists often use a chemostat, which slowly pumps resource-rich media into a flask containing bacteria, simultaneously draining its contents to keep the volume constant. Unfortunately, even in these model systems, ecosystems can evolve to be more complex than is experimentally tractable and understanding their formation remains difficult [33, 34, 42].

We set up Avida experiments based on this chemostat model [10] wherein nine resources flow into the population, and 1% of unused resources flow out. We used populations with 2500 organisms, each of which absorbed a small portion of an available resource whenever they performed the corresponding task. If too many organisms focus on the same resource, it will no longer be plentiful enough to encourage additional use.

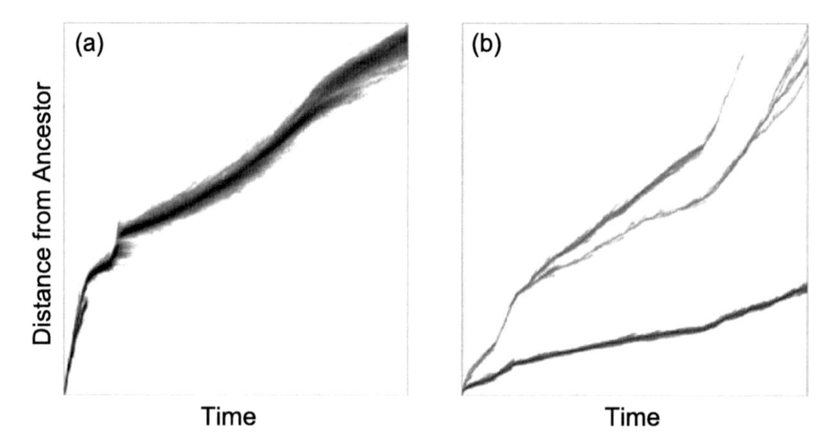

Fig. 1.5 Visualizations of phylogenies from the evolution of (a) a single-niche population, and (b) a population with limited resources. The x-axis represents time, and the y-axis is depth in the phylogeny (distance from the original ancestor). Intensity at each position indicates the number of organisms alive at a time point, at a particular depth in the tree.

Theory predicts that an environment with either a single resource or with resources in unlimited quantities is capable of supporting only one species [47], and this is exactly what we see in the standard Avida experiments. It is the competition over multiple, limited resources that is believed to play a key role in the structuring of communities [45, 49]. In 30 trials under the chemostat regime in Avida, a variety of distinct community structures developed [10]. Some evolved nine stably coexisting specialists, one per resource, while others had just a couple of generalists that divided the resources between them. Others still mixed both generalists and specialists. In all cases, the ecosystems proved to be stable because they persisted after all mutations were shut off in the system, and if any abundant phenotype were removed, it would consistently reinvade.

Phylogeny visualizations provide a striking demonstration of the differences between populations that evolved in a single niche and those from ecosystems, as displayed in Fig. 1.5. Single-niche populations can have branching events that persist for a short time, but in the long term, one species will out compete the others or simply drift to dominance if the fitness values are truly identical. By contrast, in ecosystems with multiple resources, the branches that correspond to speciation events persist.

We also studied the number of stably coexisting species as a function of resource availability [7]. We varied the inflow rate of resources over six orders of magnitude and found that multi-species communities evolved at intermediate resource abundance, but not at very high or very low resource abundance. The reason for this observation is that if resources are too scarce, they cannot provide much value to the organisms and the base metabolic rate dominates, while if resources are too abundant, then they are no longer a limiting factor, which means that space becomes the primary limit. In both cases, the system reduces down to only a single niche that organisms can take advantage of.

1.5 Outlook

Digital organisms are a powerful research tool that has opened up methods to experimentally study evolution in ways that have never before been possible. We have explained the capabilities of the Avida system and detailed the methods by which researchers can make use of them. We must be careful, however, not to be lured into the trap of thinking that because these systems can be set up and examined so easily that any experiment will be possible. There are definite limits on the questions that can be answered.

Using digital organisms, we cannot learn anything about physical structures evolved in the natural world, nor the specifics of an evolutionary event in our own history; the questions we ask must be about *how* evolution works in general and how we can harness it. Even for the latter type of questions, it is sometimes difficult to set up experiments in such a way that they give meaningful results. We must always remember that we are working with an arguably living system that will evolve to survive as best it can, not always in the direction that we intended. Avida has become, in many ways, its own bug tester. If we make a mistake, the organisms will exploit it. For example, we originally had only 16-bit inputs for the organisms to process; they quickly discovered that random guessing often took less time than actually performing the computation. In this case, the organisms indeed found the most efficient way to solve the problem we gave them, only that it was not the problem we had thought we were giving. This error happened to be easy to find and easy to fix – now all inputs are 32 bits long – but not all "cheating" will be so simple to identify and prevent. When performing an Avida experiment, it is always important that we step through the population and try to understand how some of the organisms are functioning. More often than not, they will surprise us with the cleverness of the survival strategies that they are using. Also, sometimes they will even make us step back to rethink our experiments.

Many possible future directions exist in the development of Avida. Ongoing efforts include (among others) the implementation new CPU models that are more powerful and general purpose, an overhaul of the user interface framework that will support enhanced visualization and cross-platform graphical interaction with all of the analysis tools, continued refinement of deme features and organism communication, and the implementation of more complex and realistic topologies, including physical environments in which organisms move and grow. Finally, a major new feature in development for Avida 3.0 will be a full-featured scripting language. The new AvidaScript will expose all of the power of populations and analysis tools, allowing researchers to create vastly more sophisticated experiments right out of the box.

References

1. Avida packages. URL `http://sourceforge.net/projects/avida`
2. Avida project. URL `http://avida.devosoft.org`
3. Adami, C.: Digital genetics: unravelling the genetic basis of evolution. Nature Reviews Genetics **7**(2), 109–118 (2006). DOI 10.1038/nrg1771
4. Adami, C., Brown, C.T., Haggerty, M.R.: Abundance-distributions in artificial life and stochastic models: Age and area revisited. In: F. Morán, A. Moreno, J.J. Morelo, P. Chacón (eds.) Proceedings of the Third European Conference on Advances in Artificial Life, Lecture Notes in Computer Science, pp. 503–514. Springer-Verlag, London, UK (1995)
5. Adami, C., Ofria, C., Collier, T.C.: Evolution of biological complexity. Proceedings of the National Academy of Sciences **97**, 4463–4468 (2000)
6. Barton, N., Zuidema, W.: Evolution: the erratic path towards complexity. Current Biology **13**(16), R649–R651 (2003)
7. Chow, S.S., Wilke, C.O., Ofria, C., Lenski, R.E., Adami, C.: Adaptive radiation from resource competition in digital organisms. Science **305**(5680), 84–86 (2004). DOI 10.1126/science.1096307
8. Chu, J., Adami, C.: Propagation of information in populations of self-replicating code. In: C.G. Langton, T. Shimohara (eds.) Artificial Life V: Proceedings of the Fifth International Workshop on the Synthesis and Simulation of Living Systems, pp. 462–469. International Society of Artificial Life, MIT Press, Cambridge, MA (1997)

9. Comas, I., Moya, A., Gonzalez-Candelas, F.: Validating viral quasispecies with digital organisms: A re-examination of the critical mutation rate. BMC Evolutionary Biology **5**(1), 5 (2005). DOI 10.1186/1471-2148-5-5
10. Cooper, T.F., Ofria, C.: Evolution of stable ecosystems in populations of digital organisms. In: R.K. Standish, M.A. Bedau, H.A. Abbass (eds.) Artificial Life VIII: Proceedings of the Eighth International Conference on Artificial life, pp. 227–232. International Society of Artificial Life, MIT Press, Cambridge, MA (2003)
11. Darwin, C.: On the Origin of Species by Means of Natural Selection. Murray, London (1859)
12. Dawkins, R.: The Blind Watchmaker, 2nd edn. W. W. Norton & Company, New York (1996)
13. Dennett, D.C.: The new replicators. In: M. Pagel (ed.) Encyclopedia of Evolution. Oxford University Press, Oxford (2002)
14. Dewdney, A.K.: In a game called core war hostile programs engage in a battle of bits. Scientific American **250**(5), 14–22 (1984)
15. Domingo, E., Beibricher, C.K., Eigen, M., Holland, J.J.: Quasispecies and RNA Virus Evolution: Priciples and Consequences. Landes Bioscience, Georgetown, TX (2001)
16. Drake, J.W., Holland, J.J.: Mutation rates among rna viruses. Proceedings of the National Academy of Sciences **96**(24), 13,910–13,913 (1999). DOI 10.1073/pnas.96.24.13910
17. Egri-Nagy, A., Nehaniv, C.L.: Evolvability of the genotype-phenotype relation in populations of self-replicating digital organisms in a tierra-like system. In: Proceedings of the 7th European Conferance on Artificial Life, Lecture Notes in Computer Science, pp. 238–247. Springer, Berlin (2003)
18. Elena, S.F., Lenski, R.E.: Evolution experiments with microorganisms: The dynamics and genetic bases of adaptation. Nature Reviews Genetics **4**(6), 457–469 (2003). DOI 10.1038/nrg1088
19. Gerlee, P., Lundh, T.: The genetic coding style of digital organisms. In: Proceedings of the 8th European Conference on Artificial Life, *Lecture Notes in Computer Science*, vol. 3630, pp. 854–863. Springer, Berlin (2005). DOI 10.1007/11553090_86
20. Goldberg, D.E.: The Design of Innovation: Lessons from and for Competent Genetic Algorithms. Kluwer Academic Publishers, Boston (2002)
21. Hartl, D.L., Clark, A.G.: Principles of Population Genetics. Sinauer Associates, Sunderland, MA (2006)
22. Kim, Y., Stephan, W.: Selective sweeps in the presence of interference among partially linked loci. Genetics **164**(1), 389–398 (2003)
23. Koza, J.R.: Genetic Programming IV: Routine Human-Competitive Machine Intelligence. Kluwer Academic Publishers, Norwell, MA (2003)
24. Lenski, R., Ofria, C., Pennock, R.T., Adami, C.: The evolutionary origin of complex features. Nature **423**, 139–144 (2003)
25. Lenski, R.E.: Phenotypic and genomic evolution during a 20,000-generation experiment with the bacterium, *Escherichia coli*. Plant Breeding Reviews **24**, 225–265 (2004)
26. Lenski, R.E., Ofria, C., Collier, T.C., Adami, C.: Genome complexity, robustness and genetic interactions in digital organisms. Nature **400**(6745), 661–664 (1999). DOI 10.1038/23245
27. Maynard Smith, J.: Byte-sized evolution. Nature **355**, 772–773 (1992). DOI 10.1038/355772a0
28. McVean, G.A.T., Charlesworth, B.: The effects of hill-robertson interference between weakly selected mutations on patterns of molecular evolution and variation. Genetics **155**(2), 929–944 (2000)
29. Misevic, D., Ofria, C., Lenski, R.E.: Sexual reproduction shapes the genetic architecture of digital organisms. Proceedings of the Royal Society of London: Biological Sciences **273**, 457–464 (2006)
30. Morin, P.J.: Biodiversity's ups and downs. Nature **406**(6795), 463–464 (2000)

31. Newcomb, R.D., Campbell, P.M., Ollis, D.L., Cheah, E., Russell, R.J., Oakeshott, J.G.: A single amino acid substitution converts a carboxylesterase to an organophosphorus hydrolase and confers insecticide resistance on a blowfly. Proceedings of the National Academy of Sciences **94**(14), 7464–7468 (1997). DOI 10.1073/pnas.94.14.7464
32. Nilsson, D.E., Pelger, S.: A pessimistic estimate of the time required for an eye to evolve. Proceedings of the Royal Society of London: Biological Sciences **256**(1345), 53–58 (1994)
33. Notley-McRobb, L., Ferenci, T.: Adaptive mgl-regulatory mutations and genetic diversity evolving in glucose-limited *Escherichia coli* populations. Environmental Microbiology **1**(1), 33–43 (1999). DOI 10.1046/j.1462-2920.1999.00002.x
34. Notley-McRobb, L., Ferenci, T.: The generation of multiple co-existing mal-regulatory mutations through polygenic evolution in glucose-limited populations of *Escherichia coli*. Environmental Microbiology **1**(1), 45–52 (1999). DOI 10.1046/j.1462-2920.1999.00003.x
35. Nowak, M.A.: Evolutionary Dynamics: Exploring the Equations of Life. Belknap Press of Harvard University Press, Cambridge, MA (2006)
36. Ofria, C., Adami, C., Collier, T.: Design of evolvable computer languages. IEEE Transactions on Evolutionary Computation **6**(4), 420–424 (2002). DOI 10.1109/TEVC.2002.802442
37. Ofria, C., Bryson, D.M., Baer, B., Nanlohy, K.G., Lenski, R.E., Adami, C.: The Avida User's Guide. Michigan State University, East Lansing, MI (2008)
38. Ofria, C., Wilke, C.: Avida: A software platform for research in computational evolutionary biology. Artificial Life **10**, 191–229 (2004). DOI 10.1162/106454604773563612
39. O'Neill, B.: Digital evolution. PLoS Biology **1**(1), 011–014 (2003). DOI 10.1371/journal.pbio.0000018
40. Orr, H.A.: The rate of adaptation in asexuals. Genetics **155**(2), 961–968 (2000)
41. Pennock, R.T.: Avida-ED website. URL `http://avida-ed.msu.edu/`
42. Rainey, P.B., Travisano, M.: Adaptive radiation in a heterogeneous environment. Nature **394**, 69–72 (1998). DOI 10.1038/27900
43. Rasmussen, S., Knudsen, C., Feldberg, P., Hindsholm, M.: The coreworld: Emergence and evolution of cooperative structures in a computational chemistry. Physica D **42**(1–3), 111–134 (1990). DOI 10.1016/0167-2789(90)90070-6
44. Ray, T.S.: An approach to the synthesis of life. In: C.G. Langton, C. Taylor, J.D. Farmer, S. Rasmussen (eds.) Artificial Life II, vol. XI, pp. 371–408. Addison-Wesley, Redwood City, CA (1991)
45. Schluter, D.: Ecological causes of adaptive radiation. American Naturalist **148**, S40–S64 (1996)
46. Schluter, D.: Ecology and the origin of species. Trends in Ecology & Evolution **16**(7), 372–380 (2001)
47. Tilman, D.: Resource Competition and Community Structure. Princeton University Press, Princeton, NJ (1982)
48. Tilman, D.: Causes, consequences and ethics of biodiversity. Nature **405**(6783), 208–211 (2000)
49. Travisano, M., Rainey, P.B.: Studies of adaptive radiation using model microbial systems. The American Naturalist **156**, S35–S44 (2000). DOI 10.1086/303414
50. Wilke, C.O.: Maternal effects in molecular evolution. Physical Review Letters **88**(7), 078,101 (2002). DOI 10.1103/PhysRevLett.88.078101
51. Wilke, C.O., Adami, C.: The biology of digital organisms. Trends in Ecology & Evolution **17**(11), 528–532 (2002). DOI 10.1016/S0169-5347(02)02612-5
52. Wilke, C.O., Adami, C.: Evolution of mutational robustness. Mutation Research/Fundamental and Molecular Mechanisms of Mutagenesis **522**(1–2), 3–11 (2003)
53. Wilke, C.O., Wang, J.L., Ofria, C., Lenski, R.E., Adami, C.: Evolution of digital organisms at high mutation rates leads to survival of the flattest. Nature **412**(6844), 331–333 (2001). DOI 10.1038/35085569

54. Wilkins, A.S.: The Evolution of Developmental Pathways. Sinauer Associates, Sunderland, MA (2002)
55. Yedid, G., Bell, G.: Microevolution in an electronic microcosm. The American Naturalist **157**(5), 465–487 (2001). DOI 10.1086/319928
56. Yedid, G., Bell, G.: Macroevolution simulated with autonomously replicating computer programs. Nature **420**(6917), 810–812 (2002). DOI 10.1038/nature01151
57. Zhang, H., Travisano, M.: Predicting fitness effects of beneficial mutations in digital organisms. Artificial Life, 2007. ALIFE '07. IEEE Symposium on pp. 39–46 (2007). DOI 10.1109/ALIFE.2007.367656

Chapter 2
Foundations of and Recent Advances in Artificial Life Modeling with Repast 3 and Repast Simphony

Michael J. North and Charles M. Macal

2.1 Introduction

Artificial life focuses on synthesizing "life-like behaviors from scratch in computers, machines, molecules, and other alternative media" [24]. Artificial life expands the "horizons of empirical research in biology beyond the territory currently circumscribed by life-as-we-know-it" to provide "access to the domain of life-as-it-could-be" [24]. Agent-based modeling and simulation (ABMS) are used to create computational laboratories that replicate real or potential behaviors of actual or possible complex adaptive systems (CAS). The goal of agent modeling is to allow experimentation with simulated complex systems. To achieve this, agent-based modeling uses sets of agents and frameworks for simulating the agent's decisions and interactions. Agent models show how complex adaptive systems may evolve through time in a way that is difficult to predict from knowledge of the behaviors of the individual agents alone. Agent-based modeling thus provides a natural framework in which to perform artificial life experiments. The free and open source Recursive Porous Agent Simulation Toolkit (Repast) family of tools consists of several advanced agent-based modeling toolkits.

2.1.1 *Artificial Life*

The discipline of artificial life studies the synthesis of forms and functions that appear alive. Artificial life allows scientific studies of biological systems outside the currently observable accidents of history. According to Langton [24]:

> Biology is the scientific study of life – in principle, anyway. In practice, biology is the scientific study of life on Earth based on carbon-chain chemistry. There is nothing in its charter that restricts biology to carbon-based life; it is simply that this is the only kind of life that has been available to study. Thus, theoretical biology has long

> faced the fundamental obstacle that it is impossible to derive general principles from single examples.
>
> Without other examples, it is difficult to distinguish essential properties of life – properties that would be shared by any living system – from properties that may be incidental to life in principle, but which happen to be universal to life on Earth due solely to a combination of local historical accident and common genetic descent.
>
> In order to derive general theories about life, we need an ensemble of instances to generalize over. Since it is quite unlikely that alien life-forms will present themselves to us for study in the near future, our only option is to try to create alternative life-forms ourselves – artificial life – literally "life made by Man rather than by Nature."

Langton's description of artificial life indicates the depth but belies the age of the discipline. According to Di Paolo [12]:

> To say that artificial life is a young discipline in name only is to exaggerate, but it would be mistaken to think that its goals are new. The marriage of synthetic scientific aims with computational techniques makes artificial life a product of the last fifteen years, but its motivations have much deeper roots in cybernetics, theoretical biology, and the age-old drive to comprehend the mysteries of life and mind. Little wonder that a good part of the work in this field has been one of rediscovery and renewal of hard questions. Other disciplines have sidestepped such questions, often for very valid reasons, or have put them out of the focus of everyday research; yet these questions are particularly amenable to be treated with novel techniques such as computational modeling and other synthetic methodologies. What is an organism? What is cognition? Where do purposes come from?

2.1.2 Agent-Based Modeling for Artificial Life

Agent-based modeling and simulation are used to create computational laboratories that replicate selected real or potential behaviors of actual or possible complex adaptive systems. A complex adaptive system is made up of agents that interact, mutate, replicate, and die while adapting to a changing environment. Holland has identified the three properties and four mechanisms that are common to all complex adaptive systems [19]:

1. The nonlinearity property occurs when components or agents exchange resources or information in ways that are not simply additive. An example is a photosynthetic cell agent that returns 1 calorie of energy when 1 calorie is requested, 2 calories of energy when 2 calories are requested, and 3 calories of energy when 10 calories are requested.
2. The diversity property is observed when agents or groups of agents differentiate from one another over time. An example is the evolutionary emergence of new species.
3. The aggregation property occurs when a group of agents is treated as a single agent at a higher level. An example is the ants in an ant colony.
4. The flows mechanism involves exchange of resources or information between agents such that the resources or information can be repeatedly

forwarded from agent to agent. An example is the flow of energy between agents in an ecosystem.
5. The tagging mechanism involves the presence of identifiable flags that let agents identify the traits of other agents. An example is the use of formal titles such as "Dr." in a social system.
6. The internal models mechanism involves formal, informal, or implicit representations of the world embedded within agents. An example is a predator's evolving view of the directions prey are likely to flee during pursuit.
7. The building blocks mechanism is used when an agent participates in more than one kind of interaction. An example is a predator agent that can also be prey for larger predators.

Of course, these properties and mechanisms are interrelated. For example, with aggregation, many agents can act as one. With building blocks, one agent in some sense can act as many. Agent-based models normally incorporate some or all of the properties and mechanisms of complex adaptive systems.

The goal of agent modeling is to allow experimentation with simulated complex systems. To achieve this, agent-based modeling uses sets of agents and frameworks for simulating the agent's decisions and interactions. Agent models can show how complex adaptive systems can evolve through time in a way that is difficult to predict from knowledge of the behaviors of the individual agents alone. Agent modeling focuses on individual behavior. The agent rules are often based on theories of the individual such as rational individual behavior, bounded rationality, or satisficing [39]. Based on these simple types of rules, agent models can be used to study how patterns emerge. Agent modeling may reveal behavioral patterns at the macro or system level that are not obvious from an examination of the underlying agent rules alone: These patterns are called emergent behavior. Agent-based modeling and simulation thus provide a natural framework in which to perform artificial life experiments.[1]

Agent-based modeling and simulation are closely related to the field of Multi-agent Systems (MAS). Both fields concentrate on the creation of computational complex adaptive systems. However, agent simulation models the real or potential behaviors of complex adaptive systems and MAS often focuses on applications of artificial intelligence to robotic systems, interactive systems, and proxy systems.

[1] ABMS, agent-based modeling (ABM), agent-based simulation (ABS), and individual-based modeling (IBM) are all synonymous. ABMS is used here since ABM can be confused with anti-ballistic missile, ABS can be confused with anti-lock brakes, and IBM can be confused with International Business Machines Corporation.

2.1.3 Chapter Organization

This chapter provides an overview of the Repast agent modeling toolkit from the perspective of artificial life. This chapter is organized into four parts. The introduction describes artificial life and agent-based modeling and simulation. The second section discusses the Repast agent modeling toolkit's development ecosystem and underlying concepts. The third section reviews a series of Repast artificial life models of artificial evolution and ecosystems, artificial societies, and artificial biological systems. The final section presents a summary and conclusions.

2.2 REPAST

The Recursive Porous Agent Simulation Toolkit (Repast) family of tools is one of several agent modeling toolkits available. Repast borrows many concepts from the Swarm agent-based modeling toolkit [40]. Repast is differentiated from Swarm in several respects that are discussed later.

Two generations of the Repast toolkit family are in widespread use. The first, Repast 3, is available in pure Java and pure Microsoft.Net forms and uses a library-oriented approach to model development [33]. The second, Repast Simphony, is available in pure Java. It can be used as a library, but the preferred model of development is to use its point-and-click wizard-based system to create models, execute them, and analyze the results [21, 36].

Repast is a free open-source toolkit that was originally developed by Sallach, Collier, Howe, North, and others [9]. Repast was created at the University of Chicago. Subsequently, it has been maintained by organizations such as Argonne National Laboratory. Repast is now managed by the non-profit volunteer Repast Organization for Architecture and Design (ROAD). ROAD is led by a board of directors that includes members from a wide range of government, academic, and industrial organizations. The Repast system, including the source code, is available directly from the web [38].

Repast seeks to support the development of extremely flexible models of living social agents, but it is not limited to modeling living social entities alone. From the ROAD home page [38]:

> Our goal with Repast is to move beyond the representation of agents as discrete, self-contained entities in favor of a view of social actors as permeable, interleaved, and mutually defining; with cascading and recombinant motives... We intend to support the modeling of belief systems, agents, organizations, and institutions as recursive social constructions.

2.2.1 Repast 3

At its heart, Repast toolkit version 3 can be thought of as a specification for agent-based modeling services or functions. There are three concrete implementations of this conceptual specification. Naturally, all of these versions have the same core services that constitute the Repast system. The implementations differ in their underlying platform and model development languages. The three implementations are Repast for Java (RepastJ), Repast for the Microsoft.Net framework (Repast.Net), and Repast for Python Scripting (RepastPy). RepastJ is the reference implementation that defines the core services. The fourth version of Repast, namely Repast for Oz/Mozart (RepastOz), is an experimental system that partially implements the Repast conceptual specification while adding advanced new features [30, 42]. An example Repast model user interface is shown in Fig. 2.1.

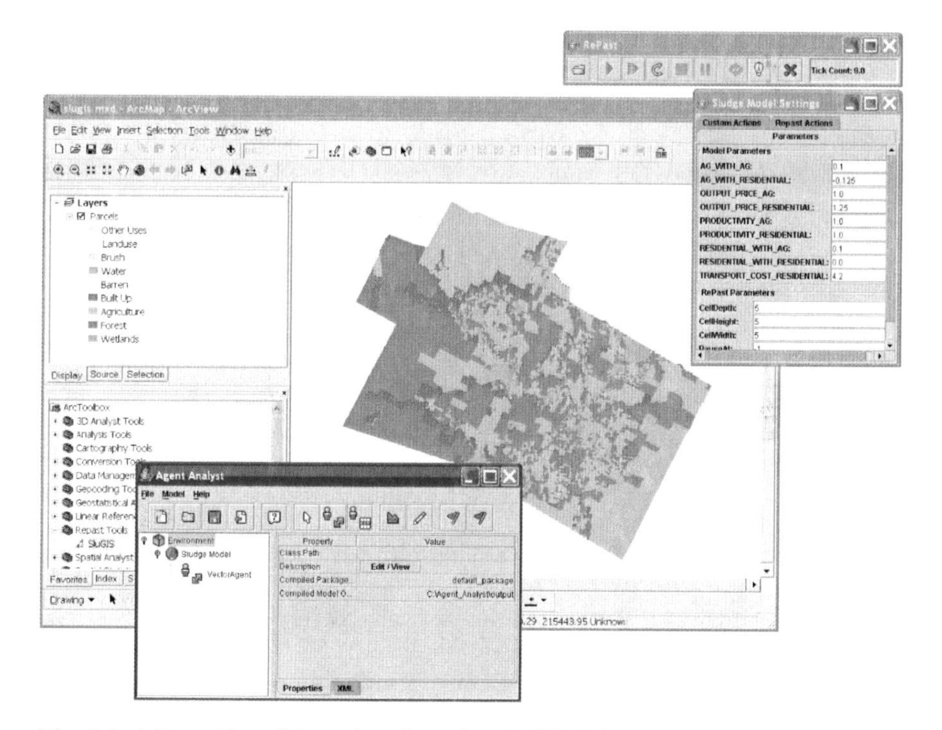

Fig. 2.1 A Repast 3 model user interface using the ESRI ArcGIS agent analyst extension.

Repast 3 has a variety of features, including the following:

- Repast 3 includes a variety of agent templates and examples. However, the toolkit gives users complete flexibility as to how they specify the properties and behaviors of agents.

- Repast 3 is fully object-oriented.
- Repast 3 includes a fully concurrent discrete event scheduler. This scheduler supports both sequential and parallel discrete event operations.
- Repast 3 offers built-in simulation results logging and graphing tools.
- Repast 3 has an automated Monte Carlo simulation framework.
- Repast 3 provides a range of two-dimensional agent environments and visualizations.
- Repast 3 allows users to dynamically access and modify agent properties, agent behavioral equations, and model properties at run time.
- Repast 3 includes libraries for genetic algorithms, neural networks, random number generation, and specialized mathematics.
- Repast 3 includes built-in Systems Dynamics modeling.
- Repast 3 has social network modeling support tools.
- Repast 3 has integrated geographical information systems (GIS) support.
- Repast 3 is fully implemented in a variety of languages including Java and C#.
- Repast 3 models can be developed in many languages including Java, C#, Managed C++, Visual Basic.Net, Managed Lisp, Managed Prolog, and Python scripting.
- Repast 3 is available on virtually all modern computing platforms, including Windows, Mac OS, and Linux. The platform support includes both personal computers and large-scale scientific computing clusters.

Repast 3's features directly support the implementation of models with Holland's three properties and four mechanisms [19]:

1. Repast 3 allows nonlinearity in agents since their behaviors are completely designed by users. Repast's Systems Dynamics, genetic algorithms, neural networks, random number generation, and social networks libraries make this process easy.
2. Repast 3 supports diversity by giving users complete control over the way their agents are defined and initialized. Again, the Repast libraries simplify the specification of diversity.
3. Repast 3 allows the aggregation property by allowing users to specify and maintain groups of agents.
4. Repast 3 supports the flows mechanism with features such as its Systems Dynamics tools and social network library.
5. Repast 3 provides for the tagging mechanism by allowing agents to present arbitrary markers.
6. Repast 3 makes the internal models mechanism available through both its flexible definition of agents and its many behavioral libraries.
7. Repast 3 supports the building blocks mechanism through its object-oriented polymorphism.

Repast 3 for Python Scripting (RepastPy) enables visual model construction with agent behaviors defined in Python [27]. RepastPy models can be automatically converted to RepastJ models using RepastPy's export option.

RepastPy users work with the interface shown in the upper left-hand window of Fig. 2.2 to add the components to their models. RepastPy users then employ Python to script the behaviors of their agents, as shown in the lower right-hand window of Fig. 2.2.

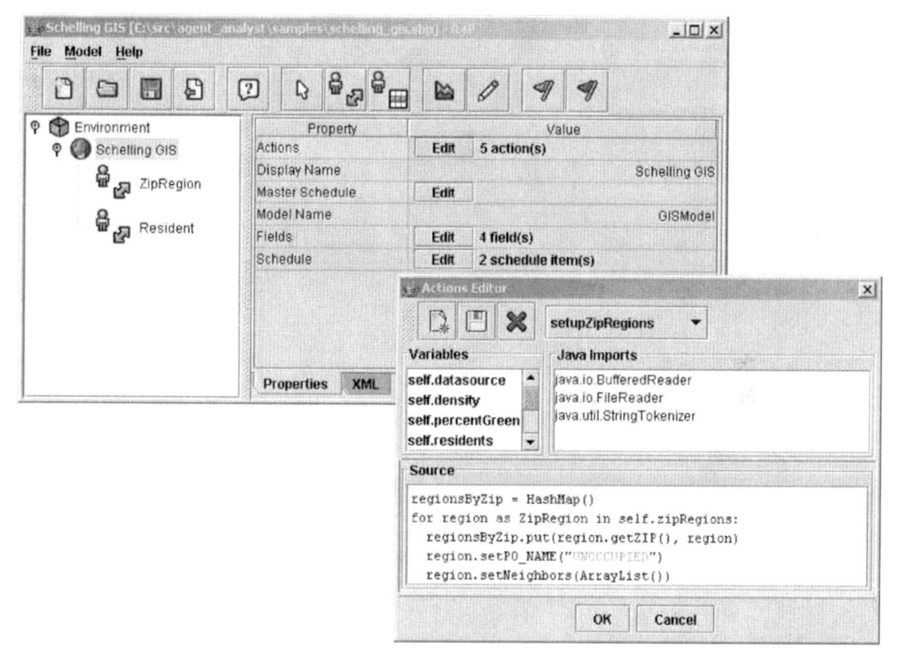

Fig. 2.2 The RepastPy interface.

The components in the example model are shown on the left-hand side of the upper window of Fig. 2.2. These components include the simulation environment specification, the model specification ("Schelling GIS"), the ZIP code region agent specification ("ZipRegion"), and the residential agent specification ("Resident"). Properties for the model specification such as the "Actions," "Display Name," and "Master Schedule" are shown on the right-hand side of the upper window in the figure. The Actions "Edit" button is used to access the Python scripting for the agent behaviors.

The Python scripting window for the example model is shown in the lower window of Fig. 2.2. The agent properties ("Variables"), the agent behavior libraries ("Java Imports"), and behavior code ("Source") can be seen in this window.

There is a special version of RepastPy known as the Agent Analyst that is an extension to the ESRI ArcGIS geographical information systems platform. ESRI ArcGIS is the leading commercial GIS, with well over one million users. Agent Analyst is a fully integrated ArcGIS Model Tool. This means that Agent Analyst has drag-and-drop integration with ArcGIS. Agent

Analyst users can create RepastPy models from within ArcGIS with a few mouse clicks. Fig. 2.1 shows the SLUDGE Geographical Information System (SluGIS) Agent Analyst model running within ArcGIS. SluGIS is described in the section on artificial societies.

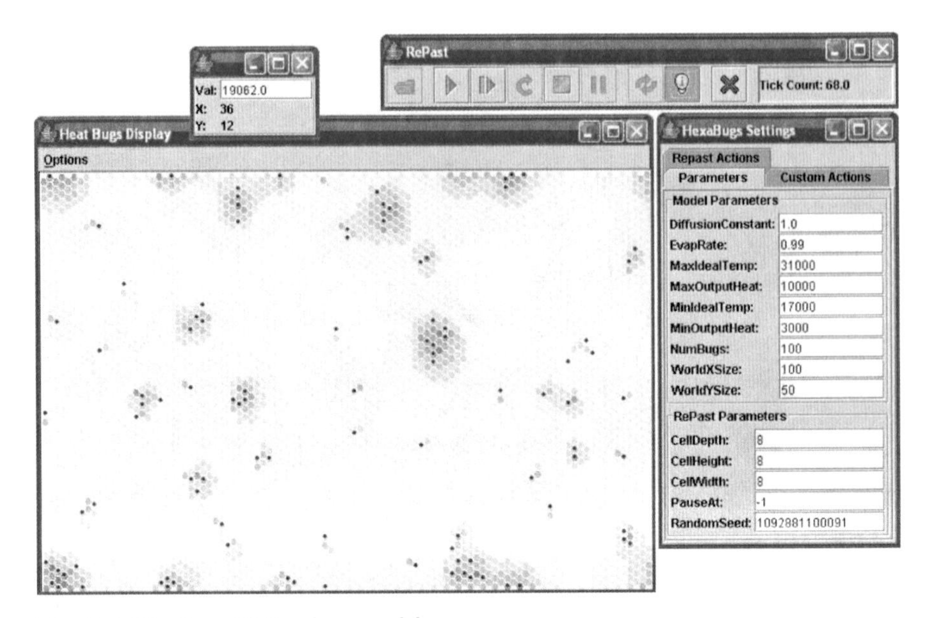

Fig. 2.3 The RepastJ Hexabugs model.

RepastJ is written entirely in Java [15]. An example RepastJ model, Hexabugs, is shown in Fig. 2.3. The Hexabugs model is discussed in the section on artificial biological systems. Since RepastJ is pure Java, any development environment that supports Java can be used. The free and open-source Eclipse development environment is recommended [13]. Eclipse provides a variety of powerful editing and development features, including code authoring wizards, automatic code restructuring tools, design analysis tools, Unified Modeling Language (UML) tools, extensible markup language (XML) tools, and integration with version control systems. Fig. 2.4 shows part of the RepastJ AgentCell model in Eclipse. The AgentCell modules are shown in the upper left "Package Explorer" tab. The cell agent component is highlighted in this tab. Part of the cell agent code is shown in the upper middle "Cell.java" tab. Some of the cell agent properties and methods can be seen in the "Outline" tab on the far right. Part of the cell agent documentation is shown in the button right tab. Additionally, a code module dependency graph can be seen in the lower left "Dependency Graph View" tab. This graph shows the connections between some of the main AgentCell modules.

Both RepastJ and RepastPy models can be developed and executed on nearly any modern computing platform. This is particularly beneficial for

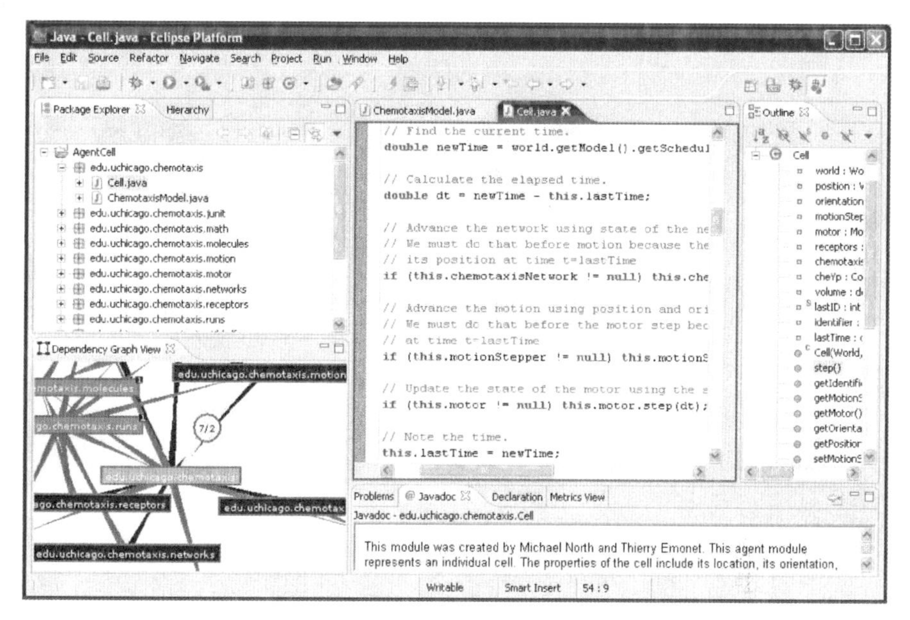

Fig. 2.4 RepastJ in the Eclipse development environment.

artificial life researchers since models can be constructed on readily available workstations and then executed on large-scale clusters without changing code. An example of this will be provided along with the description of the AgentCell model.

Repast for the Microsoft.Net framework (Repast.Net) is written entirely in C# [2]. An example Repast.Net named Rocket Bugs is shown in Fig. 2.5. The Rocket Bugs model is a Cartesian elaboration of the Hexabugs model in which some of the agents herd the other agents in the system. Some of the code for this model is displayed in the Visual Studio Environment in Fig. 2.6. It can be seen in the three windows on the left in the figure that the Rocket Bugs simulation uses a combination of Managed C++, C#, and Visual Basic.Net, all in a single seamless model. Additionally, note in the lower right that Repast.Net comes with a full set of specialized Visual Studio templates. These templates automate the initial creation of both Repast.Net models and model components such as agents.

All three versions of Repast are designed to work well with other software development tools. For example, all three versions are integrated with GISs such as the RepastPy Agent Analyst example shown in Fig. 2.1. However, since RepastJ is the most widely used version of Repast, the integration examples will focus on Java. See ROAD for many other examples [38].

RepastJ easily permits aspect-oriented software development. Aspects implement cross-cutting concerns that allow software idioms repeated throughout a model to be factored to reduce redundancy [14]. See Walker, Baniassad,

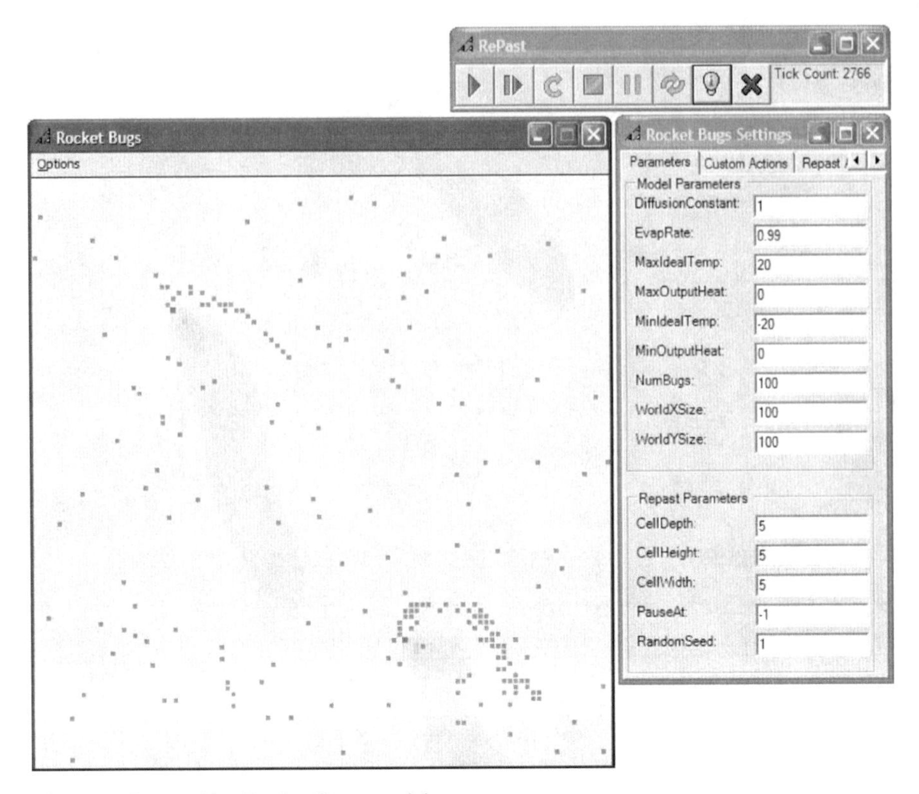

Fig. 2.5 Repast.Net Rocket Bugs model.

and Murphy for a discussion of the use of Aspects for software development [43].

RepastJ includes its own built-in logging facilities but also works with the high-performance Log4j system and also with the National Center for Supercomputing Applications' (NCSA) Hierarchical Data Format 5 (HDF5) data storage system [18, 32]. The use of Log4j, among other logging tools, in conjunction with AspectJ is discussed briefly by Cloyer et al. [8].

RepastJ unit testing is performed with JUnit as outlined in Beck and Gamma [3]. Unit testing allows software to be tested on an incremental modular level. The combination of these and other tools with RepastJ allows sophisticated models to be constructed reliably and efficiently.

The Repast 3 system has two layers. The core layer runs general-purpose simulation code written in Java or C#. This component handles most of the behind-the-scenes details. Repast users do not normally need to work with this layer directly. The external layer runs user-specific simulation code written in Java, C#, Python, Managed C++, Managed Lisp, Managed Prolog, Visual Basic.Net, or other languages. This component handles most of the center stage work. Repast users regularly work with this layer.

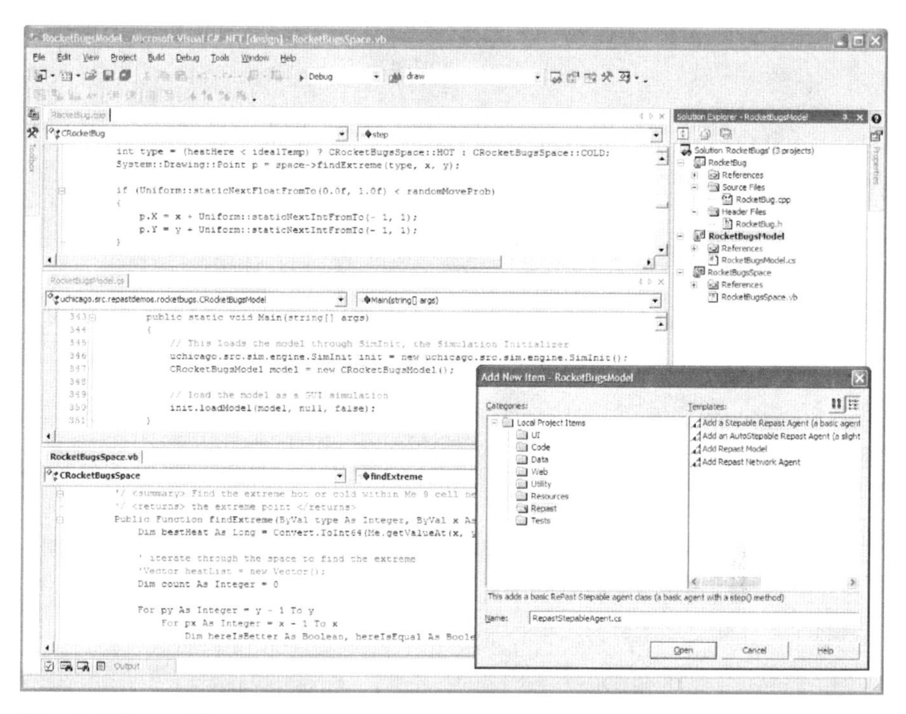

Fig. 2.6 Repast.Net in Microsoft Visual Studio.Net.

The Repast 3 system has four fundamental components, as shown in Fig. 2.7. The components are the simulation engine, the input/output (I/O) system, the user interface, and the support libraries. Each of these components is implemented in the core layer and is accessed by the user in the external layer. A Unified Modeling Language (UML) diagram showing the relationships between these components is presented in Fig. 2.8. Information on UML notation can be found in Booch [5].

The Repast 3 simulation engine is responsible for executing simulations. The Repast engine has four main parts, namely the scheduler, the model, the controller, and the agents. The relationship among these components is indicated in Figs. 2.7 and 2.8 and is discussed later in this section.

The Repast 3 scheduler is a full-featured discrete event scheduler. Simulations proceed by popping events or "actions," as they are called in Repast, off an event queue and executing them. These actions are such things as "move all agents one cell to the left," "form a link with your neighbor's neighbor," or "update the display window." The model developer determines the order in which these actions execute relative to each other using ticks. As such, each tick acts as a temporal index for the execution of actions. For example, if event X is scheduled for tick 3, event Y for tick 4, and event Z for tick 5, then event Y will execute after event X and before event Z. Actions scheduled for execution at the same tick will be executed with a simulated concurrency. In

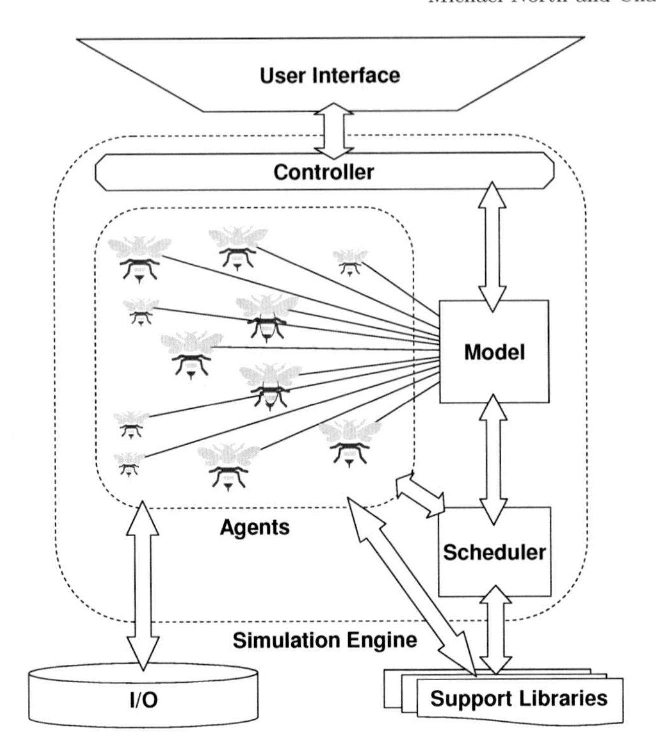

Fig. 2.7 Repast overview diagram.

this way, the progression of time in a simulation can be seen as an increase in the tick count.

The Repast 3 scheduler includes full support for concurrent task execution. Tasks become concurrent when actions are given both a starting time and duration. When durations are specified, actions that can be started in the background are run concurrently. Actions with nonzero durations will run concurrently with other actions with compatible tick counts as well as block the execution of other actions with higher tick counts until the current action is completed. For example, consider a process that contains some long-running and complicated behavior that can be started at time t with results needed at $t + 5$. Imagine that there are actions that can be run concurrently over time t to $t+5$. This behavior can be modeled as an action with a five-tick duration. In terms of implementation, this action will run in its own thread that is amenable to being run on a separate processor or even on another computer. This allows the natural introduction of complex concurrent and parallel task execution into Repast simulations. Since durations are optional, modelers can begin by creating sequential simulations and then introduce concurrency as needed.

Repast 3 schedulers are themselves actions that can be recursively nested following the composite design pattern [16]. This allows a Repast action to be

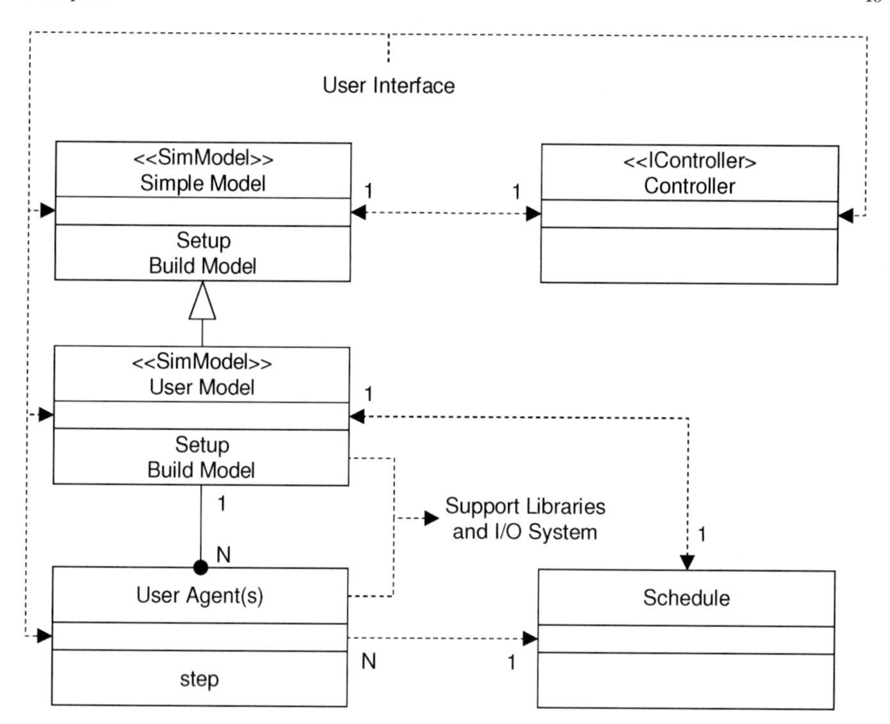

Fig. 2.8 Core Repast UML diagram.

as complex as needed for a given application. It even allows advanced multi-scale simulations to be constructed by combining existing models such that the full schedules of lower-level models run as simple actions in higher-level models.

Repast 3 models contain the definition of the simulation to be run by the scheduler. Repast models include the list of agents to be executed, the simulation initialization instructions, and the user interface specification.

Repast 3 controllers connect models and schedulers. They activate the selected model and then manage the interactions between the user or batch execution system and the model.

Repast 3 agents are created by users from components within Repast. A variety of options are available, including geographically situated agents and network-aware agents. Agents receive data from, and provide results to, the Repast I/O system.

The Repast 3 I/O system allows agents to be created based on input properties. It also can store data from both agents and overall models. Repast includes a set of results loggers that support a range of storage formats.

The Repast 3 user interface supports the display of model results and allows user to interact with running models. Repast user interface examples are shown in Figs. 2.1, 2.3, 2.5 and 2.9. Model user interfaces can include graphical outputs or maps of the agent states as well as interactive probes that allow users to view and modify agent states. An agent map is shown in the lower left of Fig. 2.3 and an agent probe is shown in the upper left. Users have full control over what is available through both maps and probes.

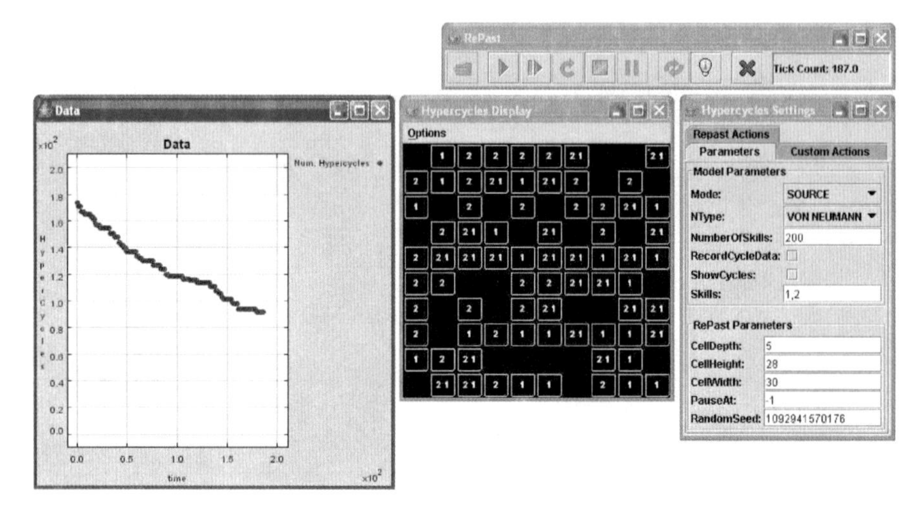

Fig. 2.9 Padgett, Lee, and Collier's RepastJ Hypercycle model.

The Repast 3 support libraries include a variety of tools for both mathematics and modeling. The mathematics support includes a range of random distribution generators and statistical aggregation tools commonly found in all kinds of simulation toolkits [25]. The modeling support includes genetic algorithms and neural network tools among other features [22].

2.2.2 Repast Simphony

Repast Simphony (Repast S) represents a substantial advance in artificial life modeling, and agent-based modeling in general, compared to previous technologies. As described in [21], Repast S builds upon the Repast 3 model development approach by introducing the following model creation process:

- The modeler creates model pieces, as needed, in the form of plain old Java objects (POJOs), often using automated tools or scripting languages such as Groovy [23]. An example agent behavior flowchart is shown in Fig. 2.10. The contents of the flowchart are automatically compiled to Groovy source code and then to Java bytecode.
- The modeler uses declarative configuration settings to pass the model pieces and legacy software connections to the Repast S runtime system.

- The modeler uses the Repast S runtime system to declaratively tell Repast S how to instantiate and connect model components.
- Repast S automatically manages the model pieces based on both interactive user input and declarative or imperative requests from the components themselves.

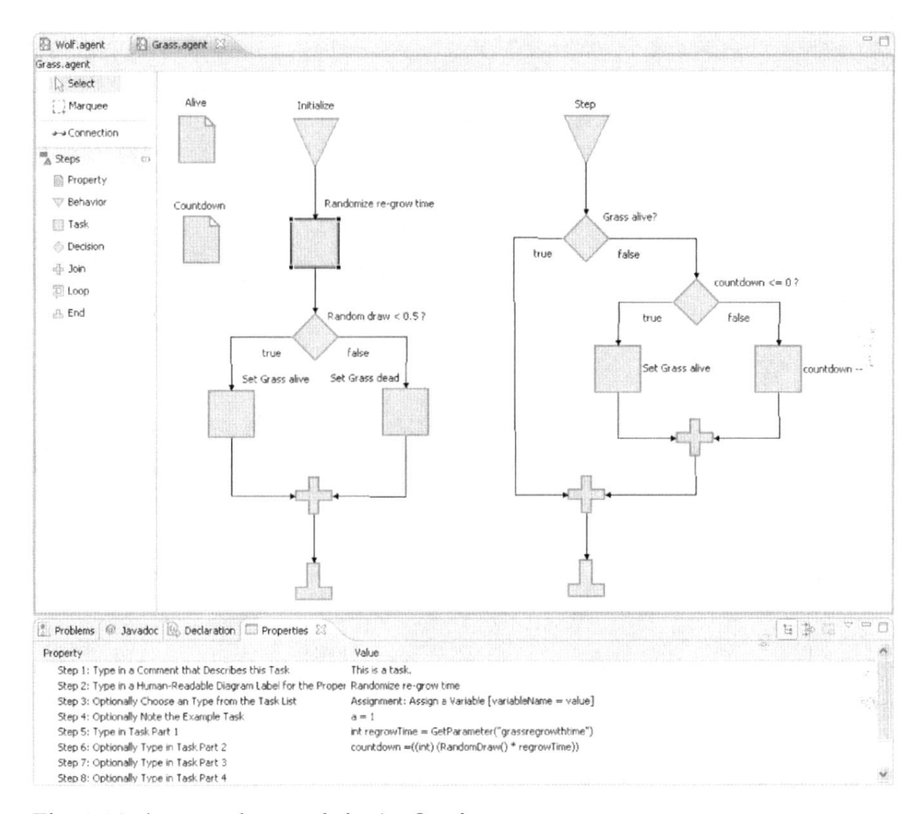

Fig. 2.10 An example agent behavior flowchart.

The POJO model components can represent anything but are most commonly used to represent the agents in the model. Although the POJOs can be created by using any method, this chapter discusses one powerful way to create POJOs for Repast S: the Repast S development environment. However, modelers can use any method – from hand coding to wrapping binary legacy models to connecting into enterprise information systems – to create the Repast S POJO model components.

Regardless of the source of the POJOs, the Repast S runtime system is used to configure and execute Repast S models. The Repast S runtime system includes the following:

- Point-and-click model configuration and operation;

- Integrated two-dimensional, three-dimensional (3D), and other views (see Fig. 2.11 for an example 3D GIS view);
- Automated connections to enterprise data sources;
- Automated connections to powerful external programs for conducting statistical analysis and visualizing model results.

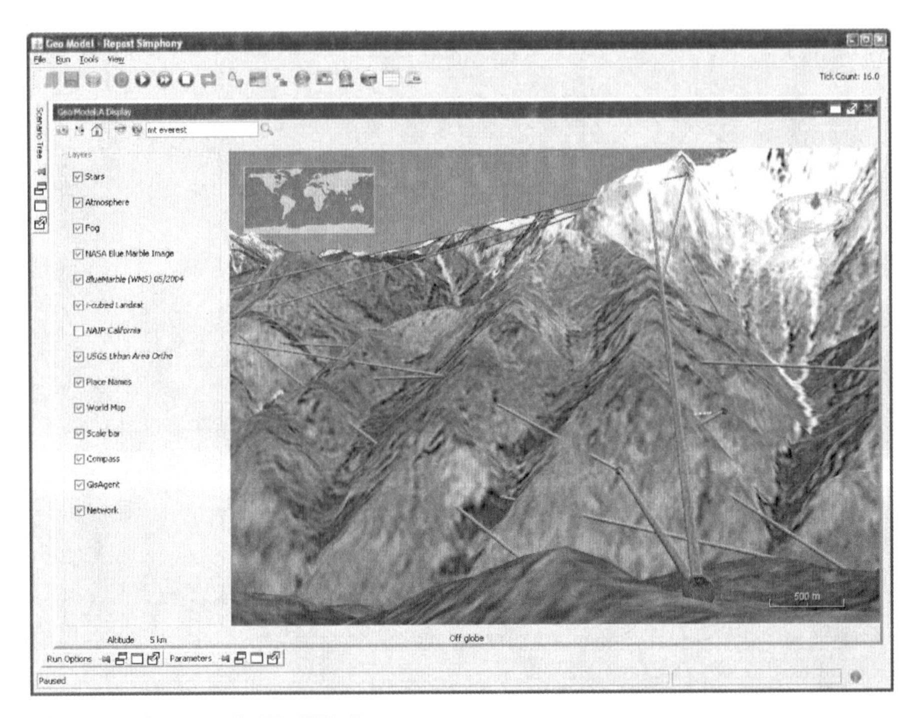

Fig. 2.11 An example 3D GIS view.

2.2.3 Using Repast 3 and Repast Simphony

As previously mentioned, all versions of Repast are distributed under a variation of the BSD license [38]. This license states that Repast can be used for virtually any purpose without fees and without a requirement to release propriety model source code. See ROAD for details [38]. This license allows Repast to be freely used in education, research, and entertainment by non-profit, government, and commercial organizations.

Many educational institutions are now using or have used Repast for either education or research. These institutions include the University of Chicago, the University of Michigan, Iowa State University, the Swiss Federal Institute of Technology Zurich, the Illinois Institute of Technology, and Harvard University. In particular, the University of Chicago is the birthplace of Repast. The educational uses generally focus on providing students with a laboratory environment for experiments with complex systems and for instructing students on agent-based modeling concepts. The research work includes the development of models in a variety of domains as well as model-theoretic studies. Several of the models are discussed in the following sections. The model-theoretic work mostly involves additions to and extensions of the Repast framework itself. This list of educational institutions using Repast is rapidly growing.

A significant number of U.S. federal government agencies and other organizations are using or have used various versions of Repast. These users have Repast models that focus on a range of mission-critical applications such as infrastructure security and network communications planning.

Several commercial organizations are working with Repast. These organizations include software developers such as ESRI and other private organizations. These corporations are using Repast for several purposes, including strategic planning and commercial software enhancement.

2.3 Repast Artificial Life Models

Agent-based modeling has been used in an enormous variety of applications to social systems, covering human as well as nonhuman systems. Applications range from modeling ancient civilizations that have been gone for hundreds of years, to modeling how to design new markets that do not currently exist, such as space tourism and digital news. Selected applications are listed in Table 2.1 for the period 2004 to 2008. Earlier models can be found in the first edition of this book [35]. All of the applications listed acknowledge using Repast as the underlying agent-based modeling toolkit.

Several of the works contend that using agent-based modeling versus other modeling techniques is necessary because agent-based models can explicitly model the complexity arising from individual actions and interactions that exist in the real world. All of the works support the use of Repast as the tool of choice based on reasons having to do with usability, ease of learning, cross-platform compatibility, and its sophisticated capabilities to connect to databases, graphical user interfaces, and GISs.

Griffin and Stanish [17] developed an agent-based model, using Repast, for the Lake Titicaca basin of Peru and Bolivia covering the late prehistoric period, 2500 BC to AD 1000. The model was used to study hypotheses for the causal variables affecting prehistoric settlement patterns and political consolidations. The model's geo-spatial structure consists of a 50,000 km^2 grid composed of 1.5-km square cells. Each cell modeled the geography, hydrology, and agricultural potential. Agents consist of settlements, peoples,

Application area	Model description and reference
Air Traffic Control	An agent-based model of air traffic control to analyze control policies and performance of a capacity constrained air traffic management facility (Conway [10]).
Anthropology	An agent-based model of prehistoric settlement patterns and political consolidation in the Lake Titicaca basin of Peru and Bolivia (Griffin and Stanish [17]). An agent-based model of linguistic diversity (de Bie and de Boer [4]).
Ecology	Agent-based model of predator–prey relationships between transient killer whales and other marine mammals (Mock and Testa [29]). Aphid population dynamics in agricultural landscapes using agent-based modeling (Parry, et al. [37]).
Energy Analysis	An agent-based model for scenario development of offshore wind energy (Mast et al. [28]). Agent-based model of residential energy generation (Houwing and Bouwmans [20]).
Epidemics	An agent-based computer model to retrospectively simulate the spread of the 1918–1919 influenza epidemic through the small fur-trapping community of Norway House in Manitoba, Canada (Carpenter [6]).
Marketing	An agent-based simulation to model the possibilities for a future market in sub-orbital space tourism (Charania et al. [7]). A multi-agent based simulation of news digital markets (López-Sánchez et al. [26]). An agent-based model of Rocky Mountain tourism (Yin [45]). An agent-based computational economics model using Repast to study market mechanisms for the secondary use of the radio spectrum (Tonmukayakul [41]).
Organizational Decision Making	Agent based modeling approach to allow negotiations in order to achieve a global objective, specifically for planning the location of intermodal freight hubs (van Dam et al. [11]). An evaluation framework for supply chains based on corporate culture compatibility using agent-based modeling (Al-Mutawah and Lee [1]). Emergency response (Narzisi et al. [31]).
Social Science	Simulating the process of social influence within a population, using dynamic social impact theory (Wragg [44]).

Table 2.1 Selected recent Repast applications

political entities, and leaders that interact with each other and with the environment (grid). Agent behavior is modeled as a set of condition-action rules that are based on hypothesized causal factors affecting agriculture, migration, competition, and trade processes. The authors report that through a series of simulation runs, the model produced a range of alternative political prehistories and the emergence of macro-level patterns that corresponded to observed patterns in the archaeological record. Simulation results provided insights into region-wide political consolidation.

De Bie and de Boer [4] developed an agent-based model of linguistic diversity implemented in Repast. The authors model language diversity as the result of language mutation. Agents adopt mutations from other agents based on social impact theory, in which less common varieties of language have a relatively greater influence per individual speaking the variety. Using the model, they demonstrate how different language patterns can exist at the same time.

Carpenter [6] constructed an agent-based model with Repast to retrospectively simulate the spread of influenza in the 1918–1919 pandemic through the small fur-trapping community of Norway House in Manitoba, Canada. According to the author, historical information about the influenza pandemic was used to create a computer simulation that could be manipulated in ways that would not be possible, or ethical, in real life. Carpenter contends that by using agent-based modeling, an artificial landscape can be populated with heterogeneous agents who move and interact in ways that more closely resemble human behavior than is possible to do using other modeling techniques. The model was used to address research questions on the influence of changes in population movement patterns on the transmission of disease, specifically whether the seasonal population movements could influence the spread of the influenza through the community and whether a winter epidemic could differ from a summer epidemic predominantly due to changes in seasonal population movement.

Charania et al. [7] used agent-based simulation and Repast to model possible futures for a market in sub-orbital space tourism. Each agent is a representation of an entity within the space industry, such as consumers, producers, the government, etc., that provides or demands different products and services. Tourism companies seek to maximize profits while they compete with other companies for sales. Individual companies decide the price they will charge for a flight aboard their vehicle. Customers evaluate the products offered by the companies according to their individual tastes and preferences.

López-Sánchez et al. [26] developed a multi-agent-based simulation of news digital markets called SimwebAB, using Repast. Their approach is to adapt traditional business models to the new market. They use the model to investigate the dynamics of the new market and gain insights into how to exploit the impending paradigm shift in news contents, marketing, and distribution. The authors contend their model is useful for informing business strategy decisions.

Yin [45] used Repast to develop an agent-based model of Rocky Mountain tourism and applied it to the town of Breckenridge, Colorado. The model was used to explore how homeowners' investment and reinvestment decisions are influenced by the level of investment and amenities available in their neighborhoods. The dynamics and indirect spatial impacts of amenity-led mountain tourism on development were explored. The author found that individual levels of appreciation of amenities and continuing investment in a neighborhood attracted investment and reinvestment and created pressure for high-density resort housing development at the aggregate level.

Tonmukayakul [41] developed an agent-based computational economics model using Repast to study market mechanisms for the secondary use of the radio spectrum. Secondary use is the temporary access of the existing licensed spectrum by users who do not own a spectrum license. Using transaction cost economics as the theoretical framework, the objective of the model was to identify the preconditions for when and why the secondary use market could emerge from the repeated interactions of agents in the simulated market and what form it might take. Understanding the dynamics for this hypothetical market could lead to policy instruments to more effectively manage the spectrum.

Al-Mutawah and Lee [1] developed an agent-based supply chain model for evaluating supply chain corporate culture compatibility, using Repast. The model focuses on a three-level supply chain. The model integrates the framework for cultural learning to evaluate the management performance of the supply chain under different scenarios and assumptions.

Wragg [44] used agent-based modeling and Repast to simulate the process of social influence within a population, using dynamic social impact theory. The motivation was to understand the social processes and power dynamics of local populations in response to recent military operations in Afghanistan and Iraq. The author contends that the simulations reproduced the expected characteristics of social influence, such as opinion clustering, opinion polarization, minority opinion decay, and, more generally, the nonlinearity of public opinion change. The author highlights the need for accurate data concerning a population's social hierarchy, social networks, behavior patterns, and human geography. These data are essential for determining the impacts of word-of-mouth and mass-media-driven information campaigns on the population.

Narzisi et al. [31] developed a agent-based disaster simulation framework, using Repast, to simulate catastrophic, emergency scenarios that required optimized emergency response. Incidents included the release of chemical agents, bomb explosions, food poisonings, and small pox outbreaks. The model includes large numbers of agents in several categories, including individuals in the population, hospitals, ambulances and on-site responders.

Conway [10] used Repast to build an agent-based model of air traffic control. The model was used to analyze the effectiveness of control policies for a capacity-constrained air traffic management facility. The model is used to address situations where capacity is overburdened and there is a potential for resultant delays to propagate throughout the flight schedule. The model includes representations of air traffic system attributes such as system capacity, demand, airline schedules and strategy, and aircraft capability.

Van Dam et al. [11] applied an agent-based modeling approach to modeling the negotiation process. The objective was to investigate the conditions under which a global objective could be achieved by individual decision makers acting in their own interests. The model incorporates the agent's decision-making process for planning the location of intermodal freight hubs.

Mast et al. [28] developed an agent-based model using Repast for investigating scenarios for developing offshore wind energy resources in the Netherlands. A simple model was developed that includes the actors involved in the supply chain, represented at a relatively high level of aggregation. The model was used to investigate opportunities and threats to this emerging industry.

Houwing and Bouwmans [20] developed an agent-based model of residential energy generation to investigate how distributed energy resources will contribute to the European Union's stated policy goals of (1) energy market liberalization and (2) decreasing environmental impacts from energy use. The modeling study focused on residential power and heat generation through the use of micro-combined heat and power units; options for heat storage were modeled. The authors estimated the impact of residential power units on household energy flows, energy costs, and CO_2 emissions. The model results show that the operational impacts are highly dependent on the control mode adopted for heating and power units. Agent-based modeling proved useful in modeling both the individual technology units and the individual decision making behaviors of the consumers to operate the units.

Mock and Testa [29] use Repast to develop an agent-based model of predator–prey relationships between transient killer whales and threatened marine mammal species in Alaska. Threatened species include sea lions and sea otters. The authors state that previously only simplistic, static models of killer whale consumption had been constructed due in part to the fact that the interactions between transient killer whales and their marine mammal prey are poorly suited to classical predator–prey modeling approaches such as those based on the Lotka–Volterra differential equation framework. This killer whale model is an agent-based model at both the individual and hunting group levels. Individual agents eat, grow, reproduce, and die. Hunting groups change in size and composition while encountering other marine mammals.

Parry et al. [37] modeled the dynamics of aphid populations in agricultural landscapes using a spatially explicit agent-based simulation model and Repast. Aphid agents interact with one another and with the landscape environment over time. The model is heavily parameterized and coupled to a GIS for geo-spatial realism. The authors demonstrate that a spatial model that explicitly considers environmental factors (e.g., landscape properties, wind speed and direction, etc.) provides greater insight into aphid population dynamics over spatial and temporal dimensions than other modeling approaches.

2.4 Conclusions

Artificial life focuses on synthesizing "life-like behaviors from scratch in computers, machines, molecules, and other alternative media" [24]. Artificial life expands the "horizons of empirical research in biology beyond the territory currently circumscribed by life-as-we-know-it" to provide "access to the domain of life-as-it-could-be" [24]. Agent-based modeling and simulation are used to create computational laboratories that replicate selected real or potential behaviors of actual or possible complex adaptive systems. Agent-based models can be used to escape the accident of history in the form of "life-as-we-know-it" by revealing alterative forms of "life-as-it-could-be."

Repast is a family of free and open-source agent modeling toolkits. Repast's features directly support the implementation of models with Holland's three properties and four mechanisms of complex adaptive systems [19]. As such, Repast 3 and Repast Simphony are natural frameworks in which to perform artificial life experiments. More information on the use of Repast and other modeling tools can be found in North and Macal [34].

Repast has many academic, government, and industry users. These users are involved in a variety of application areas, including educational, research, and commercial uses. In particular, there are many examples in which Repast has been used extensively for artificial life applications in topical areas such as artificial evolution and ecosystems, artificial societies, and artificial biological systems.

References

1. Al-Mutawah K and Lee V (2008) An Evaluation Framework for Supply Chains Based on Corporate Culture Compatibility. In: Supply Chain, Theory and Applications, Kordic V (ed.) pp. 59–72, I-Tech Education and Publishing, Vienna, Austria.
2. Archer T (2001) Inside C#. Microsoft Press, Redmond, Washington.
3. Beck K and Gamma E (1998) Test infected: Programmers love writing tests. Java Report 3:37–50.
4. de Bie P and de Boer B (2007) An Agent-Based Model of Linguistic Diversity. In: Proc. ESSLLI 2007 Workshop on Language, Games, and Evolution, Benz A, Ebert C and van Rooij R (eds.), pp. 1–8, Available online at http://frim.frim.nl/Dublin.pdf.
5. Booch G (1993) Object-oriented Design with Applications. Addison-Wesley, Reading, MA.
6. Carpenter, C., 2004, Agent-Based Modeling of Seasonal Population Movement and the Spread of the 1918–1919 Flu: The Effect on a Small Community, University of Missouri-Columbia, Master's Thesis, Department of Anthropology.
7. Charania AC, Olds JR, and DePasquale D (2006) Sub-Orbital Space Tourism Market: Predictions of the Future Marketplace Using Agent-Based Modeling, SpaceWorks Engineering, Inc., Atlanta, GA, Available online at http://www.sei.aero/uploads/archive/IAC-06-E3.4.pdf.
8. Cloyer A, Clement A, Bodkin R, and Hugunin J (2003) Practitioners report: Using aspectJ for component integration in middleware. In: Companion of the 18th Annual ACM SIGPLAN Conference on Object-Oriented Programming, Systems, Languages, and Applications. R. Crocker and G. Steele Jr. (eds.) ACM, New York.
9. Collier N, Howe T, and North M (2003) Onward and upward: The transition to Repast 2.0. In: Proc. of the 1st Annual North American Association for Computational Social and Organizational Science Conference, Electronic Proceedings. Pittsburgh, PA.
10. Conway SR (2006) An Agent-Based Model for Analyzing Control Policies and the Dynamic Service-Time Performance of a Capacity-Constrained Air Traffic

Management Facility, ICAS 2006 – 25th Congress of the International Council of the Aeronautical Sciences Hamburg, Germany, 3–8 September 2006. Available online at http://ntrs.nasa.gov/archive/nasa/casi.ntrs.nasa.gov/20060048296_2006250468.pdf.
11. van Dam KH, Lukszo Z, Ferreira L, and Sirikijpanichkul A (2007) Planning the Location of Intermodal Freight Hubs: An Agent Based Approach, In: Proceedings of the 2007 IEEE International Conference on Networking, Sensing and Control, pp. 187–192, London, UK, 15–17 April 2007.
12. Di Paolo E (2004) Unbinding biological autonomy: Francisco Varela's contributions to artificial Life. Journal of Artificial Life, Vol. 10, Issue 3, 231–234.
13. Eclipse Home Page (2008) http://www.eclipse.org/.
14. Elrad T, Filman R, and Bader A (2001) Aspect-oriented programming: Introduction. Communications of ACM 44:29–32.
15. Foxwell H (1999) Java 2 Software Development Kit. Linux Journal. Specialized Systems Consultants, Seattle, Washington.
16. Gamma E, Helm R, Johnson R, and Vlissides J (1995) Design Patterns: Elements of Reusable Object-Oriented Software. Addison-Wesley, Reading, MA.
17. Griffin AF and Stanish C (2007) An agent-based model of prehistoric settlement patterns and political consolidation in the Lake Titicaca Basin of Peru and Bolivia, structure and dynamics: eJournal of Anthropological and Related Sciences, 2(2). Available online at http://repositories.cdlib.org/imbs/socdyn/sdeas/vol2/iss2/art2.
18. Gülcü C (2003) The Complete Log4j Manual: The Reliable, Fast, and Flexible Logging Framework for Java. QOS.ch, Lausanne, Switzerland
19. Holland J (1996) Hidden Order: How Adaptation Builds Complexity. Addison-Wesley, Reading, MA.
20. Houwing M and Bouwmans I (2007) Agent-Based Modelling of Residential Energy Generation with Micro-CHP, Delft University of Technology, Available online at http://wiki.smartpowersystem.nl/images/d/dc/M_Houwing&I_Bouwmans_Napa2006_FIN.pdf.
21. Howe TR, Collier NT, North MJ, Parker BT, and Vos JR (2006) Containing Agents: Contexts, Projections, and Agents. In: Proceedings of the Agent 2006 Conference on Social Agents: Results and Prospects, Argonne National Laboratory, Argonne, IL.
22. Java Object Oriented Neural Engine (Joone) Home Page (2004) http://www.jooneworld.com/
23. König D, Glover A, King P, Laforge G, and Skeet J (2007) Groovy in Action. Manning Publications, Greenwich, CT.
24. Langton C (1994) What is Artificial Life?, The Digital Biology Project. Available at http://www.biota.org/papers/cglalife.html.
25. Law AA (2007) Simulation Modeling and Analysis. 4th ed. McGraw-Hill, New York.
26. López-Sánchez M, Noria X, Rodriguez JA, and Gilbert N (2005) Multi-agent based simulation of news digital markets. International Journal of Computer Science & Applications, II(I). Available online at http://www.tmrfindia.org/ijcsa/v21.html.
27. Lutz M and Ascher D (1999) Learning Python. O'Reilly Press, Sebastopol, CA.
28. Mast EH, van Kuik GAM, and van Bussel GJW (2007) Agent-Based Modelling for Scenario Development of Offshore Wind Energy, T. Chaviaropoulos (ed.), Proceedings of the 2007 European Wind Energy Conference & Exhibition in Milan, pp. 1–4, Brussels, EWEA.
29. Mock KJ and Testa JW (2007) An Agent-Based Model of Predator–Prey Relationships between Transient Killer Whales and Other Marine Mammals, University of Alaska Anchorage, Anchorage, AK, May 31, 2007. Available online at http://www.math.uaa.alaska.edu/~orca/.
30. Mozart Consortium: Mozart Programming System 1.3.1 (2004). Available online at http://www.mozart-oz.org/.

31. Narzisi GV, Mishra B (2006) Multi-Objective Evolutionary Optimization of Agent-Based Models: An Application to Emergency Response Planning, New York University, Available online at http://www.cs.nyu.edu/mishra/PUBLICATIONS/06.ci06PlanC.pdf.
32. NCSA, HDF 5 Home Page (2004) http://hdf.ncsa.uiuc.edu/HDF5/.
33. North M, Collier N, and Vos R (2006) Experiences creating three implementations of the Repast agent modeling toolkit. ACM Transactions on Modeling and Computer Simulation, 16(1):1–25.
34. North M and Macal C (2007) Managing Business Complexity: Discovering Strategic Solutions with Agent-Based Modeling and Simulation, Oxford University Press, New York.
35. North MJ and Macal CM (2005) Escaping the Accidents of History: An Overview of Artificial Life Modeling with Repast, In: Adamatzky A and Komosinski M (eds.). Artificial Life Models in Software, 1st ed., pp. 115–141, Springer, Heidelberg.
36. North MJ, Tatara E, Collier NT, Ozik J (2007) Visual Agent-based Model Development with Repast Simphony. In: Proceedings of the Agent 2007 Conference on Complex Interaction and Social Emergence, Argonne National Laboratory, Argonne, IL.
37. Parry H, Evans AJ and Morgan D (2004) Aphid Population Dynamics in Agricultural Landscapes: An Agent-Based Simulation Model, International Environmental Modelling and Software Society iEMSs 2004 International Conference University of Osnabrück, Germany, 14–17 June 2004. Available online at http://www.iemss.org/iemss2004/pdf/landscape/parraphi.pdf.
38. ROAD: Repast 3.0 (2004) http://repast.sourceforge.net/.
39. Sandler T (2001) Economic Concepts for the Social Sciences. Cambridge University Press, Cambridge.
40. Swarm Development Group: Swarm 2.2 (2004) http://wiki.swarm.org/.
41. Tonmukayakul A (2007) An Agent-Based Model for Secondary Use of Radio Spectrum. Ph.D. thesis, University of Pittsburgh, School of Information Sciences.
42. Van Roy P and Haridi S (2004) Concepts, Techniques, and Models of Computer Programming. MIT Press, Boston, MA.
43. Walker R, Baniassad E, and Murphy G (1999) An initial assessment of aspect-oriented programming. In: Proc. 1999 Int. Conf. Software Engineering. IEEE, Piscataway, NJ, pp. 120–135.
44. Wragg T (2006) Modelling the Effects of Information Campaigns Using Agent-Based Simulation, DSTO Defence Science and Technology Organisation, Edinburgh South Australia, DSTO-TR-1853.
45. Yin L (2007) Assessing indirect spatial effects of mountain tourism development: An application of agent-based spatial modeling. Journal of Regional Analysis & Policy 37(3):257–265. Available online at http://www.jrap-journal.org/pastvolumes/2000/v37/F37-3-8.pdf.

Chapter 3
Sodarace: Continuing Adventures in Artificial Life

Peter W. McOwan and Edward J. Burton

3.1 The Sodarace Project

Much like life, the Sodarace project [8] continues to defy easy description; see [9] and [15]. As the project matures with a new raft of software, the story develops further fascinating twists. Originally developed as an online Olympics pitching human against machine intelligence, the project has expanded to incorporate an impressive range of science- and arts-based activities. Sodarace uses creative play as a bridge to foster dialogue and shared awareness between two very different audiences: a broad public of learners, both in and out of school, and the artificial intelligence and artificial life research communities. The Sodarace project is an extension of the Sodaconstructor software, an online construction kit comprising masses, springs, and muscles, providing tools for open-ended discovery and exploration. Although the mechanics are simple, the forms created often have a very life-like appearance and gait and, as such, are frequently anthropomorphized by users.

For Sodarace, the most recent focus has been to create new software that makes available powerful, but simple to use, applications allowing users to actively engage with, and experience themselves, the process of digital evolution. In addition, central to this new framework is a continued wish to foster every opportunity for cross-fertilisation between disciplines, combined with applications supporting a multitude of potential user interests. This is achieved by placing together, for the first time on a single web site, all of the various incarnations of the Soda created family of free-to-use physics and artificial intelligence simulation software. Access to this software is combined with the user support forums and an easy to use and learn from repository containing the myriad creations built and explored by using these applications. This new centralised software resource [7] draws together all of the activities and enhances the educational goals and future sustainability of the project.

3.2 Introduction

Sodarace comprises a flexible environment for races between virtual robots, either built by the public with the Sodaconstructor interface and learning support from the Sodarace community forums or created by artificial intelligences using a new suite of software that focuses on evolutionary algorithms to introduce the public to the basics of Artificial Intelligence. These intuitive tools combined with an active user forum and the provocative Humans vs. Machines narrative have proved to be both popular, educational, and fun [9].

3.2.1 Sodarace: The Story Begins

Officially, the Sodaconstructor was released in March 2000. It started 10 years earlier with a simple version, programmed in Basic, and developed by Burton, which was later updated to Java when he joined the team of digital artists at Soda Creative in London. The applet was placed online as an element of the Play section content on Soda web site, where over a short period of time its clear design, high quality interactivity, and addictive open-ended problem-solving play made it extremely popular. Sodaconstructor exploded across the Internet through e-mails, newsgroup postings, and other such informal "word-of-mouth" advertising, the perfect example of viral marketing. By the end of the month there were one million constructors using the applet each week.

In September 2000, the applet was redeveloped in order to allow Sodaconstructions to be saved, and from this the Sodazoo was developed, where creations could be sent and the best archived for all to see and learn from. It was at this stage that the educational merit of the project and its natural symbiosis with artificial life research began to crystallize, as a number of research groups around the world contacted Soda, interested in exploring this application of the software.

In April 2002, the Sodarace project was launched, supported by funds from the UK Engineering and Physical Sciences Research Council (EPSRC). Core to Sodarace's construction and subsequent development were the Sodarace community forums, where constructors and artificial intelligence researchers could interact and get involved in developing the specifications for the new software. This proved an extremely worthwhile experience for those involved, as it gave a transparent view into the development of a major software project while also ensuring that user requirements were met. The first human versus machine races occurred in December 2003. It took the form of a race over flat terrain with simple wheel-like creations, named amoebas by the community. The artificial intelligence technique of genetic algorithms was used to optimise the parameters of the wheel to cover the racetrack in a competitive time. For the record, the humans won the first race, but only just!

Since then the project has grown both in profile and, perhaps as importantly, in accessibility and future sustainability. The new framework was designed to focus on the specific theme of digital evolution, as this had proved to be the most popular and accessible form of artificial intelligence algorithm explored in the earlier stages. It also moved to develop a centralised 'one-stop shop', to facilitate more of the interdisciplinary and peer-to-peer interactions that the project had shown to be successful in the past. This new software was launched in 2005 by an event at the UK Royal Society Summer Exhibition. There followed a UK-wide tour of science centres and festivals of new stand-alone kiosk software, which encapsulated the interactive, educational and playful nature of the Sodarace project and greatly enhanced the national profile of the project.

3.2.2 Previous Work

Sodarace makes no attempt to simulate more than two-dimensional (2D) physical interactions between the mass, spring, and muscle racers and the terrain. It is within this world that both artificial intelligence and humans compete and evolve. However, as will be discussed later, these constraints do not significantly hinder user creativity. An obvious related artificial life project is Framsticks [6] described in Chapter 5, which allows the study of evolution capabilities of three-dimensional (3D) creatures in simplified Earth-like conditions. The creatures in this case also have genotype representations of their physical body, sensors, neural network, and effectors.

Karl Sims' earlier work in visualising simulated block creatures performing evolved behaviours was also an inspiration for the Sodarace project [12, 11]. Sims' research project involved simulated Darwinian evolutions, using genetic algorithms, to develop again 3D virtual block creatures. An initial random population of several hundred creatures is created, and each creature is then tested for its ability to perform a given task, such as the ability to swim in a simulated water environment. The fittest creatures for the task are then passed into the next generation and mutations are applied to produce a new population. The new creatures are again tested, and through the mutation process, some may be conferred with improvements on their parents' abilities. As this iterative cycle of variation and selection continues, creatures with more and more successful niche behaviours can emerge.

Sodarace builds on these previous works but reduces the dimensionality and complexity of the simulation in an effort to increase interaction and ease of learning. In addition, the central driving presence of the discussion forums provides both an evolving archive of human and artificial intelligence design activity and educational support.

3.3 Sodarace, the Scientific Background

3.3.1 Sodaconstructor: The Physics Engine of Sodarace

Sodarace is based on the popular Sodaconstructor software [9], which allows the simulation of structures comprised of a linked set of node masses and interconnecting springs, simulated using a simple implementation of Newtonian and Hook's law mechanics. To provide movement, constructors may select springs to drive in simple harmonic motion (SHM). These are called 'muscles'. The interface also provides an intuitive graphical representation of the phase relations of the driven muscle elements through a moving sine wave. The muscles are mapped to appropriate positions on the wave generator to fix their relative oscillation phases; see Fig. 3.1. By connecting both static and appropriately phased SHM-driven elements together, it is possible to create a whole menagerie of lively perambulating creations. The amplitude, initial wave phase offset, and speed of the driving sinusoidal wave may also be controlled. The new upgrade of the Constructor software, finally released in 2007 after extensive discussions with the user community, was an enhanced redesign that builds upon the previous functionality but, in addition, allows users to copy, cut, and paste elements of their creations and to accurately position masses, and set spring rest lengths and muscle phase differences. See Fig. 3.2.

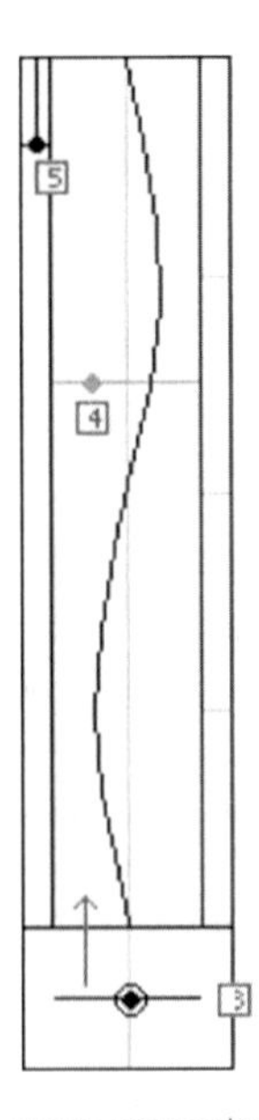

Fig. 3.1 The format of the 'muscle' wave generator, used to drive the creatures.

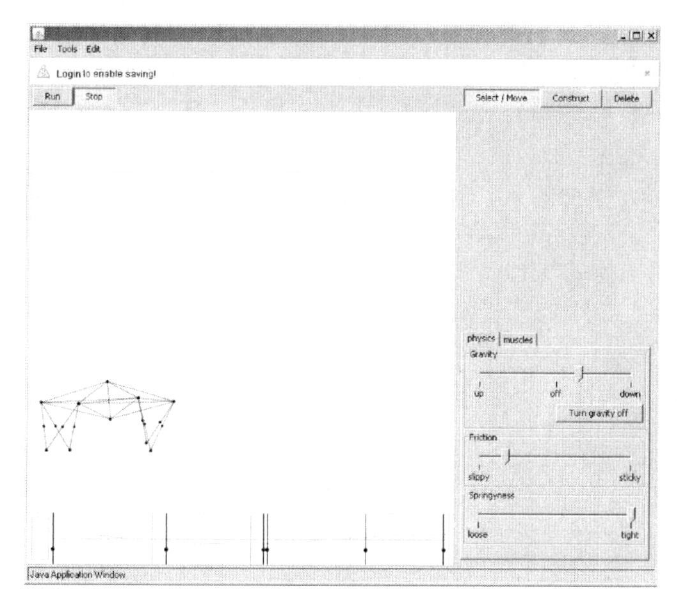

Fig. 3.2 The new Sodaconstructor interface, displaying Sodarace mascot Daintywalker, showing the additional design functionality.

3.3.2 Sodarace: The Racing Environmental Variables

To enable terrain construction, essentially the ground for the racers to race on, a simple geometric collision detection algorithm is used. In effect, these Sodarace 'bar springs' do not allow other structures to pass through them – they can be used to construct terrain that models can walk over; by default, these collisions are 'sticky,' meaning they lose energy to friction, and this can be changed to 'slippery' for friction-free collisions. The racers are held in the form of XML files (Fig. 3.3), which allows the manual editing of components but also allows the computer generation of these files.

Access to the XML also allows interested users to make the connection between a stored data structure and what is seen on the screen. Often the inner workings of software seems mysterious to a general audience. Allowing users this direct examination of 'computer code', and the ability to play and explore with it, facilitates awareness of one of the basic elements of computer programming.

One user writes in the forum

> I was doing some experiments with the xml.

Another user takes their exploration further:

```
<?xml version="1.0" encoding="ISO-8859-1" standalone="no"?>
<!DOCTYPE sodaconstructor>
<model>
 <comment></comment>
 <container width="651" height="422"/>
 <environment gravity="0.3" friction="0.05" springyness="0.2"/>
 <collisions surface_friction="0.1" surface_reflection="-0.75"/>
 <wave amplitude="0.5" phase="0.0" speed="0.001"/>
 <settings gravitydirection="down" wavedirection="forward"
  autoreverse="on"/>
<nodes>
 <mass id="m0" x="254.0" y="291.0" vx="0.0" vy="0.0"/>
 <mass id="m1" x="332.0" y="194.0" vx="0.0" vy="0.0"/>
 <mass id="m2" x="259.0" y="157.0" vx="0.0" vy="0.0"/>
</nodes>
<links>
 <spring a="m0" b="m1" restlength="124.0"/>
 <muscle a="m1" b="m2" restlength="82.0" amplitude="0.5"
  phase="0.5"/>
</links>
</model>
```

Fig. 3.3 An example of the structure of an XML file.

> I have just done a few experiments in XML. I found the “speed limit” to be around 10. Also, I found that you can break the speed limit by setting friction to a negative number. I made a mass go around with this method. Also, I found that by setting K to a negative number, springs avoid their rest length. If the spring is compressed below its restlength, it will act similar to a zipspring in 0F, vibrating at a high rate of speed. If it is expanded past its restlength, it will basically explode. At very low K settings, springs will only take a frame to turn into a box across the screen, then be pulled into various shapes, depending on the starting position.

The environmental variables that can be set globally are length and height of the environment, gravity (both strength and direction up/down), spring damping friction, and spring constant. For the terrain there are two parameters – the surface friction and the surface reflectance – which are related to the portion of velocity parallel to the surface and perpendicular to the surface, respectively, which is lost in a collision. These parameters are set by graphics sliders.

One happy user comments,

> This site is way better that the old one.

3.3.3 Taking Sodarace Out and About

The software allows the users to save their creatures and to promote their creativity through the range of social networking sites, such as Facebook, which have recently come to prominence. Storage of a creature or a race can now be promulgated through a feature that allows image-based links back to the main sodaplay.com storage site. Specifically, the new framework allows users to use the BBCode snipplet to add a thumb nail image to most forums, and to use the HTML snipplet to add a thumb nail image to most sites.

3.4 Software for Artificial Life in Sodarace

In the earlier stage of the project there were a number of approaches to developing artificial creatures to compete in the Sodaraces. The accessibility of the XML import/export format plus timing information from the original race application itself naturally allowed the application of optimization algorithms such as genetic algorithms [2, 1] and simulated annealing [5] to the task of building better racers.

It is interesting to note how the audience reacted to the issues raised by simply using an algorithm to optimize a previous human-designed racer, that we specifically chose Daintywalker, a well-beloved early creation of Burton that had, in effect, become the Sodarace mascot. The initial feeling in the forum was that this was somehow 'cheating.' However, on later reflection, the general consensus was that nature and human creativity frequently work by modifying and improving on earlier designs, which, in turn, led to some deep discussions on evolution.

Using the previous race software Wodka, an Austrian AI group, created a program that generates robots from scratch instead of optimizing existing robots. Wodka used a genetic algorithm that generates a random set of muscles, masses, and springs, forming multiple Sodarace creatures. These creatures are then loaded and raced, and those that finish quickest breed, and their fastest offspring breed, and so on. Initially, they started off as simple segmented sticks that bounced across the screen. However, later the best creatures start to develop structure and look as if they have powered 'flippers' in the front. This software was made open source and was used by others in the community to experiment with digital evolution.

3.5 Lessons Learned on Approaches for Artificial Life in Sodarace

Although a range of algorithms was originally applied to the process of creating creatures, it was clear from the user forum responses that there were two significant lessons to learn. First, to engage the broader user community in the Artificial Intelligence theme, the race software and ability to interact directly with AI elements needed to be more accessible. Second, the biological evolution analogy of trial and errors combined with survival of the fittest was the most accessible approach to pull new users into the educational experience. It was also clear that the initial Humans vs. Machines narrative was giving way to a broader view on cooperation. Artificial Intelligence, rather than being a competitive threat, was seen as a tool that could help enhance human creativity. With these views in mind, the new framework was developed.

3.5.1 Easy-to-Use Point-and-Click Racing

The Sodarace application, rather than requiring computer programming skills, now allows users to easily create terrains for the race using a point-and-click system. The terrain starts as a flat race, with start and finish flags, but by selecting and moving sections of the ground, this can be reconfigured into an appropriately challenging race course. These new challenges can then be saved to allow others to see how their racers do over the course. New creatures can easily be added to the race; they can be creatures created directly or selected by the user scouring the online archive of previously stored creations. In both cases, all that is required is that the user enters the URL for the stored creature, and this then loads that creature into the race software; see Fig. 3.4.

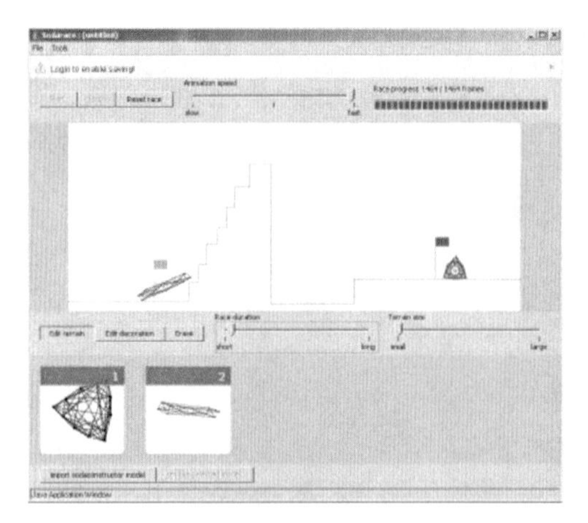

Fig. 3.4 The new Sodarace interface, allowing users to easily load racers and design their own terrains.

In selecting creatures for a race and experimenting on the creatures' viability over particular race terrains, users are pulled into a covert understanding of both physical principles and the ethos of adaptation of form (creature) to function (performance over the terrain). Trial-and-error engineering approaches then often ensue, where users optimise their or another's racing creatures to complete the pre-defined course and come in first:

> New sodarace is great, I like that it is possible to change the race track without opening the constructor first, and it is a lot easier to load models! I've been waiting for this.

3.5.2 Amoebamatic: An Easy Start for the New Engineer

Amoebamatic is a new tool for automatically generating a variety of rolling constructor models known informally as 'Amoebas.' This form of racers first named in the forums are robust wheel-like creatures, which can survive and prosper over a number of different terrain types. The Amoebamatic application, based on the idea first put forward on a fan web site, allows users to select the number of masses, number of interconnecting muscles, and amplitude and frequency of these drivers to create amoebas with a range of differing properties. Users can then tune their creatures by hand, via a range of graphics sliders and a repression of relative movement over the racetrack, to automatically create viable creatures. This both engenders the beginnings of engineering principles, the interaction of physical parameters giving rise to differing creature performance, and allows new users an easy entry into building a racer that is both stable and viable.

> The amoebamatic is perfect for this. I love breeding faster amoebas with it.

3.5.3 AI and the Travelling Kiosk Software

Examination of digital evolution is facilitated with new kiosk software; see Fig. 3.5. This software formed the basis of the activity during the UK tour of Science Centres. The application is deliberately designed to be stand-alone. Users are presented with three pre-created amoebas, one of which they select. They then select a race terrain from the three pre-defined races presented. Their selected amoeba then runs the selected race against a group of other differing types of soda creatures. The software then displays the race and the first three racers across the finish line are shown on a podium graphic. The user is challenged to ensure that the selected racer comes in at the number one position.

Users may chose to try and improve their racer by hand (see Fig. 3.6) or automatically. If they select by hand, they are presented with an Amoebamatic-like interface allowing them to manipulate the amoeba's parameters, masses, lines, amplitude, and frequency. They can then race again to see if their changes have improved performance.

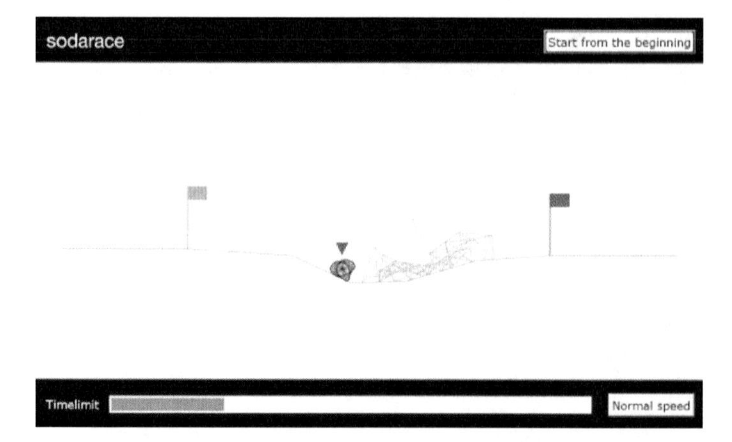

Fig. 3.5 The new stand-alone educational Sodarace kiosk application, used on the UK tour and now available for download.

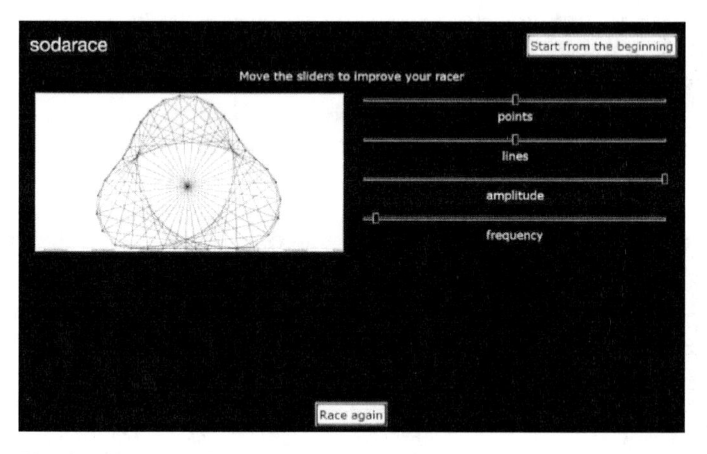

Fig. 3.6 The Amoebamatic-like user interface for manual enhancement of racers.

This trial-and-error approach exposes users to a fun incarnation of the basis of engineering experimentation: What changes will improve their racer to work best on the selected racetrack terrain? However, in addition, users can select at any time to allow the software to assist and to automatically try to improve their racer. If selected, the users view a screen showing multiple instances of their racers with small random changes of parameters applied; see Fig. 3.7. This population of potentially improved racers is then raced 'behind the scenes' and the best performing solution is highlighted. This random mutation, followed by survival of the fittest cull, continues and a graph showing users the race time for the best current racer is displayed. When the best racer has evolved to users' satisfaction, they may then watch it race by returning to the main race screen .

We deliberately selected a very reduced implementation of genetic algorithms for this process; the inclusion of crossover was felt to complicate the story for this application. The random mutation stage, however, means that each run of the software has the potential to create differing solution to the race-winning problem, which successfully engages users.

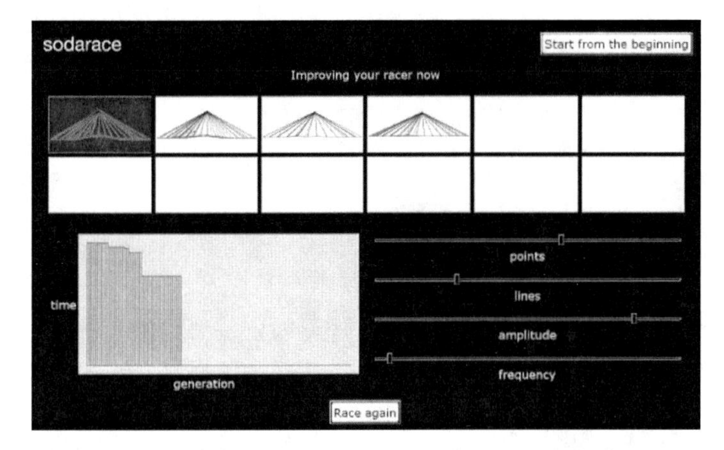

Fig. 3.7 The digital evolution user interface for automatic enhancement of racers.

Users have found that blending their by-hand optimisation with automatic computer optimisation, in effect cooperative turn-taking, has led to enhanced solutions. The direct analogy of human trial and error and computer optimisation as an approach to artificial intelligence has helped to de-mystify the field. Users feel they now understand a basic optimisation technique, because they have been an active part of that process. The parallels of the automated computer optimisation to biological evolution are also obvious and frequently commented on:

> It was interesting how evolution worked.

> This rocks, it is really fun and a nice challenge, I am going to do this at home.

And perhaps the best of all,

> I've got an idea.

3.6 Interactions in Sodarace: The Evolution of the Forums

Interactions via the user forums drive the project. It is fair to say that there is no typical usage or user for Sodarace; the range and diversity of usage are among the great strengths of the 'tools not rules' philosophy. The following section highlights some of the fascinating stories that have evolved through the project. The Sodarace forums realise the audience's own potential to create, contribute, and share significant ideas and to interact with the Artificial Intelligence and Artificial Life research communities. To date, highlights have included the spontaneous emergence amongst users of peer-to peer learning, mentoring, an embryonic scientific research process of hypothesis, experiment, theory development and subsequent racers exploitation of developed technologies, and a surprisingly seamless integration of creative and technical dialogues. There has also been a spontaneous expansion of community learning on the subject of artificial intelligence programming, with tutorial web sites and open-source code to allow interested users to construct and experiment with their own AI systems.

3.6.1 Community Development of Peer-to-Peer Learning Web Sites

Independent superuser sites have been created by model-making masters within the community. With step-by-step tutorials for beginners and expert engineering tips, they prove an invaluable resource for the model-making community and support peer-to-peer learning. Of particular interest was the spontaneous emergence of peer review of tutorials and other learning materials by accepted 'Sodarace experts' to merit inclusion on these sites.

3.6.2 The Pandora's Box: An Example of the Spontaneous Development of Scientific Method

A deceptively simple yet mind-bending model was created. As community members collaborated to study its mysterious properties, hypothesize about how it works, and construct experiments to prove their theories, we witnessed a remarkable and spontaneous microcosm of scientific method emerge.

- **The Initial Discovery** – A user in the forum writes

> Have you ever made something and then just sat and stared at the screen and said "Hummmm..." Just mess with it for a little bit.

- **The Experiments** – A short time later another user posts

> Think I figured out this phenomena. Step1: in Sodaconstructor, value of spring force has relation with only extension of the spring proportional maybe, and does not have any relation with the length at rest. See this. these are a variety length of springs. left-side one is a zero-length spring, 2nd one is 1-pixel-length, 3rd is 2-pixel, and so on. Besides, they are located as the free masses line up horizontally,

> try drag G value up slowly. Then, you will see all of them are elongated similarly, and not related to their length at rest. Step2: think about a rectangular with diagonal lines. as we see in step1, the pulling force of spring is related to its extension. Now, let's assume that the force is proportional to extension. so, for example, if the ratio of two side lines = 2 : 3, the forces acting on the springs = 2 : 3 also. so, the direction of the resultant of these two vectors overlaps the diagonal line. the resultant of these two vectors and the expanding force acting on the diagonal line are completely on a straight line and the direction of the force is opposite. So, the length of these lines come to settle in a fixed value. This happens only the shape is rectangle and the side springs are zero-length.

- **The Pandora's Calculus: Engineering Formalism and Commercial Exploitation** – The development of an empirical understanding and experimental validation of the by then named 'Pandora' effect was followed by a formalization into the Pandora's Calculus,

  > I wanted to share something I figured out in case people wanted to learn a new technique. Basically I came up with a Pandora Calculus that is similar to the way Tension Spring Calculus works. It allows you to: 1. Create Pandoras to fit any specific size you need. 2. Create multiple Pandoras on the same base. 3. Embed any Pandora on a tension spring. This means that Pandoras can be used like hubless bearings to mount rotors or turbines on, and who knows what else.

  This technique is now widely used in the community to develop numerous virtual motors to drive Sodaracers. Another user develops 'motors' for use in others racers and writes

  > So far i found out that motor 4 is just 2 RAM's connected in the middle with some crossbeams..but because of the fact that the crossbeams get compressed they make the 2 RAM's much sturdier..and better..1 and 3 look mesmerizing. but aren't really stable... but ill work on that... PLZ DONT JUST READ OVER THIS.... I WOULD REALLY LIKE TO WORK WITH SOME OF THE GREAT CONSTRUCTORS... TO IMPROVE MOTORS.

3.6.3 Interdisciplinary Interaction: Art and Music Meet Science and Engineering in Sodarace

Sodarace also raises awareness of the important interactions between science, design, and engineering, and the arts. For over 10 years artist Theo Jansen [3] has been building incredibly engineered walking constructions on the beaches of Holland that feed on the wind [4]. It looks uncannily as if some of the most complex Sodaconstructor models have left the screen and become real. In the forum, the artist agreed to discuss his work and explain the mechanisms by which his creations move. A teacher commented in the forum: 'I am also an art teacher who tries to convince colleagues of the cross-curricular potential of sodaplay. I really want to bring interactive media into the art-classroom – even if only to showcase it as an example of emergent creative culture.'

The Sodarace project has also been part of a retrospective at the UK's Institute of Contemporary Arts, archived in the Texas Museum of Digital Art and used as part of the content to the Arnold Schwarzenegger T3: Rise of the Machines movie web site.

3.6.4 Sodarace in Schools

The Sodarace project is widely used in schools around the world. Forum users often mention the education benefit that they find over a number of science and technology curricula. Specialist education resources are also being developed for release soon. Here are some representative samples taken from the forums; there are many more:

> I found out about sodaplay at school, from a friend. At the moment, I am using it to help me survive Physics, particularly waves.

> I am a High School student. I use sodaplay for geometry classes. It is easier to understand the properties of shapes and how different support loads can be dispersed. I enjoy working with the program and it helped me achieve an above average grade for the class.

> My whole 8th grade class has been using sodaplay. They all love it. We have learned ton of physics, I did a project for my physics class on your Sodaconstructor. I recreated Galileo's "Leaning Tower of Pisa" experiment and made a simple pendulum.

The project has also impacted university students.

> SodaSumo is a Java application I have been writing as my university third year project.

The discussion forums is also supporting a growing teachers' user community to develop allowing a peer-to-peer interchange of teaching ideas and materials. Younger users not yet competent or confident enough to develop racers themselves still have an active voice in the community, often creating animated artworks or cartoons, designing terrain racetrack as challenges to others, or undertaking to build interactive 'games'. All of these projects require the user to covertly develop an understanding of simple physical processes and principles of the Sodarace world.

3.7 Experiments with Sodarace

One of the significant findings during both human- and machine-generated racer construction is the ability to effectively exploit the limitations of the simulator physics to solve particular problems. Examples abound of creatures exploiting so-called 'holes' in the simulation to allow creations to fly. This

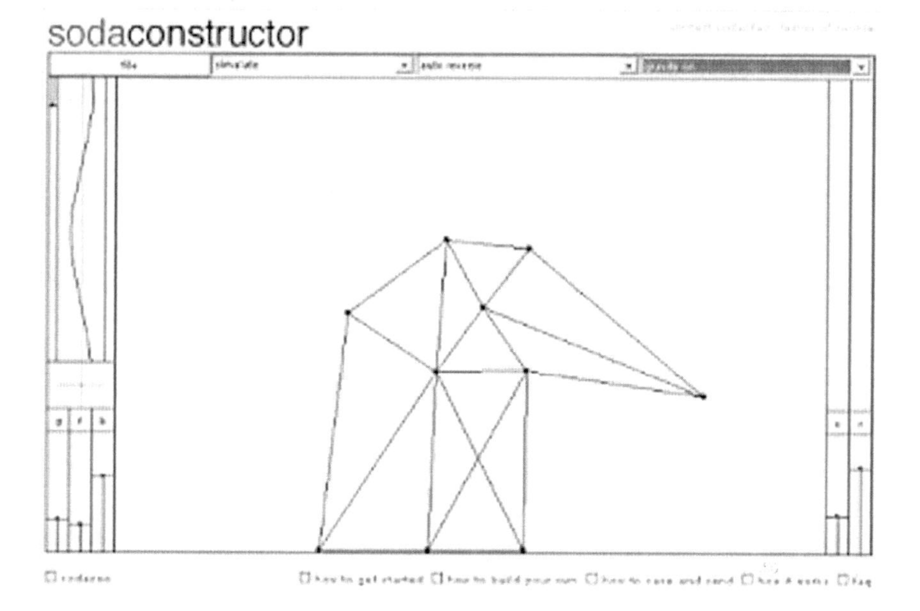

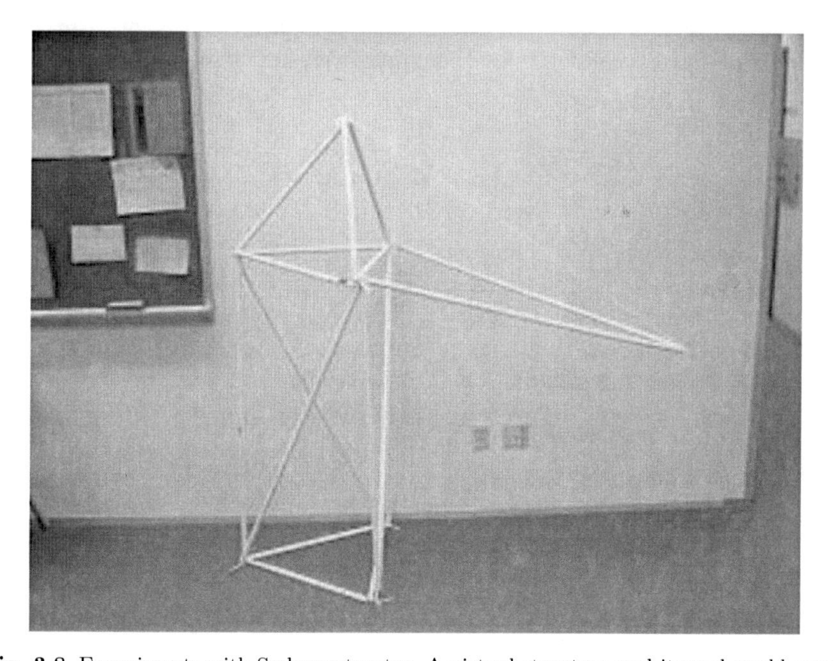

Fig. 3.8 Experiments with Sodaconstructor. A virtual structure and its real-world counterpart.

ability to capitalize on particular simulation-world niches has produced a rich range of unexpected behaviours, which still also manifest themselves in the kiosk software. In the software upgrade, we left these digital loop holes, as they were responsible for generating a whole ecosystem of creatures.

> Please do not remove these "bugs," as I think they will be a very interesting area of study.

Another study of particular interest was the Berta's Tower [14] study (see Fig. 3.8) conducted at the University of Wisconsin–Madison Center for Education. Using the theory of pedagogical praxis [10], which argues that professional practices are useful models for technology-supported school learning environments, the researchers designed and implemented workshops for middle school students based on the professional practices of engineers. During the experiment, the participants engaged in a series of engineering design challenges on Sodarace as a pathway for learning concepts in physics. The students demonstrated a statistically significant gain in understanding about the concept of center of mass and a high level of interest and engagement when using the tool [13]. The data further suggest that the short, rapid iterations of the engineering design-build-test cycle on Sodarace simultaneously increased their motivation and understanding. The study offers an example of how the use of Sodarace in a rich activity system promotes scientific learning and thus provides a potential new vision for physics education.

3.8 Summary: The Future of Sodarace

The community forums of Sodarace, central to its creation and success, show no signs of running out of new and creative ideas. The new Sodarace software, which is more powerful and easier to use, has been developed to effectively reduce the technical hurdles to entry. We have also elected to simplify the artificial intelligence story and parcel it up for wider audience who just want to play with the concepts. We have therefore shifted the focus more onto exploration of evolutionary approaches to Artificial Intelligence, as this algorithmic approach provides a clearer analogy for a wider public. The Amoebamatic design software, and the race kiosk software allow in-depth explorations of the advantages and limits of this form of optimisation. Bringing together on a single site the forums and the full range of free to use software from the Soda family such as Moovil, software aimed at engaging younger constructors, and Newtoon, easy to use software to develop physics-based games for mobile phones, will increase the opportunities for interplay, interaction and new creative explorations. This central storage of all materials on a single site [7] also enhances future sustainability for the project. The Sodaplay site is now self-supporting in terms of finance through the inclusion of appropriately filtered advertisements. As ever-engaging play will be central to user learning, and we believe the new Sodarace project framework will ensure this serious fun continues into the foreseeable future.

Acknowledgements The authors are grateful to EPSRC for funding support for the Sodarace project, to Dan Burton, Jonathan Jones-Morris, Fiddian Warman, and all at Soda for assistance in development, and to Gina Svarovsvky for her work on Berta's Tower. And most of all, thanks again go to the community of model makers, mentors, and programmers who contribute to the forum and whose creativity never, ever, ceases to amaze us.

References

1. Goldberg, D.E.: Genetic Algorithms in Search, Optimization and Machine Learning. Addison-Wesley Longman Publishing Co., Inc., Boston, MA (1989)
2. Holland, J.H.: Adaptation in natural and artificial systems. MIT Press, Cambridge, MA (1992)
3. Jansen, T.: Strandbeest website. URL `http://www.strandbeest.com/`
4. Jansen, T.: The Great Pretender. 010 Uitgeverij, Rotterdam, Holland (2007)
5. Kirkpatrick, S., Gelatt, C.D., Vecchi, M.P.: Optimization by simulated annealing. Science **220**, 671–680 (1983)
6. Komosinski, M.: The Framsticks system: versatile simulator of 3D agents and their evolution. Kybernetes: The International Journal of Systems & Cybernetics **32**(1/2; Special Issue on Artificial Life Software), 156–173 (2003)
7. McOwan, P., Burton, E.: Sodaplay website. URL `http://sodaplay.com/`
8. McOwan, P., Burton, E.: Sodarace website. URL `http://sodarace.net/`
9. McOwan, P.W., Burton, E.J.: Sodarace adventures in artificial life. In: A. Adamatzky, M. Komosinski (eds.) Artificial Life Models in Software, pp. 97–111. Springer-Verlag, New York (2005)
10. Shaffer, D.W.: Pedagogical praxis: using technology to build professional communities of practice. SIGGROUP Bulletin **24**(3), 39–43 (2003)
11. Sims, K.: Evolving 3D morphology and behavior by competition. Artificial Life **1**(4), 353–372 (1994)
12. Sims, K.: Evolving virtual creatures. In: SIGGRAPH '94: Proceedings of the 21st annual conference on Computer graphics and interactive techniques, pp. 15–22. ACM, New York (1994)
13. Svarovsky, G., Shaffer, D.: Sodaconstructing knowledge through exploratoids. Journal of Research in Science Teaching **44**, 133 – 153 (2006)
14. Svarovsky, G., Shaffer, D.W.: Berta's tower: understanding physics through virtual engineering. In: ICLS '04: Proceedings of the 6th international conference on Learning sciences, pp. 639–639. International Society of the Learning Sciences (2004)
15. Voth, D.: When virtual roberts race, science wins. Intelligent Systems **19**(2), 4–7 (2004)

Chapter 4
3D Multi-Agent Simulations in the breve Simulation Environment

Jon Klein and Lee Spector

4.1 Overview

Artificial life is often described as the study of life "as it could be," meaning that it involves the exploration of life-like interactions of agents with novel behaviors in novel environments. Although these behaviors and environments can be explored using mathematical models or even real-world experiments, one of the most commonly employed techniques is multi-agent simulation. Using multi-agent simulation software, users can define the rules of an environment and the behaviors of individual agents in order to examine how they behave collectively, their so-called "emergent behaviors."

One of the more challenging and time-consuming tasks in exploring multi-agent systems is developing the software infrastructure for modeling and simulation. Even models of conceptually simple interactions between agents often require a great deal of software infrastructure to manage and coordinate agent behaviors.

Many artificial life and multi-agent system models rely on realistic three-dimensional (3D) spaces, especially those that aim to produce results relevant in explaining real-world behaviors (such as simulations that give insights into biology) or those that may eventually be implemented in the real world (such as simulations of traffic or sensor networks). Moreover, some simulations, such as those dealing with robotics, require realistic simulation of physical interactions between agents and their environments. These applications present a particular challenge in simulation development because they require sophisticated spatial and physical simulation algorithms in order to produce accurate simulation results.

Breve is a free, open-source software package that aims to simplify the creation of 3D simulations of multi-agent systems and artificial life.[1] Breve

[1] Breve was developed and is maintained by Jon Klein. The breve-based research described in this chapter is the product of both authors.

frees simulation authors to focus on the definition of agent behaviors and on descriptions of the simulated environment, whereas the breve engine manages simulation data and algorithms. Agents' behaviors can be written in Python, or using a simple scripting language called "steve." Breve includes support for physical simulation and collision detection for the simulation of realistic creatures or robotics and an OpenGL display engine for the visualization of simulated worlds. Breve runs on a number of platforms and is distributed as a pre-built package for Mac OS X, Linux, and Windows from the breve website [7].

Breve is analogous in many ways to 3D game engines that allow for the rapid development of games while abstracting away many of the details of their implementation. Breve aims to do the same for multi-agent simulations. The similarity is more than conceptual – breve actually preforms many of the same types of computation that 3D game engines do, such as 3D rendering, management of agent interaction, and collision detection. Unlike game engines, however, breve is designed for simulation applications, especially those relating to artificial life and artificial intelligence. To this end, breve includes a rich library of features suited to simulation applications.

4.1.1 Agent-Based Modeling Paradigms

Aside from breve, there are several other multi-agent simulation software packages that can be used to study artificial life. These environments vary greatly in their representations of space and time, and these representations strongly influence the types of simulations to which an agent-based modeling system is best suited.

Some simulation environments have no explicit modeling of space. In these environments, abstract spatial structures (such as networks) can be achieved via connections between individual agents. Another popular representation of space in agent-based modeling is a discrete two-dimensional (2D) or 3D grid. The grid introduces a realistic spatial structure and allows agents to interact based on spatial proximity, but it somewhat limits the granularity of these spatial interactions. Breve uses a fully continuous representation of 3D space, which allows it to be used for realistic 3D simulation of phenomena such as swarms, robots, and vehicles.

Certain phenomena, such as chemical concentrations or temperature (in addition to many other kinds of environmental states), are difficult to accurately and efficiently model using individual agents and benefit greatly from the use of a discrete 2D or 3D grid. To facilitate the simulation of these phenomena, breve supports discrete 2D and 3D grids, which can be used independently or in concert with continuous spaces. This allows agents to move through continuous 3D spaces while interacting with discrete environmental states.

With respect to representation of time, the differences between the various simulation environments can be more subtle. Some systems use large, discrete time-steps to represent a period of time passing. Others, like breve, integrate arbitrarily small time-steps to simulate fully "continuous" time (although at the lowest levels, the representation of time is still discrete). In these environments, it is understood that the time-step is generally as small as possible, given the practical computation constraints of arbitrary small integration steps.

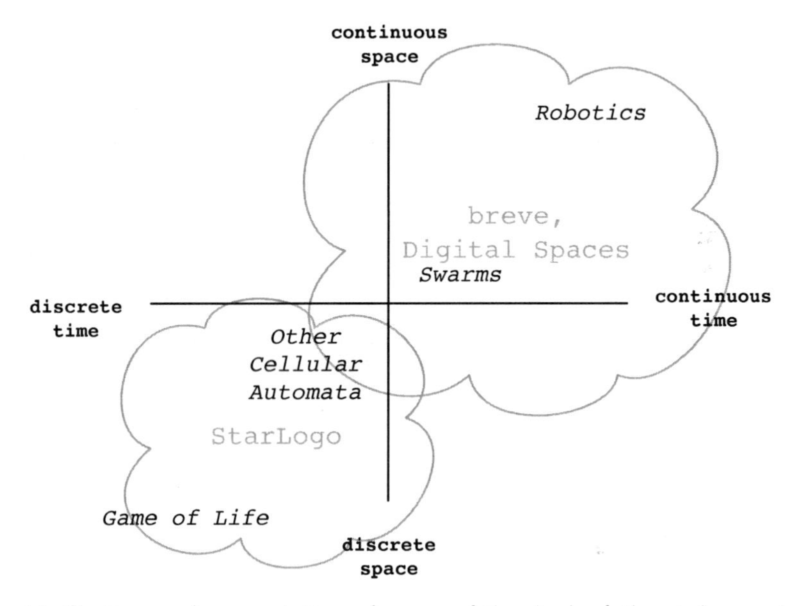

Fig. 4.1 Continuum of representations of space and time in simulation environments.

Fig. 4.1 shows a general representation of the "simulation space" defined by the spectra of discrete and continuous representations of time and space, along with the locations of common simulation paradigms and the general locations of some common agent-based modeling toolkits.

This diagram represents only a broad illustration of the distinction between discrete and continuous time and space in simulation. In reality, the distinction is often less clear, as any sufficiently extensible agent-based modeling environment is capable of modeling any representation of time or space, although doing so may require additional development effort on the part of the simulation author. In particular, some common agent-based modeling systems do not have single built-in representations of time and space. These systems may provide basic multi-agent simulation functionality such as management of agents and their interactions, while leaving the modeling of time

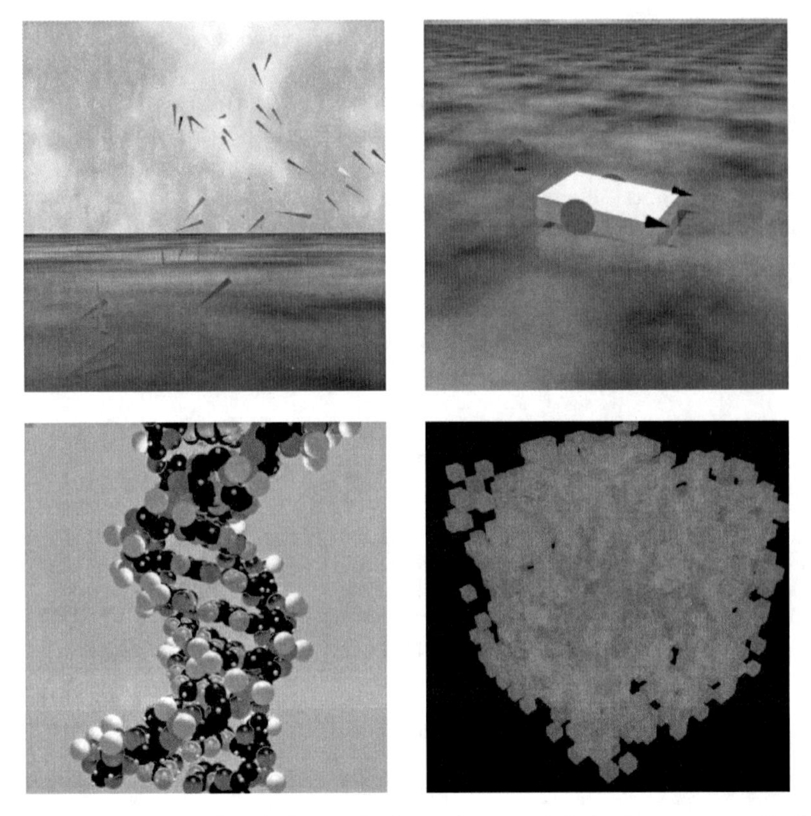

Fig. 4.2 Sample simulations in breve. Clockwise, from top left: flocking agents; a Braitenberg vehicle; a 3D game-of-life cellular automata; a DNA molecule visualized from a PDB file.

and space as a detail for the simulation author to implement according to their needs.

Fig. 4.2 shows a sample of the different simulation paradigms that can be represented in breve, including continuous non-physical simulations, continuous physical simulation, and discrete 3D cellular automata.

4.1.2 Comparison to Other Agent-Based Modeling Systems

There are several other agent-based simulation modeling environments that are useful for exploring artificial life and multi-agent systems. Some of the more notable environments are discussed here briefly.

As alluded to earlier while describing representations of space and time in simulation environments, choosing between simulation packages for a particular simulation application is, by and large, not a matter of a "laundry list" comparison of simulation features among the packages. All of the environments described here are powerful and extensible environments with support for authoring custom simulations. The important differences between the environments is generally found in the types of simulations they are best equipped to model.

- StarLogo and StarLogo TNG – StarLogo [17] is the oldest of the agent-based simulation packages described here. StarLogo allows users to model multi-agent systems through simultaneous manipulation of multiple Logo "turtles" in a discrete 2D world, using an integrated programming language derived from Logo. In addition to its multi-agent simulation applications, StarLogo (like the original Logo) is used as an educational tool to teach students and novice programmers how to program and how to construct multi-agent models based on simple agent behaviors.

  StarLogo was one of the original inspirations for breve. Although breve does not place as strong an emphasis on education applications as StarLogo, the inspiration to use an integrated, interpreted language grew directly out of StarLogo's implementation.

  StarLogo TNG ("The Next Generation") described in Chapter 6 is a new version of StarLogo currently in development that includes support for continuous 3D space and for a visual programming language.
- Repast – Repast [16] described in Chapter 2 is a Java-based agent simulation toolkit that is geared toward social simulation and that also allows for visual construction of simulations. Although Repast does include support for simulation and visualization of agents in 3D spaces, it does not provide support for physical simulation.
- MASON – MASON [12] is a Java-based simulation toolkit that supports 2D/3D simulation as well as network topologies. Like Repast, MASON supports simulation and visualization of agents in 3D spaces but does not include physical simulation functionality.
- Swarm – Swarm [13] is an Objective-C-based simulation library originally developed at the Santa Fe Institute. Compared to the other systems described here, Swarm provides support for multi-agent modeling at a relatively low level. Swarm provides a library of tools for managing agent behaviors and relationships, with less of an emphasis on features such as spatial representation and visualization. Swarm does offer a "Space" library with support for discrete 2D spaces, and sample code on the Swarm website details how Swarm can be extended to model discrete 3D spaces.
- Digital Spaces – One of the newest additions to this list is the Digital Spaces simulator, which is also the most similar to breve in terms of simulation "niche." Digital Spaces includes support for 3D physics in simulations and for importing complex 3D environments. Although Digital Spaces has not been used heavily for artificial life research, it does offer

an impressive feature set that makes it an excellent candidate for such applications.
- Framsticks – Framsticks [9] described in Chapter 5 was not originally developed as a general-purpose agent-based modeling toolkit, but it does deserve a special mention in comparison to breve because it models artificial life agents with articulated bodies in 3D physical worlds. Framsticks has traditionally taken a more specialized simulation approach with a focus on evolved virtual creatures, although versions 2 and 3 provide more general agent-based modeling capabilities in the software.

4.2 Motivations

The fundamental goal of breve as a simulation environment is to abstract away simulation implementation details as much as possible and to allow users to instead focus on constructing agent behaviors and the environment.

4.2.1 A Personal Motivation

Although there was a considerable amount of academic inspiration behind the development of breve, an equally important inspiration (for Jon Klein, breve's primary author) came on a more personal level from a college friend with a great interest and insight in artificial life and artificial intelligence, but (to put it gently) no particular talent in computer programming.

This friend would observe a simulation or other programming project, typically the product of several weeks of work, and cheerfully make an astute suggestion for a "simple" improvement that would invariably require several more weeks of work. This particular interaction was repeated so many times that, in frustration, one would eventually snap "do it yourself!" to which the friend *cheerfully* retorted "I can't!"

And so a small – but genuine – motivation for the development of breve was to silence this friend.

4.2.2 Design Principles and Goals

The guiding motivation in the implementation of breve is that experimentation with artificial life and multi-agent systems should be simple and should not be limited by programming ability. To this end, a number of design principles behind breve are described here. Given the rapid pace of evolution in the fields of multi-agent simulation and artificial life (and computing in general), these principles represent moving targets, so even after several years and many revisions of the software, these concepts still guide the continued development of breve.

- *breve should be approachable by people of all programming abilities.* Practically speaking, this means breve should use a simple language and simple interfaces to artificial life simulation features such as evolvable code, genetic algorithms, and physical simulation.
- *breve should be an integrated development environment, not a software library or application programming interface (API).* Although some powerful agent-based simulation toolkits, such as Swarm, are implemented as a software library, it can represent a considerable barrier for new users, especially those without a great deal of prior programming experience.

  The decision to implement an integrated environment instead of an external library does come with downsides. Integrating breve into other software projects, for example, is a more complicated prospect than integrating an environment that is distributed as a software library.
- *breve should allow users to immediately see the connection between code and simulation behavior.* Breve is intended not only to enable the development of of multi-agent systems but also to encourage the exploration of these systems through an integrated visualization engine and by streamlining the write/run/revise cycle of simulation development. This principle (and especially its manifestation in StarLogo) strongly influenced the decision to make an integrated scripting language a main component of the simulation environment.

Ironically, the decision to implement a custom language to make breve more accessible to new users may have had the opposite of its intended effect. Although the steve language is arguably simpler and more approachable than many other languages used for simulation software (such as Java, C, C++), the main breve developer may have overestimated the enthusiasm that potential users would feel for learning a new, proprietary language. With programming languages, familiarity may trump simplicity.

Because of this, and keeping in line with breve's development goals, breve now supports the Python language and is in the process of transitioning toward Python as the preferred simulation language.

As for the college friend who so greatly influenced the development of breve, he still does not use breve to implement his simulation ideas. Whether this is a shortcoming of breve or a character flaw is, after all these years, not entirely clear.

4.3 Writing Simulations in breve

This section provides a conceptual overview of how simulations are constructed in breve. Although breve simulations can be written in either a proprietary language called steve or in the popular Python language, this section describes general concepts which apply to writing simulations regardless of which language frontend is used. The details of the individual scripting languages supported by breve are described later in Sections 4.3.5 and 4.3.6.

4.3.1 Object Orientation and the Built-in breve Classes

The breve environment makes extensive use of object-oriented concepts both as a programming model and as a more general representation of the simulated world. Every object in a simulation corresponds directly to a programming language object in the steve or Python frontend language.

Furthermore, all programming in breve is preformed by extending the built-in classes – there is no code or data in the steve language that is not associated with an object. (Although the Python language frontend does allow code outside of classes, this code is not able to interact directly with the breve engine.) The breve distribution includes a standard library of classes that interface with the internal features of the breve engine or that themselves implement useful simulation features. These classes are then subclassed to implement custom behaviors.

The special class "Object" represents the root object class in breve, and all other objects are subclasses, either directly or indirectly, of the Object class. Practically speaking, very few classes inherit directly from Object. Instead, breve defines two Object subclasses that serve as a logical distinction of all objects in breve: the Real subclass, which includes all objects which have a physical presence in the simulated world, and Abstract, which includes those that do not. Abstract objects are most often used to encapsulate data and computation, and they more closely resemble objects in traditional programming languages than do Real objects.

4.3.2 The Controller Object

The "controller" object in breve is a special object that is created by the engine when the simulation begins. The controller object is a custom subclass of the class "Control" that the user implements to set up and manage the simulation. Because the controller object is the only object created when a simulation begins, it is responsible for creating all other objects from its *init* method.

The controller object is accessible to every object in the simulation and, because there are no global variables in the steve language, the controller plays an important role in holding global simulation state data and in facilitating communication between the agents.

4.3.3 The breve Simulation Loop

Agents in a breve simulation are simulated following a pattern called the *breve simulation loop*. The basic behavior of each agent is summarized in the following pseudocode:

```
init agent [run user defined ''init'' method]
while ( simulation running )
    foreach agent:
        iterate agent [run user defined ''iterate'' method]:
            examine internal/external states (using sensors)
            perform computation
            change behavior (using actuators)
```

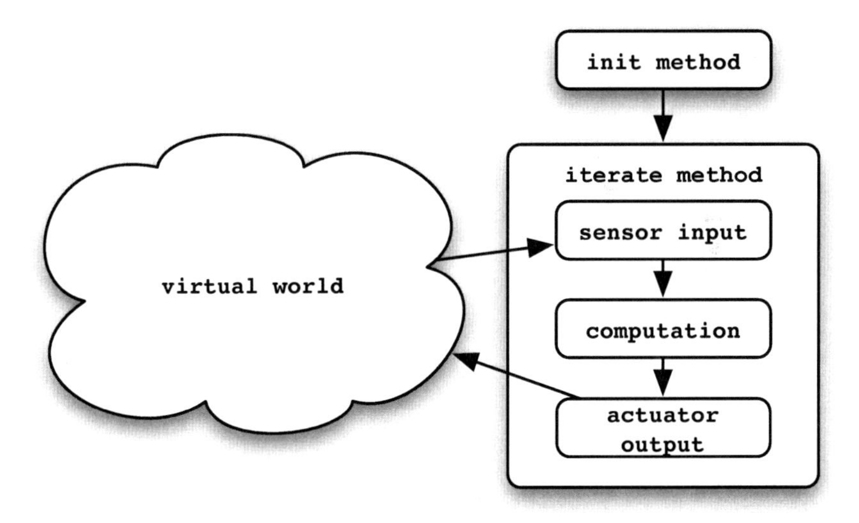

Fig. 4.3 The breve simulation loop.

An abstract diagram of this process is illustrated in Fig. 4.3, and two examples of how this process is implemented with actual simulations are shown in Fig. 4.4 and Fig. 4.5.

Fig. 4.4 shows the simulation loop for Creatures, a demo breve simulation of physically simulated creatures evolving via a genetic algorithm. In spite

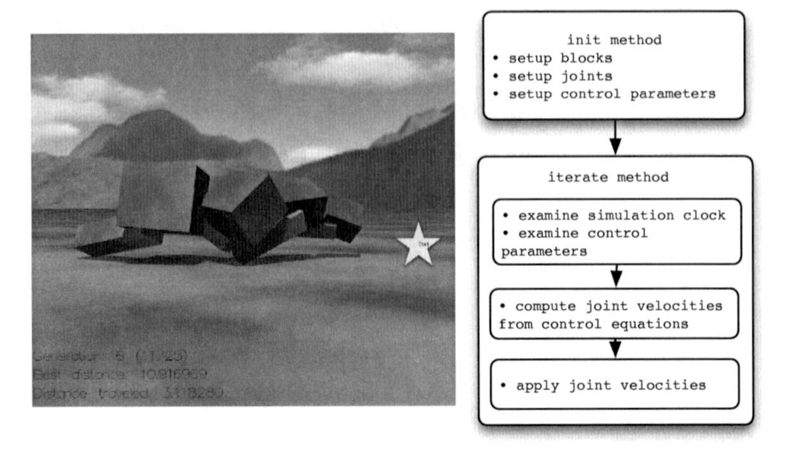

Fig. 4.4 The breve simulation loop for an agent in the "Creatures" simulation.

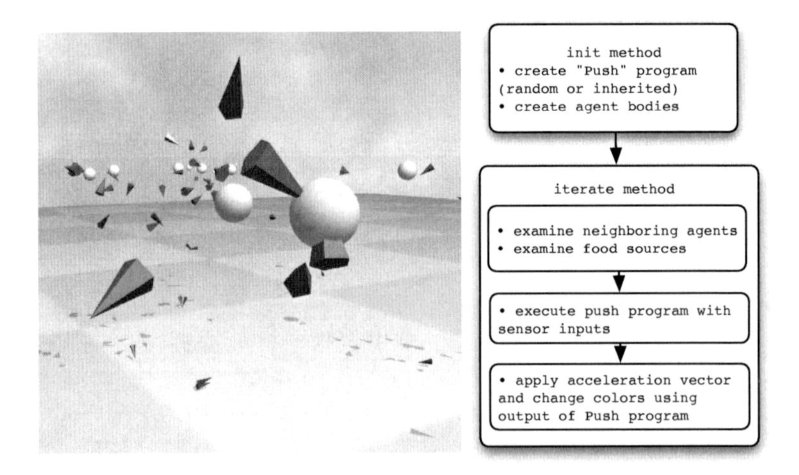

Fig. 4.5 The breve simulation loop for each agent in the "SwarmEvolve 2.0" simulation.

of the complexity of the simulation behind the scenes, the simulated agent's behavior is rather straightforward: The agent takes simple inputs from its internal state and from its genome and produces outputs in the form of forces applied to joints.

Fig. 4.5 shows a breve demo, SwarmEvolve-2.0, with more complex interactions between agents and their environment. In this simulation, agents are initialized by creating evolvable code in the Push programming language (described in detail in Sect. 4.4.4) that will control their behavior. Then at each time-step, the agents sense the world to provide input to their Push programs, execute the Push programs, and then use the results to change various behaviors, such as heading and color. This figure shows only a conceptual overview of the SwarmEvolve-2.0 agent behaviors and many of the details are omitted. The SwarmEvolve agents, for example, have access to many other types of input sensors and output actuators than those shown. A more detailed description of the simulation is described in Sect. 4.6.1.

4.3.4 Defining Callbacks and Agent Behaviors

Callback functions are used in breve to allow the simulation engine or front-end application to trigger events in simulation code. When possible, callback behaviors are automatically defined through the presence of methods with special names. In some cases, though, callbacks are defined explicitly via requests from simulation code. Some of the main callback types are described here.

- In response to collisions. Breve handles the collisions between objects at two different levels. First, if physical simulation is enabled for an object, the collision is resolved in the physics engine. Then the collision triggers a callback to one or more agents at the user simulation level. This callback can be handled by providing the breve engine with a callback method to be executed upon collisions with a certain object type.
- In response to object "announcements". Object announcements in breve provide a simple way for agents to communicate general events or state changes to all interested parties, without keeping track of message recipients or making method calls directly. With object announcements, any agent can register itself as a listener for another agent and can schedule a method call to be executed when a specific announcement is made. Agents may have any number of different announcement types identified by unique strings. Methods in the Object class are provided for both making announcements and for registering as listeners to announcements in other objects.
- In response to keyboard and mouse input. A number of special callbacks are defined that are executed automatically in the simulation controller object in response to keyboard and mouse events. Support for keyboard or mouse interaction can be added to a simulation simply by defining callback methods with the the proper names.
- In response to network events. Breve includes rudimentary network transfer support, and, as with other events, network transfers trigger method calls in the simulation's controller object. This allows users to react and respond to simulation objects sent over the network from other simulations.

4.3.5 The steve Programming Language

Breve simulations can be written in a simple object-oriented language called "steve." The steve language resembles and borrows features from languages like C, Objective-C, and SmallTalk and includes support for 3D vectors and matrices as native types.

Table 4.1 shows a simple demo simulation written in the steve language. In this simple simulation, agents are given an initial upward velocity and a downward acceleration to simulate the behavior of a fountain of particles.

```
@include "Mobile.tz"
@include "Control.tz"

Controller Fountain.

Control : Fountain {
    + to init:
        250 new Particles.

        self point-camera at (0, 9, 0) from (40.0, 2.1, 0.0).
}

Mobile : Particle (aka Particles) {
    + to iterate:
        if (self get-location)::y < -6.0:
            self reset.

    + to init:
         self set-acceleration to (0, -9.8, 0.0).
        self reset.

    + to reset:
        self set-color to random[(0, 1, 1)].
        self move to (0, 0, 0).
        self set-velocity to random[(10, 20, 10)] + (-5, 4, -5).
        self set-rotational-velocity to random[(.6, .6, .6)].
}
```

Table 4.1 A sample simulation in steve.

Method calls in steve resemble SmallTalk and Objective-C in that they use *keywords* to designate method arguments and the method arguments may be placed in any order. This feature enhances the readability of method calls for readers not familiar with the breve APIs.

Although steve is object-oriented, simple data-types such as int, float, matrix, vector, list, and hash are not first class objects in steve. Steve's type system treats these simple datatypes differently than true objects that are added to the breve engine and follow the simulation-loop pattern described above. Similar to typing in Objective-C, the simple data-types are statically typed, while the true objects follow a dynamic-typing scheme in which method calls are resolved at runtime and do not depend on the object type, so long as the object responds to the given method.

Steve also includes garbage collection via a reference counting scheme. This garbage collection is used automatically for the simple datatypes, but it must be manually enabled for other objects on a per-object basis. This is because objects in a breve simulation are referenced internally by the breve engine and may continue to play a role in a simulation (through their iteration methods) even when they are not referenced explicitly by any other simulation object. In addition, objects not explicitly referenced by others (in the classical programming sense) may reacquire references through simulation events (such as announcements or collisions) or through the steve language's "all" command, which provides a list of all objects of a requested type.

4.3.6 The Python Programming Language

As of breve version 2.6, simulations can be written in the popular Python language. Because of Python's powerful reflection and introspection features, most of the benefits of steve's tight integration with breve can be emulated with Python, so that, by and large, the user experience and APIs are identical for simulations written in both steve and Python. Table 4.2 shows the same sample simulation listed in Table 4.1 but this time implemented in Python.

The Python and steve languages can communicate seamlessly through bridge objects, allowing code from both languages to be mixed together in a single simulation. This allows even existing simulations written in steve to take advantage of the rich library of Python modules, one of Python's greatest strengths. Python objects can be instantiated and manipulated directly from steve, and vice versa.

4.4 Breve Features and Technical Details

This section describes some of the basic simulation features that breve provides and details on how they are implemented.

4.4.1 3D Spatial Simulation

The foundation of the simulation functionality provided by breve, and the one upon which the others are built, is a subsystem that manages objects and their interactions in 3D worlds. The spatial simulation system handles placement of objects, manages collisions, and integrates accelerations and velocities to compute object positions. This system preforms all of the *non-*

```
import breve

class Fountain( breve.Control ):
    def __init__( self ):
        breve.Control.__init__( self )
        Fountain.init( self )

    def init( self ):
        breve.createInstances( breve.Particle, 250 )
        self.pointCamera(
            breve.vector( 0, 9, 0 ),
            breve.vector( 40.0, 2.1, 0.0 ) )

class Particle( breve.Mobile ):
    def __init__( self ):
        breve.Mobile.__init__( self )
        self.setAcceleration( breve.vector( 0, -9.8, 0.0 ) )
        self.reset()

    def iterate( self ):
        if ( self.getLocation().y < -6.0 ):
            self.reset()

    def reset( self ):
        self.setColor( breve.randomExpression( breve.vector( 0, 1, 1 ) ) )
        self.move( breve.vector( 0, 0, 0 ) )
        self.setVelocity( (
            breve.randomExpression( breve.vector( 10, 20, 10 ) ) +
            breve.vector( -5, 4, -5 ) ) )

        self.setRotationalVelocity(
            breve.randomExpression( breve.vector( 0.6, 0.6, 0.6 ) ) )

# Create an instance of our controller object to initialize the simulation
Fountain()
```

Table 4.2 A sample simulation in Python.

physical simulation tasks, although it is tightly integrated with the physical simulation engine, which is described below.

4.4.2 Physical Simulations

Breve includes support for realistic rigid body dynamics for simulation of virtual creatures and robotics. While the core functionality and interface to the breve physics engine has stayed the same over the years, the under-the-hood implementation has evolved, piece by piece, from a home-grown implementation to its current state, which largely uses the Open Dynamics Engine (ODE) library, a physical simulation engine based on a generalized coordinate method which models articulated bodies as a series of differential equations.

The original physics engine implementation was based heavily on Brian Mirtech's work on impulse-based rigid body simulation [15]. The original release of breve included its own physics engine based on the Featherstone dynamics algorithm, a reduced coordinated algorithm that calculates the motion of articulated bodies by recursively computing force and mass properties for subtrees of jointed bodies. Physical contact in the engine was modeled using a micro-collision model in which resting contact is simulated through the use of a large number of low-velocity collisions and impulses. Collision detection was implemented as a two-pass process: The first stage, a prune-and-sweep sweep, found potential collisions by maintaining lists of sorted bounding box minima and maxima in all three dimensions; the second stage, which determined which candidate pairs were intersecting, used Mirtech's V-Clip algorithm [14], based on the Lin-Canny closest feature collision detection algorithm, with extensions to properly handle the interiors of polyhedra.

The prune-and-sweep portion of the original collision detection model is still used for collision detection and for the breve's *neighbor detection*. For this task, the prune-and-sweep algorithm efficiently discovers object pairs overlapping by a user-defined threshold, without running a second-stage collision check. This allows objects to quickly and efficiently discover nearby objects without preforming an $O(n^2)$ search of the entire simulation space – because of spatial coherence between simulation steps, this method of discovering neighbors is very close to $O(n)$.

Although the physical simulation in breve is a useful tool for developing simulations involving artificial life, evolution of creature morphologies, and robotics, it should be stressed that neither the original breve physics engine nor the ODE engine are intended to be truly predictive of real-world physical behaviors. Although the physics engine used in breve does place an emphasis on stability, it is still important for simulation developers to include sanity checks in physical simulations to avoid situations that can lead to instabilities or inaccuracies. In particular, excessively large velocity and mass values can cause simulations to "blow up" and should be avoided.

When used in conjunction with evolutionary computation, this aspect of physical simulation represents a particular challenge due to the uncanny ability of the evolutionary process to exploit any available means of improving performance, including exploiting bugs or inaccuracies in physical simulation. Karl Sims described this problem in his work on evolving virtual creatures [20]. For this reason, work with physical simulation often requires additional care and attention to detail.

4.4.3 Visualization

Breve includes an OpenGL-based visualization engine that renders 3D simulations in realtime. On several occasions (described later in more detail), breve's visualization engine has allowed for the discovery or understanding of phenomena that may have otherwise gone unnoticed.

Beyond simply rendering agents and their environment, the visualization engine can render lines, drawings, and a variety of "special effects" that may be used to enhance understanding of the simulation state. Shadows and reflections, for example, provide additional visual cues of agent location and orientation.

In addition to conveying "literal" information about agents in a simulation (size, location, orientation), the visualization engine can be used to convey arbitrary information about an agent's state by changing display qualities using a number of visualization features. High-dimensional agent states can be represented using visual qualities such as color, size, and transparency. In this way, even simulations modeling abstract agents or non-spatial simulation can benefit from 3D visualization.

The performance impact of high-quality visualization in breve is minimal. Most modern computers include powerful graphics cards that support hardware-accelerated 3D rendering, so in most cases visualization can be achieved with virtually no computational costs and thus no with impact on simulation speed.

4.4.4 The Push Programming Language

Push [40] is a stack-based programming language intended primarily for use in evolutionary computation systems. Although the steve and Python languages, in which breve simulations are written, are intended to be simple for human programmers to use, neither is well suited for use in evolutionary computation due to syntactical and semantical language constraints.

Push has an unusually simple syntax that facilitates program evolution. Despite its simple syntax, Push provides more expressive power than most other program representations that are used for program evolution. Push programs can process multiple data types (without the syntax restrictions that usually accompany this capability), and they can express and make use of arbitrary control structures (e.g., recursive subroutines and macros) through the explicit manipulation of their own code (via the "CODE" stack and data type) and through manipulation of their own execution (via the "EXEC" stack).

These features allow Push to support the automatic evolution of modular program architectures in a particularly simple way. Push can also support entirely new evolutionary computation paradigms such as "autoconstructive evolution," in which genetic operators and other components of the evolutionary system themselves evolve (as in the Pushpop [22] system, and in SwarmEvolve2, described in Sect. 4.6.1).

Push offers a powerful way to evolve open-ended agent behaviors in breve. By providing Push programs with direct access to agent methods via callback instructions into simulation code, agent behaviors can be fully controlled via evolving code.

4.5 Development History and Future Development

Breve was originally developed by Jon Klein as part of master's thesis work at Chalmers University in Göteborg, Sweden. The first version of breve was released in late 2001. The initial versions of the breve integrated development environment were released for Mac OS X only, although the breve engine was available as a command-line application for Linux and Windows platforms. Since 2002, continued development of breve has been largely supported by Hampshire College.[2]

4.5.1 Transition to Open Source

With the release of breve 1.7, in late 2003, breve made the conversion from a closed-source project to a fully open-source project with source code available under the GPL. Since opening the source code to breve, users have contributed many improvements to the code, such as an integrated graphical interface for breve on Linux and Windows platforms as well as initial revisions of code to interface with Python and other frontend languages.

4.5.2 Push3

Beginning with version 2.0, released in late 2004, breve includes built-in support for the Push programming language designed specifically for use with genetic programming. The Push programming language is described in more detail in Section 4.4.4 and its applications in breve are described in Section 4.6.

[2] This material is based upon work supported by the National Science Foundation under Grant No. 0308540 and Grant No. 0749184. Any opinions, findings, and conclusions or recommendations expressed in this publication are those of the authors and do not necessarily reflect the views of the National Science Foundation.

4.5.3 Python Integration

As of version 2.6, released in late 2007, breve features support for writing simulations and accessing all breve engine functionality through the Python scripting language. Because of its power and its popularity, Python is now the preferred language for new simulation development.

4.5.4 Future Development

Currently in development for the next release of breve is enhanced support for XML import and export of data. Simulation parameters in particular will be separated into standard XML files to be loaded when a simulation begins. This change will greatly simplify the process of exploring the parameter space of a simulation, which currently requires user interface interaction or changes to simulation files themselves.

Also in development for the next release of breve is a new graphical user interface, using the open-source Qt environment, that will unify the simulation experience on Mac OS X, Linux, and Windows. Among other improvements, the new interface will provide graphical user interfaces for managing the simulation parameter sets described above.

A final area of breve development in the longer term is providing greatly enhanced networking support. Breve currently offers support for simple network transfer of individual objects. Future enhancements will expand this functionality to support real-time interactions of agents in different simulations, similar to networking capabilities of modern game engines.

4.6 ALife/AI Research with breve

Here we demonstrate the range of breve's applicability by describing, briefly, some of the AI and ALife research that we have conducted using the software. Other individuals and research groups have been putting it to additional uses in a diverse set of areas including physically simulated evolving creatures [11], evolving ecologies [10], swarm robotics [6, 42], artificial intelligence applied to homeland security applications [43], simulations of sorting behaviors in ants [5], cognitive science research [4], and self-assembly in physical systems [2].

4.6.1 Evolving Swarms

The original distribution of breve included a Swarm demo that was essentially a re-implementation of Reynold's Boids system [18]. SwarmEvolve is an evolutionary extension of the original Swarm demo, in which the flying agents reproduce and evolve in the context of a simple energy dynamics: They collect energy from "feeders" and expend energy when they fly, collide, or crowd one another.

In the simplest, most highly constrained version of SwarmEvolve (1.0), the velocities of agents are computed as weighted combinations of environmental vectors (toward the nearest feeder, toward the closest agent, etc.), following the general scheme used in Reynold's Boids but extended for a modestly enriched environment. The weights used by each agent are derived from its genome, which is a sequence of floating-point numbers. Each agent is a member of one of three pre-defined species, and when an agent dies, it is replaced by a new agent whose genome is inherited, with mutation, from the most fit member of its own species.[3] In SwarmEvolve 2.0 there are no pre-specified species groups, behavior patterns, or reproductive regimes; an agent's genome is a program expressed in a Turing-complete language (Push), and its behavior (including perceptual, motor, and reproductive behavior) is the result of executing that program.

The SwarmEvolve systems have been used to demonstrate the emergence of several types of collective behavior, including multicellular organization, altruistic feeding behaviors, and tag-mediated cooperation [38, 37, 39]. They also produce life-like, visually compelling displays of 3D flocking behavior [30]. Screen shots of SwarmEvolve are shown in Fig. 4.6.

Fig. 4.6 A screenshot of SwarmEvolve 1.0 (left). The large pentagons are feeders from which the agents receive energy. In SwarmEvolve 2.0 (right), the feeders are spheres that shrink when they are eaten and re-grow slowly. The dark cones falling to the ground are corpses of agents that have run out of energy.

[3] Fitness is calculated as the product of age and energy.

4.6.2 Evolution of Cooperation

Using breve, we investigated the evolution of tag-mediated cooperation – altruism toward other agents that share a similar "tag" marker [19] – in multiple simulation paradigms in order to understand the phenomenon from different perspectives and in different contexts. Some of these studies used breve's 2D or 3D spatial simulation facilities to explore the interaction between neighborhood structures and the emergence of cooperation [37, 32, 33].

Breve's visualization engine played an important role in understanding the underlying mechanism of how colonies of cooperators were able to flourish in the spatial simulations. Without writing additional simulation code for visualization, we were able to step through simulations generation by generation, to observe altruistic donations as they took place, and to determine the conditions under which such cooperative activities would spread through the population.

4.6.3 Division Blocks

The Division Blocks project [34] combines several of breve's facilities, including complex physical simulation, high-quality graphical output, and support for efficient neural network processing, to explore the open-ended evolution of development, form, and behavior. Division Blocks are simulated rectangular blocks that can grow and shrink, divide and form joints, exert forces on joints, and exchange resources. They are controlled by recurrent neural networks that evolve, along with the blocks, by natural selection. Energy is approximately conserved, and all energy derives ultimately from a simulated sun via photosynthesis. A screen shot of the Division Blocks system is shown in Fig. 4.7.

4.6.4 Genetic Programming Research

Breve has proven to be useful as a vehicle for a wide range of evolutionary computation research – even research that does not use the system's core facilities for three-dimensional simulation. Breve's incorporation of the Push programming language [21, 41, 40, 36], in particular, has supported several research projects that extend or apply genetic programming techniques. Examples of such projects include:

- Automatic Quantum Computer Programming. In this project we used breve's support for PushGP in combination with its support for the QGAME quantum computer emulator to evolve quantum algorithms that

Fig. 4.7 A screen shot of the Division Blocks system.

compute functions of interest. This project has produced several human-competitive results [24, 26, 25, 1, 27, 23].

- Trivial Geography. In this project we used breve's PushGP implementation to develop and test a simple population-structuring technique that produced surprisingly dramatic improvements in problem-solving performance on a suite of test problems (ten symbolic regression problems and a quantum computing problem) [31].
- Unwitting Distributed Genetic Programming. In this project we developed a web-based system for distributed computation of genetic programming via asynchronous javascript and XML (AJAX), requiring no explicit user interaction and no installation of client-side software. Clients automatically and unknowingly participated in a distributed genetic programming run simply by visiting a webpage, thereby allowing for the solution of genetic programming problems without running a single local fitness evaluation [8]. Breve was used to handle server-side tasks such as population management and reproduction.
- Genetic Programming for Finite Algebras. In this project we applied genetic programming to problems in pure mathematics, in the study of finite algebras. We documented the production of human-competitive results in the discovery of particular algebraic terms, using both breve's implementation of PushGP and the ECJ genetic programming system [44]. We showed that GP can exceed the performance of every prior method of finding the relevant terms in either time or size by several orders of magnitude [28].

4.7 Educational Applications of breve

One of the more rewarding and unexpected uses of breve has been its application in educational settings. Because breve offers a relatively low barrier of entry to experimenting with multi-agent systems, it has been a useful tool in many educational contexts, even in those that do not deal explicitly with artificial life or simulation.

4.7.1 Artificial Life and Braitenberg Vehicles

The Braitenberg vehicles simulation included with breve has served as a useful introduction to many topics related to agent-based simulation, artificial life, artificial intelligence, and computer programming in general. It has been used for this purpose in several artificial life courses at Hampshire College.

Braitenberg vehicles, named after Valentino Braitenberg, are simple robotic designs that are capable of displaying surprisingly life-like behaviors from very simple circuitries of sensors and wheels [3]. One of the simplest creatures, dubbed "aggressor," demonstrates aggressive light-chasing behavior with a simple criss-crossed connection of light sensors and motors: Light sensed with the right sensor causes activation of the left wheel and vice versa, such that the vehicle turns toward and speeds in the direction of any lights in the environment.

Even by the standards of the rest of the breve demos, which are designed to be easy to use and to modify, the Braitenberg Vehicles simulation is exceptionally simple. A set of Braitenberg classes included with breve allows for the creation of Braitenberg vehicles and for placement of wheels and sensors in only a few lines of code. Students with no programming experience at all are quickly able to learn to design vehicles and place them in environments with lights (which the vehicles sense) and other obstacles. The students are thus able to quickly design and run simple robotics experiments with realistic physical simulation. A breve Braitenberg vehicle is shown in Fig 4.2.

4.7.2 Reactive Bouncy Balls for Kids

The "bouncy" breve simulation was developed for use in an activity with a fifth grade class at the Smith College Campus School. Like the Braitenberg vehicle simulation, the bouncy simulation allows users to create agents simply, but here the quest for simplicity has been taken to an extreme: The user can create an agent and specify its behavior with a single line of code such as

```
new bouncy with size 3.0 color (0.7, 0, 0.4) preferences (0, -1, 0.1).
```

Fig. 4.8 The breve "bouncy" simulation.

This creates a "bouncy" ball with size 3, color reddish-purple (the three numbers in parentheses specify amounts of red, green, and blue), and a preference to move quickly *away* from green and slowly *toward* blue. Users can create arbitrary numbers of such balls, with different sizes, colors, and preferences, and observe the often complex dynamics that emerge as the balls bounce around in a physically simulated arena. A slightly more complex syntax allows for the specification of balls that change their colors and preferences when they collide with other balls, producing extremely complex patterns of activity. A screen shot of the bouncy simulation is shown in Fig. 4.8.

4.7.3 Biology and SuperDuperWalker

SuperDuperWalker is a software-based framework for experiments on the evolution of locomotion. It simulates the behavior of evolving agents in a 3D physical simulation environment and displays this behavior in real time. A genetic algorithm controls the evolution of the agents. Students manipulate parameters with a graphical user interface and plot outputs using standard utilities. The software supports an inquiry cycle that has been piloted in a course titled "Biocomputational Developmental Ecology" at Hampshire College [35]. A screen shot of the SuperDuperWalker simulation is shown in Fig. 4.9.

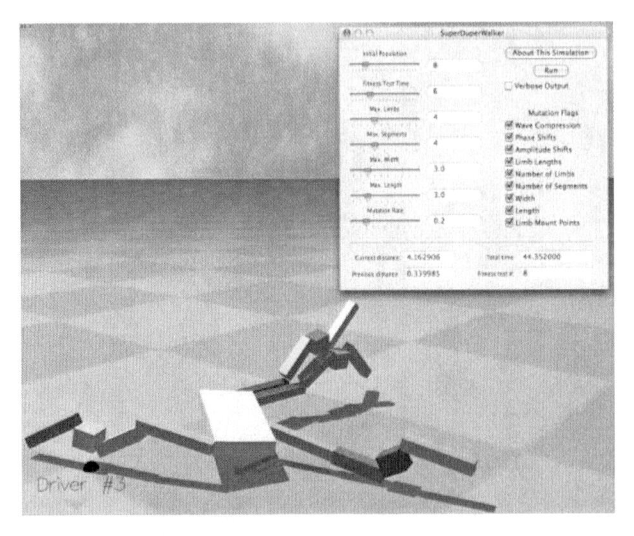

Fig. 4.9 The breve SuperDuperWalker simulation.

4.7.4 Artificial Intelligence in 3D Virtual Worlds

Breve has been used as the foundation for a new approach to the introductory artificial intelligence curriculum, which has been presented as a course twice (as of this writing) at Hampshire College under the title "Artificial Intelligence in 3D Virtual Worlds." This course covers roughly the same material as

Fig. 4.10 The breve WubWorld simulation.

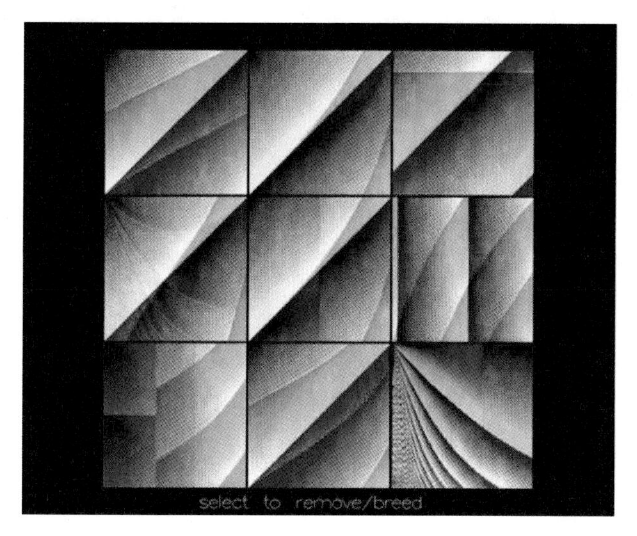

Fig. 4.11 The breve PushImageEvolve simulation.

is covered in other "agent-based" introductory AI courses, with an emphasis on reactive agents, neural networks, and evolutionary computation. Traditional AI topics such as knowledge representation, heuristic search, planning, and logic-based approaches are also covered but in less detail. The primary novelty of the course is that students work within breve, creating programs to control agents in engaging, dynamic, and visually rich virtual worlds. These virtual worlds include "WUB World," an environment inhabited by monsters (Wildly Unpredictable Biots, or WUBs), agents, obstacles, worm-holes, and energy sources, and a world in which teams of agents compete in a "capture the flag" game. The materials for this course, including the breve simulation files, are available online [29]. A screen shot of the WubWorld simulation is shown in Fig. 4.10.

4.7.5 Algorithmic Art

Breve's rich visualization subsystem is useful not only for developing and observing simulations but also for work in the computational arts. An "Algorithmic Arts" course, taught at Hampshire College (twice as of this writing), has used breve to provide students with tools for dynamically manipulating images, lines, shapes, and color in 3D space. The integration of these tools with breve allowed students to develop portfolios of algorithmic artwork that employed turtle graphics, Lindenmayer systems, iterated function systems, reaction/diffusion systems, texture mapping, 3D simulation, and interactive genetic algorithms. Fig. 4.11 shows a screen shot of the PushImageEvolve simulation for interactive image evolution, which allows the user to evolve image-generating Push programs using a simple point-and-click interface.

4.8 Conclusion

Breve is a powerful, free software package for multi-agent simulation. With features such as realistic physical simulation, real-time 3D visualization, support for evolutionary computation, and integration with powerful scripting languages, breve greatly simplifies the rapid development and exploration of advanced artificial life experiments. We have successfully applied breve to the development of multi-agent systems to study artificial life and a wide variety of other fields.

References

1. Barnum, H., Bernstein, H.J., Spector, L.: Quantum circuits for OR and AND of ORs. Journal of Physics A: Mathematical and General **33**(45), 8047–8057 (2000)
2. Bhalla, N., Bentley, P.J., Jacob, C.: Mapping virtual self-assembly rules to physical system. In: A. Adamatzky, L. Bull, B.D.L. Costello, S. Stepney, C. Teuscher (eds.) Proceedings of the 2007 Conference on Unconventional Computing. Luniver Press, Beckington (2007)
3. Braitenberg, V.: Vehicles: Experiments in Synthetic Psychology. The MIT Press, Cambridge, MA (1986)
4. Cohen, P.R., Morrison, C.T., Cannon, E.: Maps for verbs: The relation between interaction dynamics and verb use. In: L.P. Kaelbling, A. Saffiotti (eds.) Proceedings of the 19th International Joint Conference on Artificial Intelligence (IJCAI-05), pp. 1022–1027. Professional Book Center, Edinburgh (2005)
5. Don, O., Amos, M.: An ant-based algorithm for annular sorting. ArXiv Nonlinear Sciences e-prints (2005)
6. Hamann, H., Wörn, H.: An analytical and spatial model of foraging in a swarm of robots. In: E. Sahin, W.M. Spears, A.F.T. Winfield (eds.) Swarm Robotics, Second International Workshop, SAB 2006, Rome, Italy, September 30-October 1, 2006, pp. 43–55. Springer, Berlin (2006)
7. Klein, J.: breve website. URL `http://www.spiderland.org/`
8. Klein, J., Spector, L.: Unwitting distributed genetic programming via asynchronous javascript and XML. In: H. Lipson (ed.) GECCO '07: Proceedings of the 9th Annual Conference on Genetic and Evolutionary Computation, vol. 2, pp. 1628–1635. ACM Press, New York (2007)
9. Komosinski, M., Ulatowski, S.: Framsticks – artificial life. In: C. Nédellec, C. Rouveirol (eds.) ECML 98 Demonstration and Poster Papers, pp. 7–9. Chemnitzer Informatik-Berichte, Chemnitz (1998)
10. Kriplean, T.L.: Evolving an ecology of two-tiered organizations. In: F. Rothlauf (ed.) GECCO '05: Proceedings of the 2005 Workshops on Genetic and Evolutionary Computation, pp. 402–406. ACM Press, New York (2005)
11. Lassabe, N., Luga, H., Duthen, Y.: A New Step for Evolving Creatures. In: IEEE-ALife'07, Honolulu, Hawaii, 01/04/2007-05/04/2007, pp. 243–251. IEEE Press (2007)

12. Luke, S., Cioffi-Revilla, C., Panait, L., Sullivan, K., Balan, G.: Mason: A multiagent simulation environment. Simulation **81**(7), 517–527 (2005)
13. Minar, N., Burkhart, R., Langton, C., Askenazi, M.: The swarm simulation system, a toolkit for building multi-agent simulations. Tech. Rep. 96-06-042, Sante Fe Institute (1996)
14. Mirtich, B.: V-clip: Fast and robust polyhedral collision detection. ACM Transactions on Graphics **17**(3), 177–208 (1998)
15. Mirtich, B., Canny, J.F.: Impulse-based simulation of rigid bodies. In: SI3D '95: Proceedings of the 1995 Symposium on Interactive 3D Graphics, pp. 181–188, 217. ACM Press, New York (1995)
16. North, M., Macal, C.: Escaping the accidents of history: An overview of artificial life modeling with repast. In: A. Adamatzky, M. Komosinski (eds.) Artificial Life Models in Software, pp. 115–142. Springer-Verlag New York, Secaucus (2006)
17. Resnick, M.: Starlogo: an environment for decentralized modeling and decentralized thinking. In: CHI '96: Conference Companion on Human Factors in Computing Systems, pp. 11–12. ACM Press, New York (1996)
18. Reynolds, C.W.: Flocks, herds, and schools: A distributed behavioral model. Computer Graphics **24**(4), 25–34 (1987)
19. Riolo, R.L., Cohen, M.D., Axelrod, R.: Evolution of cooperation without reciprocity. Nature **414**, 441–443 (2001)
20. Sims, K.: Evolving virtual creatures. In: Proceedings of SIGGRAPH '94 (Orlando, Florida, July 1994), pp. 15–22 (1994)
21. Spector, L.: Autoconstructive evolution: Push, pushGP, and pushpop. In: Proceedings of the Genetic and Evolutionary Computation Conference (GECCO-2001), pp. 137–146. Morgan Kaufmann, San Francisco (2001)
22. Spector, L.: Adaptive populations of endogenously diversifying pushpop organisms are reliably diverse. In: R. Standish, M.A. Bedau, H.A. Abbass (eds.) Proceedings of the Eighth International Conference on Artificial Life, pp. 142–145. MIT Press, Cambridge, MA (2002)
23. Spector, L.: Automatic Quantum Computer Programming: A Genetic Programming Approach. Kluwer Academic Publishers, Boston (2004)
24. Spector, L., Barnum, H., Bernstein, H.J.: Genetic programming for quantum computers. In: Genetic Programming 1998: Proceedings of the Third Annual Conference, pp. 365–373. Morgan Kaufmann, Madison, WI (1998)
25. Spector, L., Barnum, H., Bernstein, H.J., Swamy, N.: Finding a better-than-classical quantum AND/OR algorithm using genetic programming. In: P.J. Angeline, Z. Michalewicz, M. Schoenauer, X. Yao, A. Zalzala (eds.) Proceedings of the Congress on Evolutionary Computation, vol. 3, pp. 2239–2246. IEEE Press, Washington (1999)
26. Spector, L., Barnum, H., Bernstein, H.J., Swamy, N.: Quantum computing applications of genetic programming. In: L. Spector, W.B. Langdon, U.M. O'Reilly, P.J. Angeline (eds.) Advances in Genetic Programming 3, chap. 7, pp. 135–160. MIT Press, Cambridge, MA (1999)
27. Spector, L., Bernstein, H.J.: Communication capacities of some quantum gates, discovered in part through genetic programming. In: J.H. Shapiro, O. Hirota (eds.) Proceedings of the Sixth International Conference on Quantum Communication, Measurement, and Computing (QCMC), pp. 500–503. Rinton Press, Princeton (2003)
28. Spector, L., Clark, D.M., Lindsay, I., Barr, B., Klein, J.: Genetic programming for finite algebras. In: Proceedings of the 10th Annual Conference on Genetic and Evolutionary Computation. ACM Press, London (2008)
29. Spector, L., Klein, J.: CS 263: Artificial Intelligence in 3D Virtual Worlds. URL `http://hampshire.edu/lspector/cs263/cs263s04.html`
30. Spector, L., Klein, J.: Evolutionary dynamics discovered via visualization in the breve simulation environment. In: Workshop Proceedings of the 8th International Conference on the Simulation and Synthesis of Living Systems. University of New South Wales (2002)

31. Spector, L., Klein, J.: Trivial geography in genetic programming. In: T. Yu, R.L. Riolo, B. Worzel (eds.) Genetic Programming Theory and Practice III, vol. 9, chap. 8, pp. 109–123. Springer, New York (2005)
32. Spector, L., Klein, J.: Genetic stability and territorial structure facilitate the evolution of tag-mediated altruism. Artificial Life **12**(4), 1–8 (2006)
33. Spector, L., Klein, J.: Multidimensional tags, cooperative populations, and genetic programming. In: R.L. Riolo, T. Soule, B. Worzel (eds.) Genetic Programming Theory and Practice IV, vol. 5, chap. 15, pp. 97–112. Springer, New York (2006)
34. Spector, L., Klein, J., Feinstein, M.: Division blocks and the open-ended evolution of development, form, and behavior. In: H. Lipson (ed.) Proceedings of the 9th Annual Conference on Genetic and Evolutionary Computation, pp. 316–323. ACM Press, New York (2007)
35. Spector, L., Klein, J., Harrington, K., Coppinger, R.: Teaching the evolution of behavior with superduperwalker. In: C.K. Looi, G.I. McCalla, B. Bredeweg, J. Breuker (eds.) Proceedings of the 12th International Conference on Artificial Intelligence in Education, pp. 923–925. IOS Press, Washington (2005)
36. Spector, L., Klein, J., Keijzer, M.: The push3 execution stack and the evolution of control. In: H.G. Beyer, U.M. O'Reilly (eds.) GECCO 2005: Proceedings of the 2005 Conference on Genetic and Evolutionary Computation, vol. 2, pp. 1689–1696. ACM Press, Washington (2005)
37. Spector, L., Klein, J., Perry, C.: Tags and the evolution of cooperation in complex environments. In: Proceedings of the AAAI 2004 Symposium on Artificial Multiagent Learning. AAAI Press, Melno Park, CA (2004)
38. Spector, L., Klein, J., Perry, C., Feinstein, M.: Emergence of collective behavior in evolving populations of flying agents. In: E. Cantu-Paz, J.A. Foster, K. Deb, L.D. Davis, R. Roy, U.M. O'Reilly, H.G. Beyer, R. Standish, G. Kendall, S. Wilson, M. Harman, J. Wegener, D. Dasgupta, M.A. Potter, A.C. Schultz, K.A. Dowsland, N. Jonoska, J. Miller (eds.) Proceedings of the Genetic and Evolutionary Computation Conference, vol. 2724, pp. 61–73. Springer-Verlag, Berlin (2003)
39. Spector, L., Klein, J., Perry, C., Feinstein, M.: Emergence of collective behavior in evolving populations of flying agents. Genetic Programming and Evolvable Machines **6**(1), 111–125 (2005)
40. Spector, L., Perry, C., Klein, J., Keijzer, M.: Push 3.0 programming language description. Tech. Rep. HC-CSTR-2004-02, School of Cognitive Science, Hampshire College (2004)
41. Spector, L., Robinson, A.: Genetic programming and autoconstructive evolution with the push programming language. Genetic Programming and Evolvable Machines **3**(1), 7–40 (2002)
42. Szymanski, M., Breitling, T., Seyfried, J., Wörn, H.: Distributed shortest-path finding by a micro-robot swarm. In: M. Dorigo, L.M. Gambardella, M. Birattari, A. Martinoli, R. Poli, T. Sttzle (eds.) Ant Colony Optimization and Swarm Intelligence, 5th International Workshop, ANTS 2006, Brussels, Belgium, September 4-7, 2006, Proceedings, vol. 4150, pp. 404–411. Springer, Berlin (2006)
43. Veeraswamy, A., Amavasai, B.: Optimal path planning using an improved a * algorithm for homeland security applications. In: AIA'06: Proceedings of the 24th IASTED International Conference on Artificial Intelligence and Applications, pp. 50–55. ACTA Press, Anaheim, CA (2006)
44. Wilson, G.C., McIntyre, A., Heywood, M.I.: Resource review: Three open source systems for evolving programs–lilgp, ECJ and grammatical evolution. Genetic Programming and Evolvable Machines **5**(1), 103–105 (2004)

Chapter 5
Framsticks: Creating and Understanding Complexity of Life

Maciej Komosinski and Szymon Ulatowski

Life is one of the most complex phenomena known in our world. Researchers construct various models of life that serve diverse purposes and are applied in a wide range of areas – from medicine to entertainment. A part of artificial life research focuses on designing three-dimensional (3D) models of life-forms, which are obviously appealing to observers because the world we live in is three dimensional. Thus, we can easily understand behaviors demonstrated by virtual individuals, study behavioral changes during simulated evolution, analyze dependencies between groups of creatures, and so forth. However, 3D models of life-forms are not only attractive because of their resemblance to the real-world organisms. Simulating 3D agents has practical implications: If the simulation is accurate enough, then real robots can be built based on the simulation, as in [22]. Agents can be designed, tested, and optimized in a virtual environment, and the best ones can be constructed as real robots with embedded control systems. This way artificial intelligence algorithms can be "embodied" in the 3D mechanical constructs.

Perhaps the first well-known simulation of 3D life was Karl Sims' 1994 virtual creatures [34]. Being visually attractive, it demonstrated a successful competitive coevolutionary process, complex control systems, and interesting (evolved) behaviors. However, this work did not become available for users as documented software. A number of 3D simulation engines was developed later (see [36] for a review), but most of them are either used for a specific application or experiment (and are not available as general tools for users) or focus primarily on simulation (without built-in support for genetic encodings, evolutionary processes or complex control). Notable exceptions are the breve simulation package described in Chapter 4 and a recent version of StarLogo – StarLogo TNG (Chapter 6).

Framsticks [19, 17, 18], a software platform described in this chapter, does not address a single purpose or a single research problem. On the contrary, it is built to support a wide range of experiments and to provide all of its functionality to users, who can use this system in a variety of ways. The significance of understanding is central for the development of Framsticks. Although the system is a simplified model of reality, it is easily capable of producing phenomena more complex than a human can comprehend [14]. Thus, it is essential to provide automatic analysis and support tools. Intelligible visualization is one of the most fundamental means for human understanding of artificial life-forms, and this feature is present in the software.

The Framsticks system is designed so that it does not introduce restrictions concerning complexity and size of creatures. Therefore, neural networks can have any topology and dimension, allowing for a range of complex behaviors, some described in Section 4 of [14]. Avoiding limitations is important because Framsticks is ultimately destined to experiments with open-ended evolution, where interactions between creatures and the environment are the sources of competition, cooperation, communication, and intelligence.

Further sections of this chapter focus on the following issues:

5.1 — history of Framsticks and available software
5.2 — simulation model, morphology, control system, communication, environment
5.3 — genetics, genotype–phenotype relationship, mutation and crossing over, evolution
5.4 — scripting, experiment definitions and system framework, creating custom neurons and experiments, popular experiments
5.5 — selected tools provided to support research and education
5.6 — sample experiments that have already been performed, as well as some new ideas
5.7 — entertaining Framsticks applications
5.8 — summary.

5.1 Available Software and Tools

Framsticks was first released in late 1996, but the first official releases became available on the Internet in June 1997. Versions 1.x provided numerous parameters to customize experiments, but the experiment logic, visualization, and neurons were hard-coded.

In 2002, starting with version 2.0, the scripting language *FramScript* was introduced, which allows for flexible control of most parts of the software – on both high level (adjusting parameters) and low level (creating custom procedures). Scripting is addressed in Section 5.4.

In 2004, the first official release of the Framsticks Theater (a simple-interface, attractive graphical application) took place, with unofficial releases available since 2002.

In 2008, version 3.0 introduced an accurate mechanical simulation mode. Useful features were added to the particle agents simulation mode and support for agent communication was greatly enhanced. A few applications were joined: The Framsticks Viewer was included in the Framsticks Theater, and the Framsticks Server was included in the Framsticks Command-Line Interface.

The Framsticks family of programs includes the following:

- Framsticks GUI (Graphical User Interface) – the most popular program, where simulated creatures, genotypes, and the virtual world are presented visually and allow for user interaction (dragging creatures, online genotype visualization, etc.).
- Framsticks CLI (Command-Line Interface) – a program where commands are issued using text. Useful for long, time-consuming and/or well-defined experiments, which can be performed automatically (batch processing) or remotely. This program has no overheads of the GUI and can be compiled for most operating systems. The CLI also acts as the Framsticks Network Server. The server communication protocol is published, which allows for development of third-party clients.
- Framsticks Theater – a complete Framsticks simulator with a simple menu and predefined actions ("shows"), described in more detail in Section 5.7. The Theater can also be used as a viewer of genotypes specified by users or read from files.
- Framsticks Editor (FRED) – a simple open-source graphical editor that allows users to easily design creatures without any knowledge about genetic encodings. It is mentioned in Section 5.7.
- Framsticks Network Clients – open-source programs that interact with the Framsticks Server (server(s) and client(s) can also be run on a single computer). Two basic roles of clients are (1) the GUI for the server and (2) visualization of the virtual world simulated on the server, as shown in Figs. 5.1 and 5.2. However, clients can use the server in a number of ways, including distributed evolution, modeling of ecosystems and migration, real-time interaction in mixed realities, and much more. Many clients can connect to the same server at the same time, and clients can exchange information between themselves.
- Other helper programs and scripts: brain optimizers, analyzers of experiment output data, graph generators, and so forth.

The above applications are in continuous development, with new releases coming out periodically. Framsticks genotypes and experiment proposals can be browsed, downloaded, and uploaded using the Internet database – Framsticks Experimentation Center.

Framsticks software documentation is available in many forms, including web site information [19], single-document Framsticks Manual, step-by-step tutorials (strongly recommended for beginners [20]), a web forum, and the FramScript reference that describes objects and functions useful when writing scripts.

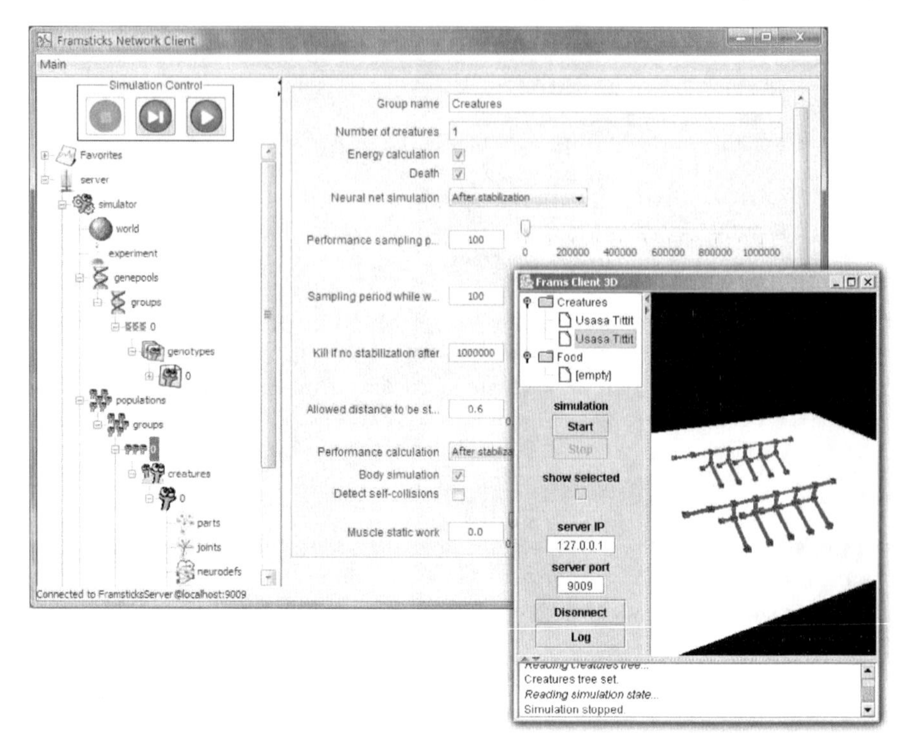

Fig. 5.1 Two sample network clients: the GUI and the virtual world viewer.

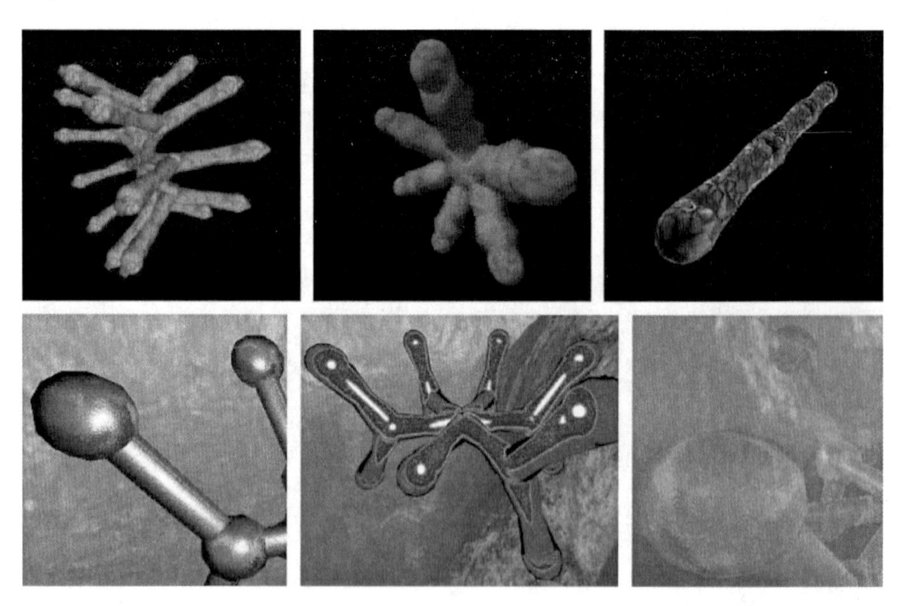

Fig. 5.2 Selected 3D real-time display styles in the Java-based network client. Top row: Basic, Fur, and Stone. Bottom row: Metal, Cartoon, and Glass.

5.2 Simulation

5.2.1 Creature Model

Creatures in Framsticks are modeled as bodies and brains. Bodies consist of body parts connected by body joints; this constitutes an undirected, spatial 3D graph structure. Brain is made of neurons (signal processing units, receptors, and effectors) and neural connections; these form a directed graph that may be connected to body where the neural units are embodied.

The four elements of the model – parts, joints, neural units, and neural connections – can be characterized by a variable number of properties like mass, position, orientation, friction, stiffness, durability, assimilation ability, weight, and neural parameters. For details on implementation of this model, refer to the GDK [37].

5.2.2 The Three Modes of Body Simulation

There is always a trade-off between simulation accuracy and simulation time. To perform evolution and evaluate thousands of individuals, fast simulation is required; on the other hand, the model should be realistic (detailed) enough to allow for realistic behaviors. When we expect emergence of more and more sophisticated phenomena, the evolution takes more time – thus simulation needs to be faster, and therefore less accurate. While some experiments are focused on realistic 3D mechanical simulations (e.g., in robotics), others are only concerned with information processing in the agent's brain (e.g., in Agent-Based Modeling applications).

In Framsticks step-by-step simulation, all major kinds of interactions between basic physical objects are considered: static and dynamic friction, damping, action and reaction forces, energy losses after deformations, gravitation, and uplift pressure – buoyancy (in a water environment). However, some experiments do not require such complexity and realism, therefore Framsticks provides three simulation modes to meet diverse requirements.

It is possible to adjust gravity force in all the three simulation modes. To improve performance, collision detection may be deactivated between agents or groups of agents (inter-creature collisions) when collisions are not of interest.

Accurate Rigid Body Dynamics

In this mode, Framsticks uses the Open Dynamics Engine [35] to simulate bodies made of boxes or cylinders that correspond to body parts and joints

Fig. 5.3 Accurate collision detection in the rigid body dynamics mode.

(Figs. 5.3 and 5.4, left). The simulation of physics is accurate enough to expect similar behavior from the real-world robotic equivalent. A number of parameters are provided to customize simulation (cf. [16]); it is also possible to deactivate collisions among creature parts (intra-creature) when these are not required.

Simplified Elastic Body Dynamics

This mode uses the native Framsticks simulation engine, Mechastick (Fig. 5.5). It provides much better performance than ODE at the cost of realism. In order to avoid computational complexity in this simulation mode, intra-creature collisions are not detected. Creatures are built of interconnected linear segments ("sticks") corresponding to body parts and joints. Articulations exist between sticks where they share an endpoint; the articulations are unrestricted in all three degrees of freedom (bending in two planes plus twisting).

In this simulation mode, bodies exhibit elastic dynamics which limits size and complexity of constructs under normal gravity conditions. On the other

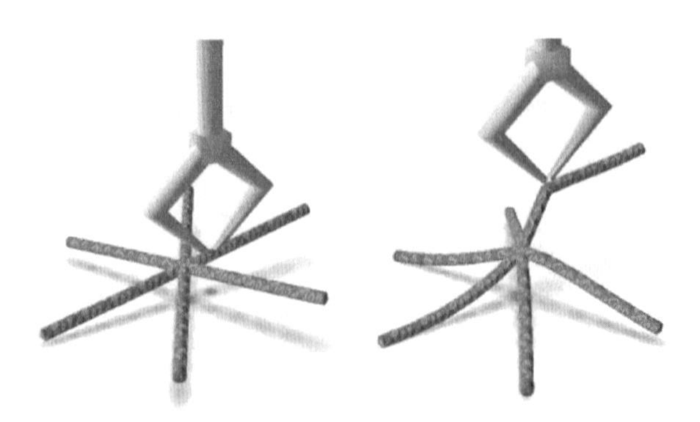

Fig. 5.4 Body dynamics simulation modes: rigid (left) and elastic (right).

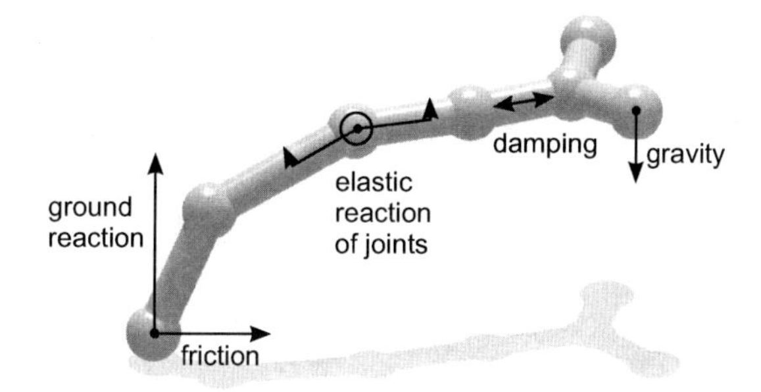

Fig. 5.5 Forces considered in Mechastick.

hand, flexibility and elastic deformations are extremely advantageous for efficient locomotion techniques (compare Fig. 5.4) and other activities. It is much harder to design or evolve naturally moving structures when rigid body dynamics is employed.

For each simulated element, Mechastick provides information about axial and angular stress values (Fig. 5.6). This information can be used to test and evolve stable, durable, and robust structures.

Particle Agents

When studying mechanics of the animal limb or robotic leg, one can spend many simulation steps to make the structure perform a single movement. However, there are many applications where morphology of agents is not considered at all, or the complete agent body is considered solid. Such experiments include navigation, communication, swarming and group behaviors, various massive multi-agent setups, and settings where agents are primarily considered as brains, not bodies.

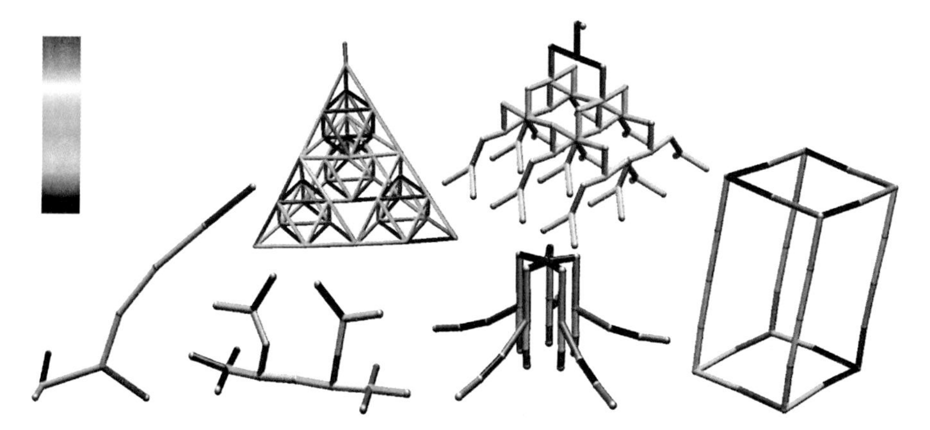

Fig. 5.6 Visualizations of axial stress in the Mechastick simulation engine.

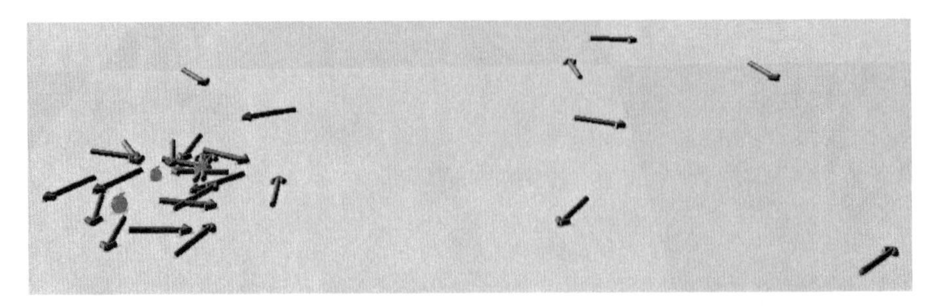

Fig. 5.7 Particle agents simulation mode. Dynamics of arrow-shaped bodies is not regarded in this experiment.

These experiments do not need true simulation of body dynamics – simple "move forward" or "turn left" commands are sufficient. In the "particle agents" mode, simple high-level methods can be used (rotating agents, setting their direction and speed) to facilitate experiment design (Figs. 5.7, 5.10, and 5.12). When bodies are not important, the simulation can be extremely simplified by making each agent a point of mass.

5.2.3 Brain

Brain (the control system) is made of neurons and their connections. A neuron may be a signal processing unit, but it may also interact with body as a receptor (sensor) or effector (actuator). There are some predefined types of neurons, for example:

- "N": the standard Framsticks neuron, which is a generalized version of the popular weighted-sum, sigmoid transfer function neuron used commonly in AI. The three additionally introduced parameters influence speed and tendency of changes of the inner neuron state and the steepness of the sigmoid transfer function. In a special case, when the three parameters are assigned specific values, the characteristics of the "N" neuron become identical to the popular, reactive AI neuron. In this case, neural output reflects instantly input signals. More information and sample neuronal runs can be found in the *simulation details* section at [19].
- "Sin": a sinusoidal generator with frequency controlled by its inputs.
- "Rnd": random noise generator.
- "Thr": thresholding neuron.
- "Delay": delaying neuron.
- "D": differentiating neuron.

It is possible to easily add custom, user-designed neurons using FramScript; an example is shown in Section 5.4.

The neural network can have any topology and complexity. Neurons can be connected with each other in any way (some may be unconnected). Inputs should be connected to outputs of another neuron (including sensors), while outputs should be connected to inputs of other neurons (including effectors). Sample control systems are shown in Fig. 5.8. Note that a single control system may contain many unconnected or independent subsystems.

Neurons can send multiple values in their output. This extension allows one to design complex neurons that provide a vector of output values, which is used, for example, in the Fuzzy and the Vector Eye neurons described later.

A section in Part I of the Framsticks tutorial [20] concerns understanding and designing control systems, and there are exercises in Part IV on the development of custom script-based neurons.

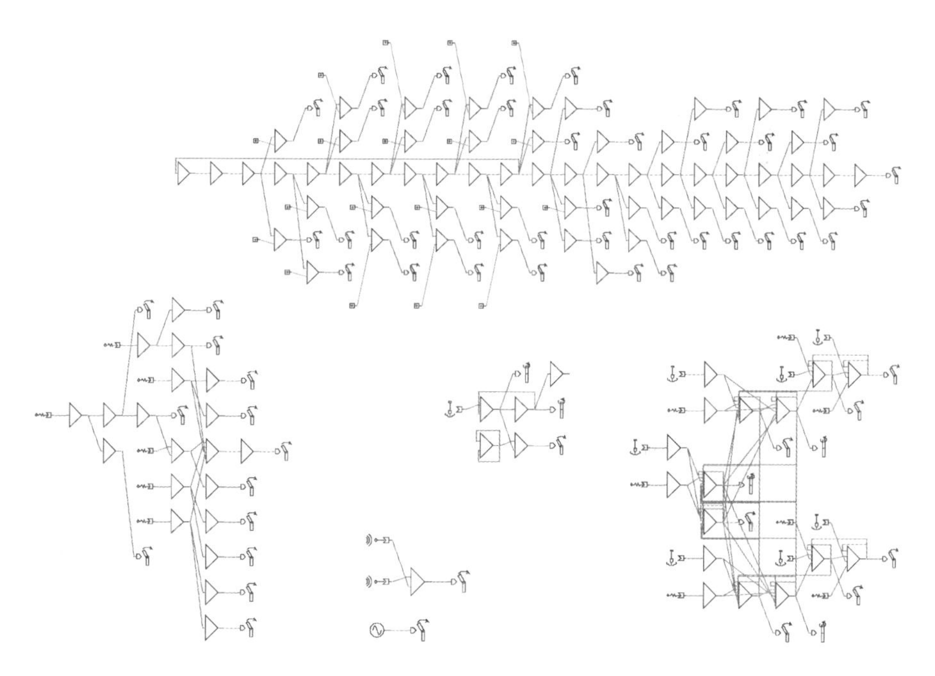

Fig. 5.8 Sample neural networks. Triangles are the standard signal-processing neurons ("N"). Receptors act as inputs (shown usually on the left side: gyroscope, touch, smell). Controlled muscles (rotating, bending) are usually on the right side. Some inputs are connected to the output of the sinus generator (⊚) or the constant signal generator (⊞). Note recurrent connections. Parallel connections are also allowed.

5.2.4 Receptors and Effectors

Receptors and effectors are interacting between body and brain. They must be connected to brain in order to be useful, but they also interact with creature's body and the world. The three basic Framsticks receptors (sensors) include "G" for orientation in space (equilibrium sense, gyroscope), "T" for detection of physical contact (touch), and "S" for detection of energy (smell); see Figs. 5.8 and 5.9.

The two basic Framsticks effectors are muscles: bending and rotating. Positive and negative changes of muscle control signal make adjacent sticks move in either direction, which is analogous to the natural systems of muscles, with flexors and extensors. The strength of a muscle determines its effective ability of movement and speed (acceleration). If energetic issues are considered in an experiment, then muscle strength affects energy consumption.

A sample framstick equipped with basic receptors and effectors is shown in Fig. 5.9. Other examples of receptors and effectors are energy level meter, water detector, vector eye, linear actuator, and thrust.

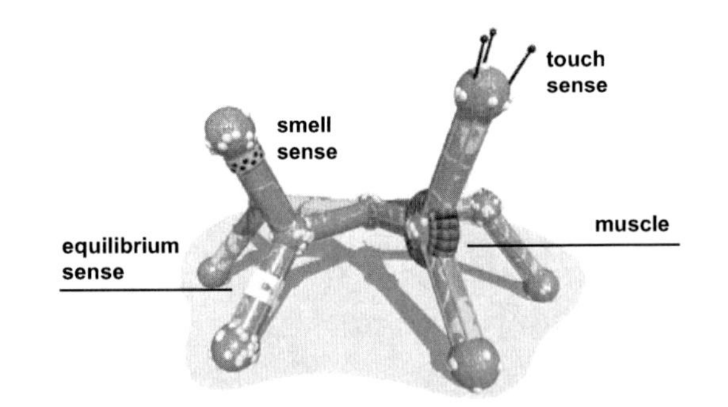

Fig. 5.9 Basic receptors (equilibrium, touch, smell) and effectors (muscles).

5.2.5 Communication

Since agents, their neurons, and the experiment logic are all script based, it is possible to implement virtually any kind of communication. Creatures, their components, and properties of these components are all accessible in script objects and can be modified by the experimenter according to their needs.

There are also dedicated objects and functions that facilitate implementation of the most common communication settings. The two basic communication objects are Channel and Signal.

- Channel objects are defined by unique names that represent either a physical medium (e.g., "light", "smell"), some abstract information ("danger", "goal"), or characterize a group of signal holders ("flock", "team_23").

- Signal objects store the value that is actually transmitted in a channel. The type of the value is arbitrary so that objects can be transmitted apart from simple values. For a signal, one can also set its power and flavor (which can be used to differentiate between multiple signals in a single channel). Signals belong to other objects – World, Creature, and Neuro – as members of their `signals` collections. World signals are stationary, while Creature signals and Neuro signals are carried by their owner.
- Functions used to receive signals can read the aggregated signal intensity from a channel (useful for physical quantities), find the strongest signal in a channel (taking into account location of transmitters and signal intensity), or enumerate all nearby signals.

Framsticks allows for visual presentation of signals in the world, which is useful both for debugging and as a part of the experiment visualization.

Communication: The Fireflies Example

In this example, Framsticks communication features are used to simulate light transmission, and two specialized neural units (the SeeLight receptor and the Light effector) are required. Agents equipped with these units are able to synchronize their flashing patterns only using local information (aggregated light intensity). See Fig. 5.10 and the Framsticks Theater show for a live demonstration.

A manually designed neural network that handles synchronization of flashing is shown in Fig. 5.11. The output of the light receptor contributes to the charging potential of neuron #6, effectively shortening the flashing cycle. Agents that flash "too late" receive most of the incoming light signal during their charging phase. This makes them charge faster and catch up with the other agents.

Fig. 5.10 The fireflies example and spatial intensity of the "light" signals.

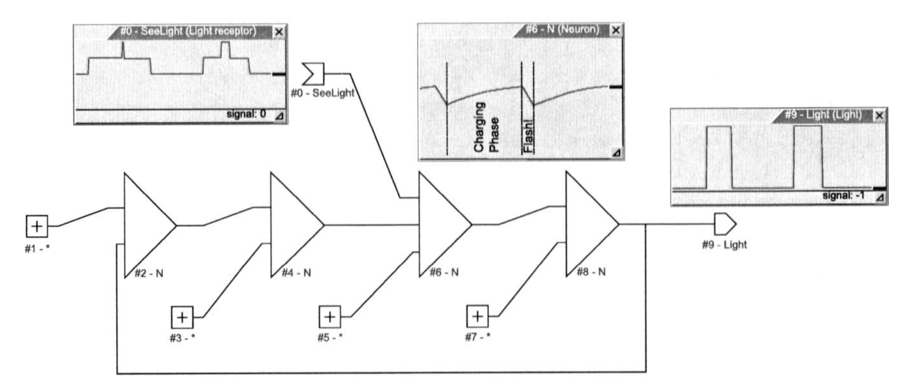

Fig. 5.11 A sample neural network that handles synchronization of flashing of a firefly.

The representation of this network in the *f1* genetic encoding (cf. Section 5.3) is as follows:

```
X [SeeLight,flavor:0] [*] [-1:2.26,6:-2,in:0.01,fo:0.01,si:1]
[*] [-2:1,-1:-0.5,si:9999,fo:1,in:0]
[*] [-2:2,-6:0.3,-1:-0.4,in:0.01,fo:0.01,si:1]
[*] [-2:1,-1:-0.5,si:9999,fo:1,in:0] [Light,-1:-1,flavor:0]
```

This genotype can be subject to evolution to obtain various flashing behaviors, depending on the fitness function used. The full source code for the the Light effector and SeeLight receptor is available in the "scripts" subdirectory of the Framsticks distribution (cf. Section 5.4 on scripting). In short, the Light effector, when created, registers one signal in the "light" channel:

```
Neuro.signals.add("light");
```

In each simulation step, this effector adjusts the power of this signal depending on its neural input value:

```
Neuro.signals[0].power=...;
```

The SeeLight receptor is very simple. In each step, it just sets its neural output to the amount of light perceived:

```
Neuro.state=Neuro.signals.receive("light");
```

Communication: The Boids Example

The implementation of boids [31] uses flexibility of the Framsticks communication to efficiently obtain the list of neighbor creatures. The data being communicated between agents are references to their own objects. The neighbor list is processed on each step by the creature handler to calculate aggregated direction of flight based on the motion of the neighbors. See Fig. 5.12 and the Framsticks Theater show for a live demonstration.

When an agent is created (the onBorn() function), it registers one signal in the "flock" channel and sets the signal value to reference its own body:

```
var signal=Creature.signals.add("flock");
signal.value=Creature.getMechPart(0);
```

To receive data, each agent in every simulation step receives a set of signals (i.e., references to neighbors within the specified range) and then can enumerate this array to perform necessary calculations:

```
var neighbors=creature.signals.receiveSet("flock",maxrange);
for(i=0;i<neighbors.size;i++) {...}
```

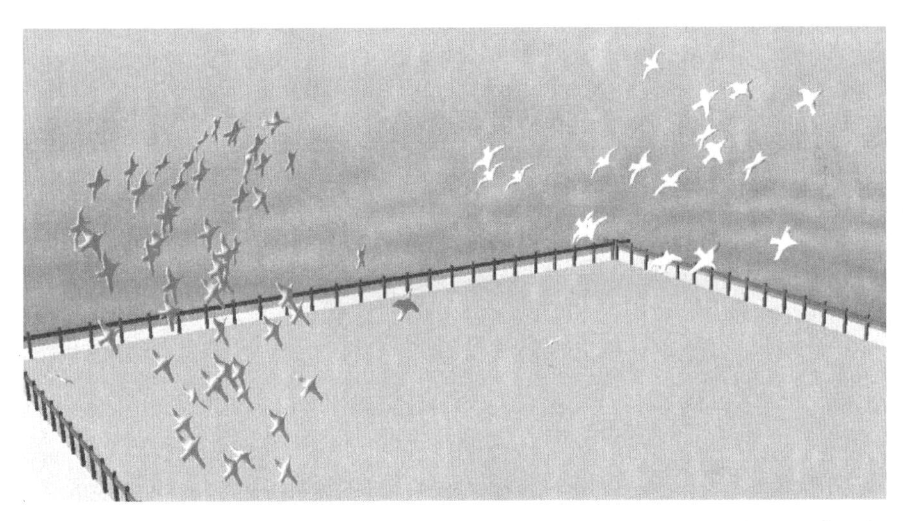

Fig. 5.12 The boids show in the Framsticks Theater. Creatures consist of one part only. They are assigned a visual style that uses a bird 3D model for display.

5.2.6 Environment

The world can be flat, built of smooth slopes, or built of blocks. It is possible to adjust the water level, so that not only walking/running/jumping creatures but also the swimming and amphibian ones are simulated. The boundaries of the virtual world can be one of three types:

- hard (surrounding wall: it is impossible to cross the boundary);
- wrap (crossing the boundary means teleportation to the other side of the world);
- none (the world is infinite).

These options are useful in various kinds of experiments and measurements of creature performance.

World heightfields are defined in a simple way, and a number of Framsticks heightfield generators exist. Real geographic information system (GIS) data from Earth or other planets can also be used: The PrettyMap conversion utilities can convert raster elevation and symbolic maps from most file formats into the Framsticks world heightfield format.

5.3 Genetics and Evolution

Framsticks supports multiple genetic encodings (also called representations or "genotype languages") [21]. The system manipulates and transforms genotype strings expressed in various representations and ultimately decodes them into the internal representation used by the simulator to construct a creature (phenotype). This means that one can describe a creature using genomes expressed in different "languages."

Any creature can be completely described using a low-level representation named *f0*, by listing all of the creature components and their properties. Other higher-level encodings convert their representations into the corresponding *f0* version (possibly through another intermediary representation), as shown in Fig. 5.13. The reverse mapping from lower-level into higher-level encodings is usually difficult or impossible to compute, which is also true in the biological realm. As a consequence, in the general case it is not possible to convert a lower-level representation into a higher-level one (or a higher-level one into another higher-level one).

Each encoding has its own genetic operators (mutation, crossover, and the optional repair) and a translation procedure that allows users to compute a phenotype from each genotype expressed in this encoding. A new encoding can be added relatively easily by implementing these components. Framsticks software is accompanied by the Genotype Development Kit (GDK)

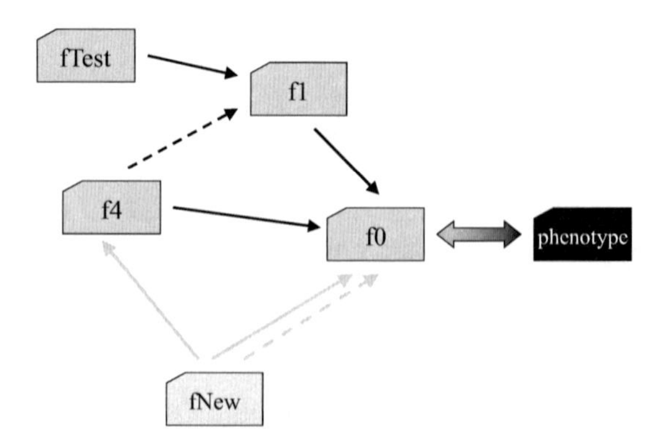

Fig. 5.13 A graph illustrating the idea of hierarchical genetic encodings (as nodes) and their translation procedures (as arcs). The dashed arrow depicts an approximate translation. Multiple, alternative translation methods may exist, as shown for the translation from *fNew* to *f0*.

which simplifies this process [37]. The most popular genetic encodings are characterized below.

The direct low-level, or *f0*, encoding describes agents exactly as they are represented in the model (Sect. 5.2.1). It does not use any higher-level features to make genotypes more compact or flexible. Its useful characteristics are that it has a minimal decoding cost and that every possible creature can be described using this encoding. Each *f0* genotype consists of a list of descriptions of all the elements a creature is composed of: parts, joints, neurons, and neural connections.

The recurrent direct encoding (labeled *f1*) also describes all the parts of the corresponding phenotype. Body properties are represented locally, so that most of the properties (and neural network connections) are maintained when a genotype section is moved to another place of the genotype. Control elements (neurons, receptors) are described near the elements under their control (muscles, sticks). Only tree-like body structures can be represented in *f1* (no cyclic bodies can be described, but arbitrary neural network topologies are possible). This concise encoding is relatively easy for humans to manipulate and manually design creatures by editing their genotypes. For example, the 'X' char means a stick, parentheses are used to branch body structure, 'r' and 'R' letters are used to rotate the branching plane, neurons are described in square brackets with their relative links and weights, and so forth.

The developmental encoding (*f4*) is development-oriented, similar to the encodings applied for evolving neural networks [6]. An interesting merit of developmental encoding is that it can incorporate symmetry and modularity, features commonly found in natural systems, yet difficult to formalize. *f4* seems similar to *f1*, but codes are interpreted as commands by cells (sticks, neurons, etc.). Cells can change their properties and divide. Each cell maintains its own pointer to the current command in the genetic code. After division, cells can execute different codes in parallel and differentiate themselves. The final body (phenotype) is the result of a development process. It starts with an undifferentiated ancestor cell and ends with a collection of interconnected differentiated cells (sticks, neurons, and connections).

Other available representations include similarity-based encoding, biological encoding (with a finite genetic alphabet and start/stop codons), chemical encoding (metabolic rules of growth), messy encoding (any sequence of symbols is valid), and a parametric Lindenmayer system encoding [9].

Each of the genetic encodings and the corresponding genetic operators have been carefully designed and tested, and each encoding was based on theoretical considerations and experimental tests. More detailed descriptions can be found in [18]. Examples of simple genotypes and corresponding phenotypes (creatures) are shown in Fig. 5.14.

The procedure of genotype translation may provide additional information regarding the relation of individual genes in the source and target encodings. If this information is available, then it is possible to track the relationship between parts of a genotype (genes) and parts of the corresponding creature

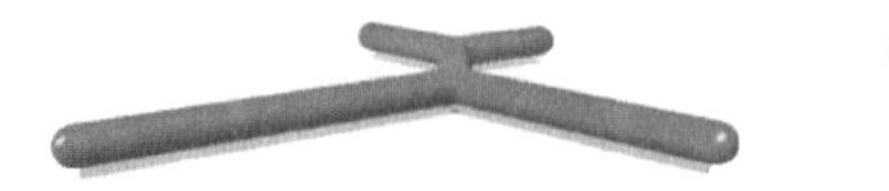

Fig. 5.14 Left: example of the *f1* genotype XXX(XX,X(X,X)). Right: example of the *f4* genotype with the repetition gene: rr<X>#5<,<X>RR< <llX>LX>LX> >X.

(phenes). Details of this process and examples are shown in [21]. Fig. 5.15 illustrates how this information can be visualized and used both ways.

In the Framsticks software, it is possible to select parts of the phenotype and genotype to get an instant visual feedback and understand their relationship; see Fig. 5.16. Users can move the cursor along the genotype to see which phenotype parts are influenced by the genotype character under the cursor. Another feature related to "genetic debugging" (a feature not yet available for biological genomes...) is the ability to modify a genotype by adding, deleting, or editing its parts while the corresponding phenotype is instantly computed and displayed. Framsticks can be used to illustrate the phenomena of polygeny and pleiotropy and to perform direct experiments with artificial genetic encodings, increasing comprehension of the genotype-to-phenotype translation process, and properties of genetic encodings – modularity, compression, redundancy, and many more.

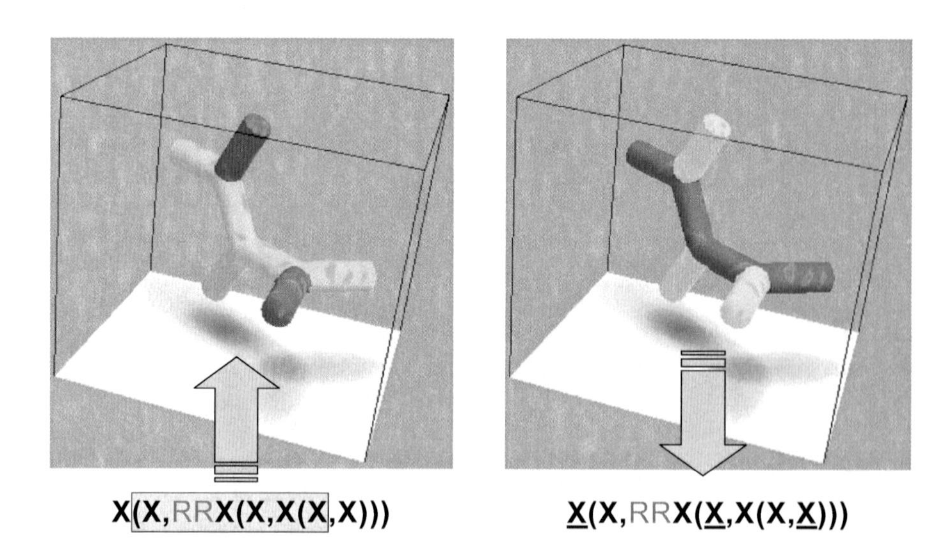

Fig. 5.15 A simple mapping between an *f1* genotype and the corresponding phenotype. Left: user selected a part of the genotype, corresponding phenes are highlighted. Right: user highlighted some parts of the body, corresponding genes are underlined.

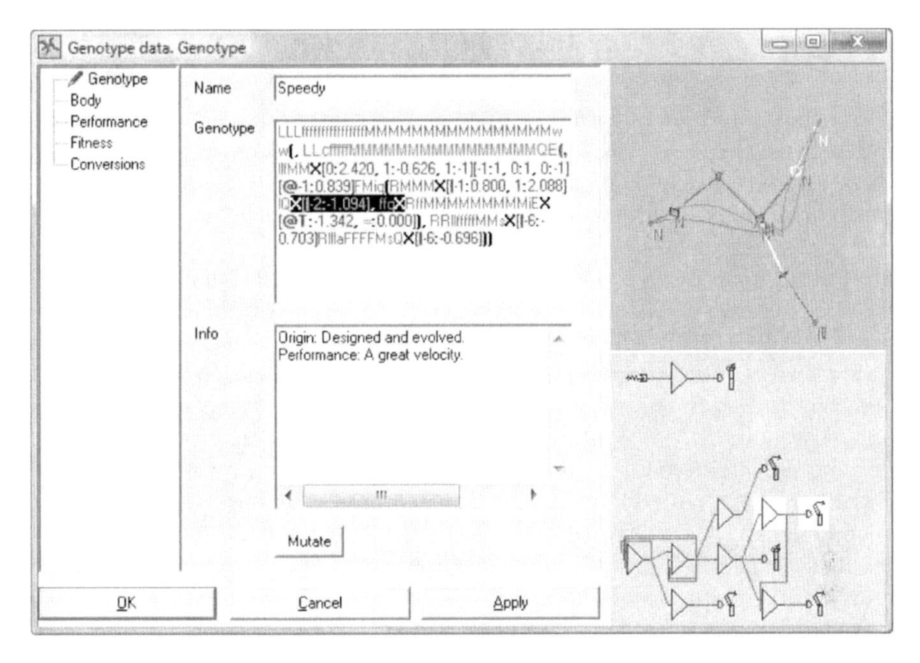

Fig. 5.16 A sample genotype and the corresponding creature (body and brain). Some genes are selected by a user, and the corresponding parts of body and brain are highlighted.

5.4 Scripting

The Framsticks environment has its own virtual machine, thus it can interpret commands written in a simple language – *FramScript*. FramScript can be used for a range of tasks, from custom fitness functions, macros, and user-defined neurons, to user-defined experiment definitions, creature behaviors, events, and even 3D visualization styles. Understanding FramScript allows one to exploit the full potential of Framsticks, because scripts control the Framsticks system.

The FramScript syntax and semantics are very similar to JavaScript, JAVA, C, or PHP. In FramScript,

- All variables are untyped and declared using *var* or *global* statements.
- Functions are defined with the *function* statement.
- References can be used for existing objects.
- No structures and no pointers can be declared.
- There is the *Vector* object which handles dynamic arrays.
- FramScript code can access Framsticks object fields as "Object.field".

5.4.1 Creating Custom Script-Based Neurons

To demonstrate how scripting can be used, we will design a "noisy" neuron which generates occasional noise (random output). Otherwise it will pass the weighted sum of inputs into its output. In neuron definitions, inputs can be read, any internal function can be defined, output can be controlled directly, and internal states can be stored using "private" neuron properties. "Public" neural properties can be used to influence the neuron behavior; genetic operators (mutation, crossing over) will by default operate on public properties.

For a neuron, two functions can be defined: the initialization function (*init*) and the working function (*go*) which is executed in each simulation step. For our noisy neuron, we do not need the initialization function – there are no internal properties to initialize. However, the public "error rate" property will be useful to control how much noise is generated. For each neuron, we first have to define its name, long name, description, and the number of preferred inputs (any number in this case) and outputs (the noisy neuron provides one meaningful output signal):

```
class:
name:Nn
longname:Noisy neuron
description:Occasionally generates a random value
prefinputs:-1
prefoutput:1
```

The error rate property ("e") will be a floating-point number (f) within the range of 0.0 and 0.1:

```
prop:
id:e
name:Error rate
type:f 0.0 0.1
```

Finally, we implement the working function, which uses the `rnd01` function of the `Math` object to obtain a random value in the range from 0.0 to 1.0:

```
function go()
{
  if (Math.rnd01 < Fields.e)
    Neuro.state = Neuro.weightedInputSum;
  else
    Neuro.state = (Math.rnd01 * 2) - 1.0;
}
```

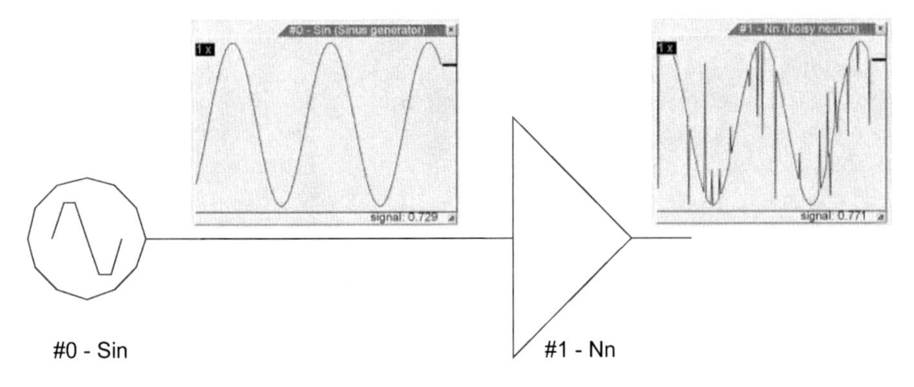

Fig. 5.17 The noisy neuron defined by the script, connected to the sinus generator. Random output values are generated with the rate of 0.1.

We join these three fragments into a single file, name it "noisy.neuro", place it in the appropriate directory, run Framsticks, build a creature that uses the noisy neuron with the error rate set to 0.1, and start the simulation to see the result shown in Fig. 5.17. Now the FramScript source of the neuron can be easily modified to extend its functionality and to exhibit more complex behaviors. The noisy neuron is ready to be used in neural networks, and even in evolution, without any additional work.

5.4.2 Experiment Definitions

A very important feature of Framsticks is that you may define custom rules for the simulator. There are no predetermined laws, but there is a script called the *experiment definition.* It is analogous to the neuron definition explained in the previous section. The experiment definition script is more complex and defines behavior of the Framsticks system in a few related areas:

- Creation of objects in the world. The script defines where, when, and how much of which objects will be created. An object is an evolved organism, food particle, or any other element of the world designed by a researcher. Thus, depending on the script, food or obstacles might appear, move, and disappear, their location might depend on where creatures are, and so forth.
- Objects interactions. Object collision (contact) is an event, which may cause some action defined by the script developer. For example, physical contact may result in energy ingestion, pushing each other, exchange of information, destruction, or reproduction.
- Evolution. Any model of evolution or coevolution can be employed, including many gene pools and many populations (generally called groups);

independent evolutionary processes can be performed under different conditions.
- Evaluation criteria. These are flexible and do not have to be as simple as the basic performances supplied by the simulator. For example, fitness may depend on time or energy required to fulfill some task, or degree of success (distance from target, number of successful actions, etc.).

The script is built of "functions" assigned to system events, which include:

- `onExpDefLoad` – occurs after experiment definition has been loaded. This procedure should prepare the environment, create necessary gene pools and populations, etc.
- `onExpInit` – occurs at the beginning of the experiment.
- `onExpSave` – occurs on "save experiment data" request.
- `onExpLoad` – occurs on "load experiment data" request.
  The script should restore the system state saved by `onExpSave`.
- `onStep` – occurs in each simulation step.
- `onBorn` – occurs when a new organism is created in the world.
- `onKill` – occurs when a creature is removed from the world.
- `onUpdate` – occurs periodically, which is useful for efficient performance evaluation.
- `on[X]Collision` – occurs when an object of population [X] has touched some other object.

Therefore, researchers may define the behavior of the system by implementing appropriate actions within these events. A single script may introduce parameters which allows users to perform a number of diversified experiments using one experiment definition.

5.4.3 Illustrative Example ("Standard Experiment" Definition)

The file "standard.expdef" contains the source for the standard experiment definition script used to optimize creatures on a steady-state (one-at-a-time) basis, with fitness defined as a weighted sum of their performances. This script is quite versatile and complex. Below, its general concept is explained, with much simplified actions assigned to events. This digest gives an idea of what components constitute a complete experiment definition.

- `onExpDefLoad`:
  - create a single gene pool named "Genotypes"
  - create two populations: "Creatures" and "Food"

- `onExpInit`:
  - empty all gene pools and populations
  - place the initial genotype in “Genotypes”
- `onStep`:
  - if too little food: create a new object in “Food”
  - if too few organisms: select a parent from “Genotypes”; mutate, crossover, or copy it. Based on the resulting genotype, create a new individual in “Creatures”
- `onBorn`:
  - move the new object into a randomly chosen place in the world
  - set its starting energy depending on the object’s type (creature or food)
- `onKill`:
  - if “Creatures” object died, save its performance in “Genotypes” (possibly creating a new genotype). If there are too many genotypes in “Genotypes”, remove one.
- `onFoodCollision`:
  - send a piece of energy from the “Food” object to the colliding “Creature” object.

5.4.4 Popular Experiment Definitions

The most popular experiment definitions are outlined below. More specific experiments are referred to in Section 5.6.7.

- **standard** is used to perform a range of common evolutionary optimization experiments. It provides one gene pool, one population for individuals, one “population” for food, steady-state evolutionary algorithm, fitness as a weighted sum of performance values, support for custom fitness formulas, fitness scaling, roulette or tournament selection, and stagnation detection. It can log fitness and automatically produce charts using gnuplot [38].
- **generational** is a simple generational optimization experiment that resembles a popular “genetic algorithm” setup. It provides two gene pools (previous and current generation), one population for creatures, generational replacement of genotypes, roulette selection, and script-defined fitness formula.
- **reproduction** models asexual reproduction in the world. Each creature with a sufficient energy level produces an offspring, which is located to its parent. Food is created at a constant rate and placed randomly (Fig. 5.29).
- **neuroanalysis** simulates all loaded creatures one by one and computes the average and standard deviation of the output signal for each neuron

in each creature. After evaluation, a simple statistics report is printed. No evolution is performed.

- **boids** models emergent, flocking behavior of birds [31]. Users can activate or deactivate individual behavior rules (separation, alignment, cohesion) and instantly see results (Fig. 5.12).
- **evolution_demo** demonstrates the process of evolution. Individuals are placed in a circle. One individual is selected and then cloned, mutated, or crossed over. It is then evaluated in the middle of the circle and, depending on its fitness, may replace a poorer, existing individual or disappear.
- **learn_food** illustrates a social phenomenon of knowledge sharing in the context of exploration of the environment and exploitation of knowledge about the environment (Fig. 5.7). When an individual encounters food, it eats a bit, remembers its location, and gets a full amount of energy. The energy of each individual provides information on how current its information about food coordinates is. When agents collide, they learn from each other where the food was (e.g., by weighted averaging their knowledge). An individual that cannot find food for a long period of time loses all energy and dies, and a newborn one is created.

  It is interesting to see how knowledge sharing (cooperation, dependence) versus no sharing (self-sufficiency, risk) influences minimal, average, and maximal life span in various scenarios of food placement (e.g., neighboring and random). The dynamics in this experiment depends on the number of individuals, size and shape of their body (affects collisions and thus chances of sharing knowledge), world size, food eating rate, food placement, learning strategy, and the behavioral movement pattern.
- **mazes** evaluates (and evolves) creatures walking between two specified points in a maze.
- **deathmatch** is an educational tool intended for use in practical courses in evolutionary computing, evolutionary robotics, artificial life, and cognitive science. Following the "education by competition" approach, it implements a tournament among teams of creatures, as well as among teams of students. To win, a team has to provide (design or evolve) a creature that stays alive longer than creatures submitted by other teams. To survive, creatures need energy which can be collected by touching energy resources, winning fights, avoiding fights, or cooperation.
- **standard-eval** evaluates loaded genotypes thoroughly one by one and produces a report of fitness averages, standard deviations, and average evaluation times. No evolution is performed.
- **standard-log** logs genetic and evaluation operations, producing a detailed history of evolutionary process.
- **standard-tricks** serves as an example of a few advanced techniques: Random force can be applied to parts of a living creature during its life span, neural property values can be used in the fitness function, and statistical data can be acquired regarding movement of simulated body parts.

5.5 Advanced Features for Research and Education

Research works in the area of artificial life often concern studies of evolutionary and coevolutionary processes – their properties, dynamics and efficiency. Various methods and measures have been developed in order to analyze evolution, complexity, and interaction in complex adaptive systems (CAS). Another line of research focuses on artificial creatures themselves, regarding them as subjects of survey rather than "black boxes" with assigned fitness, and thus helping understand their behaviors.

Artificial life systems – especially those applied to evolutionary robotics and design [3, 4, 22] – are so complex that it is difficult or impossible to fully understand behaviors of artificial agents. The only way is to observe them carefully and use human intelligence to draw conclusions. Usually, behaviors of such agents are nondeterministic, and their control systems are sophisticated, often coupled with morphology and very strongly connected functionally [24]. Using artificial life techniques, humans are able to generate successful and efficient solutions to various problems, but they are unable to comprehend their internals and logic. This is because evolved structures are not designed by a human, not documented, and inherently complex.

Therefore, to effectively study behaviors and populations of individuals, one needs high-level, intelligent support tools [17]. It is not likely that automatic tools will soon be able to produce understandable, nontrivial explanations of sophisticated artificial agents. Nonetheless, their potential in helping researchers is huge. Even simple automatic support is of great relevance to a human; this becomes obvious after spending hours on investigating relatively simple artificial creatures.

One of the purposes of Framsticks is to allow creating and testing such tools and procedures and to develop methodology needed for their use. Realistic artificial life environments are the right place for such research. In the future, some of the advanced analysis methods developed within artificial life methodology may become useful for real-life studies, biology, and medicine.

In education, it is primarily important to make complex systems attractive and easier to understand. Visualization of relations between genotypes and phenotypes described in Section 5.3 is an example of such an educational instrument.

5.5.1 Brain Analysis and Simplification

An (artificial) creature is composed of body and brain. Bodies can be easily seen, and basic statistical information is easy to obtain (the number of parts, body size, weight, density, degree of consistency, etc.). Brains are much more difficult to present and comprehend. Framsticks provides a dedicated algorithm for laying out neural networks so that their structure can be exposed.

After it is applied, complex neural networks that initially looked chaotic have their brain structures revealed; see Fig. 5.8. Additionally, if a neuron that is embodied (located in body) is selected, it is highlighted in both the body view and the brain view (Fig. 5.16).

Signal flow charts are useful to understand how the brain works. Users can open multiple views of a single brain and connect probes to neurons, as shown in Figs. 5.17 and 5.18. It is also possible to enforce states of neurons using these probes so that parts of the brain can be turned off, oscillation can be stopped, or the desired signal shape can be interactively "drawn." In this way, muscles can be directly controlled while simulation is running.

Some sensors reflect states of the body. If the body is moved, output values of these neurons change (e.g., equilibrium or touch sensors) which is immediately seen in the neural probes. Fig. 5.18 shows a creature under such analysis.

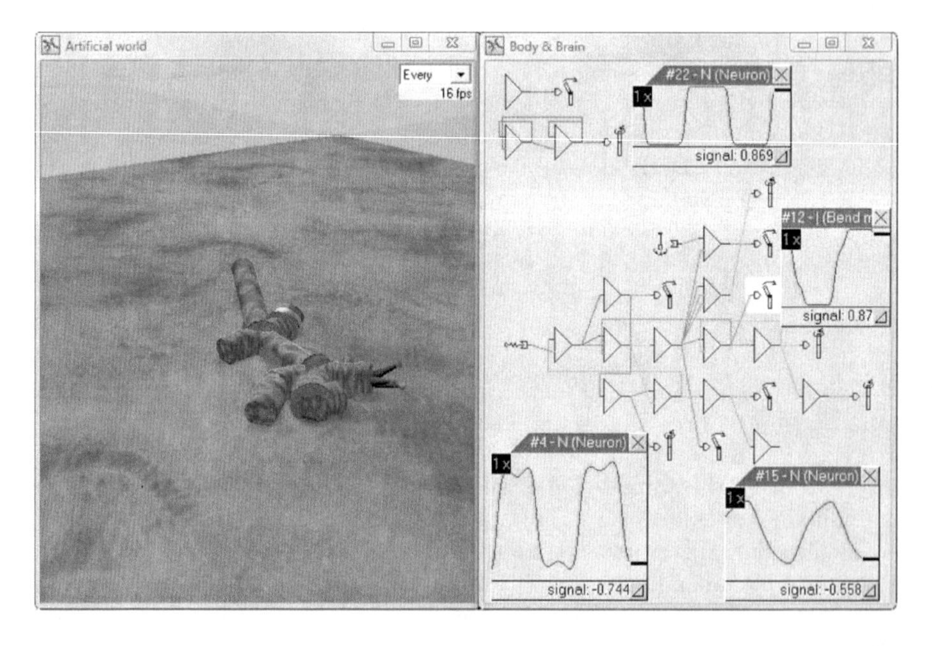

Fig. 5.18 A simulated creature with the control system under investigation. Four neural probes are attached showing signals in different locations of the neural network.

A simple automatic tool for brain analysis is the "neuroanalysis" experiment definition. During simulation, it observes each neuron in each creature and computes averages and standard deviations for all neural output signals. The final report summarizes the activity of brains and helps in the location of inactive and redundant brain areas. It also gives clues on possible ways of simplification of analyzed neural networks.

The "brain simplifier" macro is also available. It prunes neurons that do not directly or indirectly influence bodies, thus making evolved neural networks simpler and easier to apprehend.

5.5.2 Numerical Measure of Symmetry of Constructs

There is no agreement among scientists on why symmetry in biology is such a common evolutionary outcome, but this phenomenon must be certainly related to the properties of the physical world. According to one of the hypotheses [23], a bilaterally symmetrical body facilitates visual perception, as such a body is easier for the brain to recognize while in different orientations and positions. Another popular hypothesis suggests that symmetry evolved to help with mate selection. It was shown that females of some species prefer males with the most symmetrical sexual ornaments [29, 30]. For humans, there are proved positive correlations between facial symmetry and health [41] and between facial symmetry and perception of beauty [32].

It is hard to imagine the objective measure of symmetry. The only thing that can be assessed objectively is whether a construct is entirely symmetrical or not. The natural language is not sufficiently precise to express intermediate values of symmetry – we say that something is nearly symmetrical, but we are not able to the degree of symmetry numerically in the same manner as, for instance, angles can be described. This lack of expressions in natural languages describing partial symmetry is reasonable since many objects in the real world are symmetrical.

However, symmetry is not such a common concept in artificial worlds, and in order to study the phenomenon of symmetry and its implications, there was a need for defining a numerical, fully automated, and objective measure of symmetry for creatures living in artificial environments as well constructs, models, and 3D objects simulated in virtual settings. In Framsticks, the implementation of parametrized, numerical symmetry measure is provided in the Symmetry object; see [11] for a detailed description and Section 5.6.3 for a sample application.

5.5.3 Estimating Similarity of Creatures

Similarity is sometimes considered to be a basic or simple property. However, automatic measures of similarity can be extremely helpful! Similarity can be identified in many ways, including aspects of morphology (body), brain, size, function, behavior, performance, or fitness, but once a quantitative measure of similarity is established, it allows one to do the following:

- analyze structure of populations of individuals (e.g., diversity, convergence, etc.), facilitating better interpretation of experimental results,
- discover clusters in groups of organisms,
- reduce large sets or populations of creatures to small subsets of diverse representatives, thus reducing the complexity and size of experimental data and making it more comprehensible,
- infer dendrograms (and, hopefully, phylogenetic trees) based on morphological distances between organisms,
- restrict crossing over to only involve similar parents and thus avoid impaired offspring,
- introduce artificial niches, or species, by modifying fitness values in evolutionary optimization [5, 28],
- test correlation between similarity and quality of individuals, determine global convexity of the solution space [13], and develop efficient distance-preserving crossover operators [39, 27].

For the model of morphology considered in this chapter, a heuristic method was constructed that is able to estimate the degree of dissimilarity of two individuals. This method treats body as a graph (with parts as vertices and joints as edges) and then matches two body structures taking into account both body structure and body geometry. For a more detailed description of this method, see [15]; a sample application is presented in Section 5.6.4.

5.5.4 History of Evolution

In real life it is possible to trace genetic relationships within existing creatures, but we do not precisely know what happened during mutations and crossing overs of their genomes. Moreover, it is hard to trace all genetic relations over a longer timescale and in high numbers of individuals.

Framsticks as an artificial life environment allows one not only to retain all parent–child relationships but also to estimate genotype shares of related individuals (how many genes have mutated or have been exchanged). This lets users derive and render the real tree of evolution as shown in Fig. 5.19. The vertical axis is time, and the horizontal one reflects a degree of local genetic dissimilarity (between each pair of individuals). Vertices in this tree are single individuals. This way of visualizing evolution exposes milestones (genotypes with many descendants), and these can be automatically identified. The overall characteristics of the evolutionary process (convergence, high selective pressure, or random drift) can also be seen in such pictures.

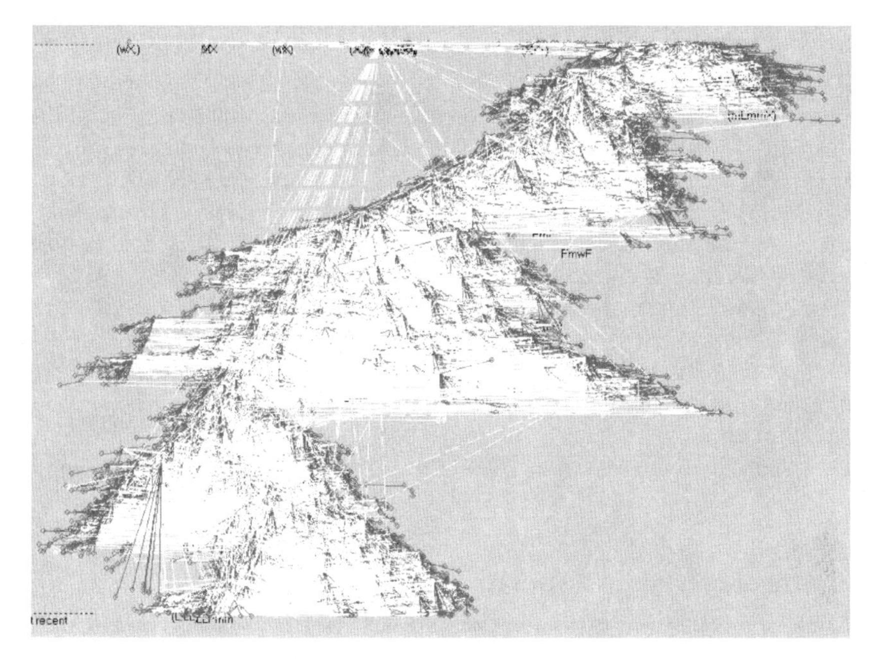

Fig. 5.19 The real tree of evolution; single ancestor and the beginning of time on top. Dark lines represent mutations, and white lines are crossovers.

5.5.5 Vector Eye

Vector Eye is a high-level sensor that provides a list of edges in the scene that are visible from some location in space. This information can be accurate – with no noise or imperfections that would exist if these edges were detected in a raster image.

Many simple sensors that are commonly used in robotics (touch, proximity, 3D orientation) provide either single-valued outputs or a constant rate of information flow (e.g., camera). Vector Eye provides a variable amount of information depending on the shape perceived and the relative position and orientation of the sensor with respect to the shape.

Vector Eye as a neuron uses multiple-valued output to send coordinates of perceived edges. This neuron is usually attached to some element of creature body, and a convex object to be watched is supplied as a parameter of this neuron; see the right part of Fig. 5.24 for illustration.

5.5.6 Fuzzy Control

Framsticks provides support for evolvable fuzzy control based on the Mamdani model: Both premises and conclusions of fuzzy rules are described by fuzzy variables [25]. The fuzzy system may be considered as a function of many variables, where input signals are first fuzzily evaluated by particular fuzzy rules, the outcomes of firing rules are then aggregated, and the resulting fuzzy set is defuzzified [33, 40].

Fuzzy sets in Framsticks are represented in a trapezoidal form, each set being defined by four real numbers within the normalized domain of $[-1, 1]$. The fuzzy rule-based control system is implemented in the Fuzzy neuron, with fuzzy sets and rules described as parameters of this neuron. Dedicated mutation and crossing-over operators have been developed; see [8, 7] for details and Section 5.6.6 for a sample application.

5.6 Research and Experiments

5.6.1 Comparison of Genotype Encodings

There are a number of studies on the evolution of simulated creatures that exhibit realistic physical behavior. In these systems, the use of a physical simulation layer implements a complex genotype–fitness relationship. Physical interactions between body parts, the coupling between control and physical body, and interactions during body development can all add a level of indirection between the genotype and its fitness. The complexity of the genotype–fitness relationship offers a potential for rich evolutionary dynamics.

The most important element of the genotype-to-fitness relationship is the genotype-to-phenotype mapping determined by the genotype encoding. There is no obvious simple way to encode a complex phenotype – which consists of a variable-size, structured body and the corresponding control system – into a simpler genotype. Moreover, performance of the evolutionary algorithm can greatly vary from encoding to encoding for reasons not immediately apparent. This makes genetic encodings and genetic operators a subject of intense research.

The flexibility of the genetic architecture in Framsticks allowed one to analyze and compare various genotype encodings in a single simulation environment. The performance of the three encodings described in detail in Section 5.3 was compared in three optimization tasks: maximization of passive height, active height, and velocity [18]. Solutions produced by evolutionary processes were examined and considered successful in these tasks for all encodings. However, there were some important differences in the degree of success. The *f0* encoding performed worse than the two higher-level encodings. The most important differences between these encodings are that *f0* has a minimal bias and is unrestrictive, while the higher-level encodings (*f1* and *f4*) restrict the search space and introduce a strong bias toward structured phenotypes. These results indicate that a more structured genotype encoding, with genetic operators working on a higher level, is beneficial in the evolution of 3D agents. The existence of a bias toward structured phenotypes can overcome the apparent limitation that entire regions of the search space are inaccessible for the optimization search. This bias may be useful in some applications (engineering and robotics, for example). The significant influence of encodings can be clearly seen in the obtained creatures: Those described by the *f0* encoding displayed neither order nor structure. The two encodings restricting morphology to a tree produced more clear constructs, with segmentation and modularity visible for developmental encoding (Fig. 5.20).

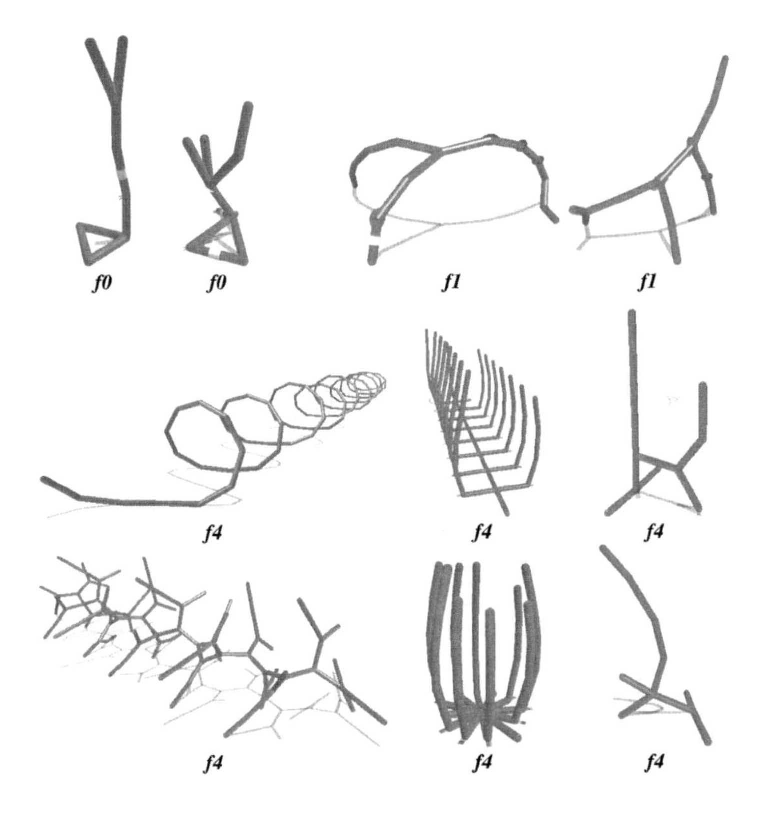

Fig. 5.20 Representative agents for three distinct genetic encodings and height maximization task.

5.6.2 Automatic Optimization Versus Human Design

Designing agents by hand is a very complex process. In professional applications, it requires extensive knowledge about the control system, sensors, actuators, mechanics, physical interactions, and the simulator employed. Designing neural control manually may be especially difficult and tedious (see for example [16]). For this reason, agents built by humans usually have lower fitness than agents generated by evolution. However, human creations are often interesting because of their explicit purpose, elegance, simplicity (a minimum of means), symmetry, and modularity. These properties are opposed to evolutionary outcomes, which are characterized by hidden purpose, complexity, implicit and very strong interdependencies between parts, as well as redundancy and randomness [18].

The difficult process of designing bodies and control systems manually can be circumvented by a hybrid solution: Bodies can be hand-constructed and control structures evolved for them. This popular approach can yield interesting creatures [14, 17, 1, 19], often resembling in behavior creatures found in nature [10].

5.6.3 Symmetry in Evolution and in Human Design

Following considerations from Section 5.5.2, the bilateral symmetry estimate allows one to compute the degree of symmetry for a construct (Fig. 5.21); it is therefore another automatic tool that helps a human examine and evaluate virtual and real creatures and designs [11].

It is also possible to rank creatures according to their symmetry value. The ranking shown in Fig. 5.22 presents 30 diversified constructs – small, big, symmetric, asymmetric, human-designed, and evolved. The horizontal axis shows values of symmetry. Creatures are oriented such that the plane of

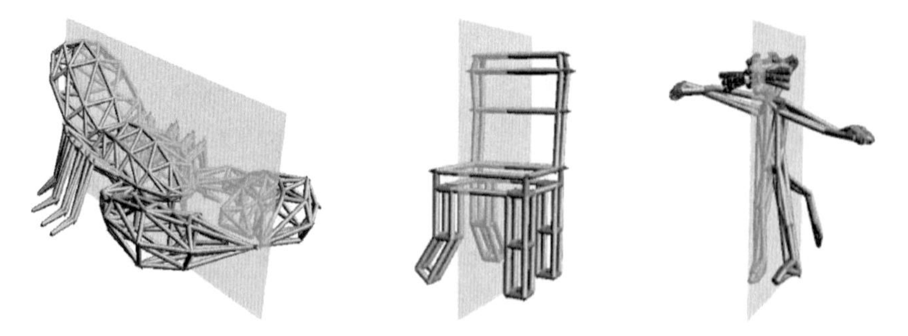

Fig. 5.21 Highest-symmetry planes for three sample constructs. Symmetry values are 1.0, 0.92, and 0.84.

symmetry for each of them is a vertical plane perpendicular to the horizontal axis. Constructs that were hand-designed and have regular shapes are the most symmetrical ones (located on the right side, with symmetry close to 1.0). On the other hand, large evolved bush-like creatures, for which symmetry planes were not obvious, are located on the far left (low values of symmetry).

Symmetry has long dominated in architecture and it is an unifying concept for all cultures of the world. Some famous examples include the Pantheon, Gothic churches, and the Sydney Opera House. A question comes up about the symmetry of human designs compared to the symmetry of evolved constructs. To investigate this issue, a set of 84 representative individuals has been examined. This set included evolved constructs originated from various evolutionary processes oriented toward speed and height and designed constructs that served various purposes – most often efficient locomotion, specific mechanical properties or aesthetic shape.

Analysis revealed that the vast majority (92%) of designed creatures appeared to be symmetrical or nearly symmetrical (symmetry higher than 0.9). Moreover, 82% of designed creatures were completely symmetrical – with the symmetry value 1.0. Clearly, human designers prefer symmetry, and there were no human designs in the set with symmetry less than 0.6.

Although half of the evolved creatures were highly symmetrical, symmetry of the rest is distributed fairly uniformly. It has to be noted that among evolved creatures with complete symmetry, many structures were very simple and symmetrical human designs were much more complex. Simple structures

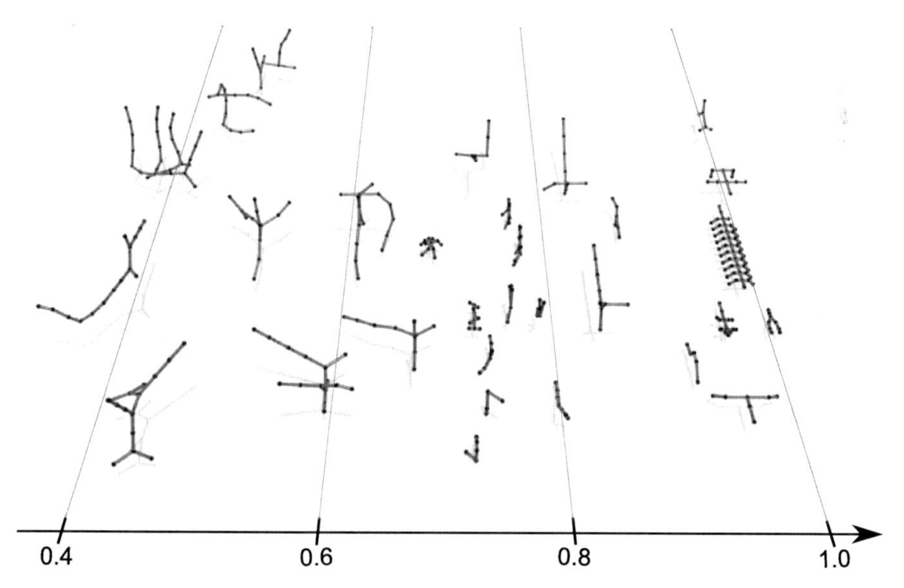

Fig. 5.22 30 diverse constructs arranged horizontally according to their values of symmetry (the most symmetrical on the right).

may exhibit symmetry by chance, while complex ones require some purpose or bias to be symmetrical.

5.6.4 Clustering with Similarity Measure

The similarity measure outlined in Section 5.5.3 allowed one to perform a number of experiments [15]; a sample clustering application is presented here. The UPGMA clustering method has been applied after computing dissimilarity for every pair of considered individuals. Fig. 5.23 shows the result of clustering of 10 individuals resulting from the experiments with maximization of body height. The clustering tree is accompanied by creature morphologies.

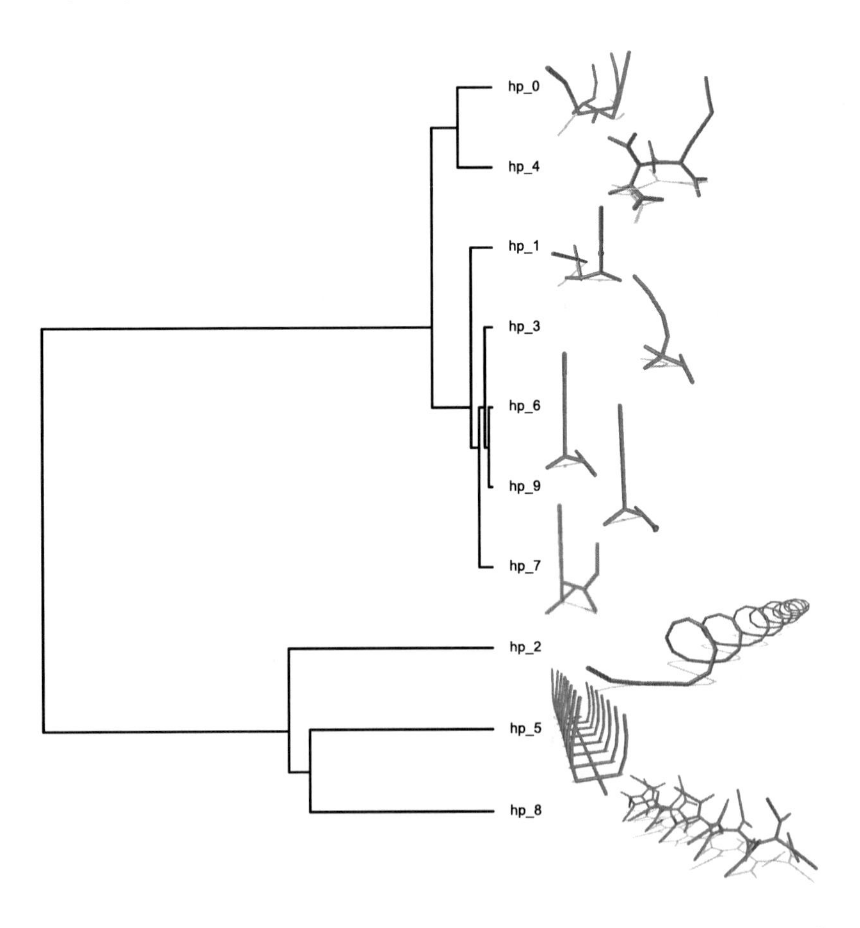

Fig. 5.23 The clustering tree for 10 best individuals in the height maximization task.

It can be seen that the three large organisms are grouped in a single, distinct cluster. They are similar in size but different in structure, so the distance in between them is high. Moreover, the measure also captured functional similarity (hp_1, 3, 6, 9, 7); all of these agents have a single stick upward and a similar base. The agents hp_0 and hp_4 are of medium size, but certainly closer to the group of small organisms than to the large ones. They are also similar in structure – this is why they constitute a separate cluster.

The similarity measure is very helpful for the study and analysis of groups of individuals. Manual work of classification of the agents shown in Fig. 5.23 yielded similar results, but it was a mundane and time-consuming process. It also lacked objectivism and accuracy, which are properties of the automatic procedure.

5.6.5 Bio-Inspired Visual Coordination

One of the factors that play an important role in the success of living organisms is the way they acquire information from the environment. Their senses are interfaces between neural systems and the outer world. Living organisms exhibit a vast number of sensor types, including olfactory, tactile, auditory, visual, electric, and magnetic ones. Among these, visual sensing provides a lot of information about the environment; it is therefore popular in natural systems and often used in artificial designs.

In the area of machine vision, considered problems usually concern object recognition and classification. The domain of artificial life adds the aspect of active exploration of the environment based on information that is perceived. The Vector Eye sensor described in Section 5.5.5 allows one to build and test models of complex cognitive systems, while Framsticks allows these systems to be embodied and situated in a virtual world. The purpose of building such biologically inspired cognitive models is twofold. First, they help understand cognitive processes in living organisms. Second, implementations of such models can cope with the complexity of real-world environments because these models are inspired by solutions that proved to be successful in nature.

A sample experiment [12] concerns a visual–motor model that facilitates stimulus–reaction behaviors, as it is the basic schema of functioning of living organisms. In this case the stimulus is visual, and the motor reaction is movement of an agent. This visual–motor system consists of three components: the Vector Eye sensor, the visual cortex, and the motor area. Vector data acquired by the sensor are transformed and aggregated by the visual cortex and fed to the motor area, which controls agent movements, as shown in Fig. 5.24. This biologically inspired visual–motor coordination model has been verified in a number of navigation experiments and perceived 3D shapes and proved to be flexible and appropriate for such purposes.

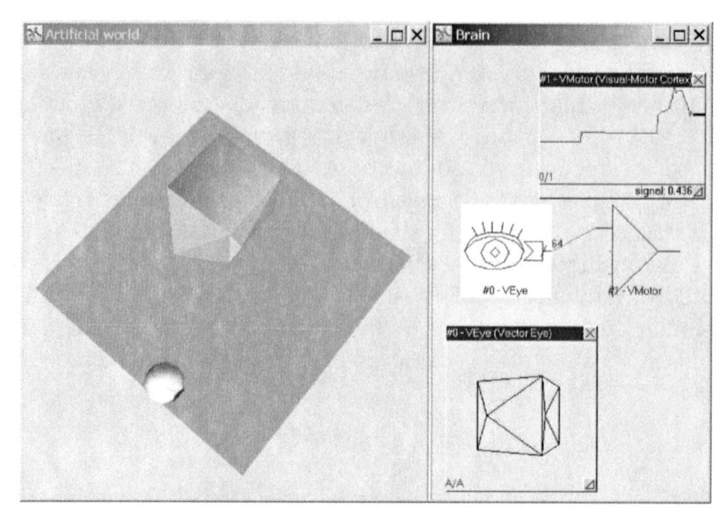

Fig. 5.24 Left: the agent equipped with the Vector Eye sensor circles around a 3D object. Right: Vector Eye perceives edges; this variable-rate information stream is processed by the visual cortex module and ultimately transformed into motor actions of the agent. Thus, agent behavior depends on what it perceives, which depends on its behavior.

5.6.6 Understanding Evolved Behaviors

Traditional neural networks with many neurons and connections are hard to understand. They are often considered "black boxes," successful but impossible to explain – and therefore not trustworthy for some applications. However, there is another popular paradigm used in control: the *fuzzy* control. It is employed in many domains of our life, including washing machines, video cameras, ABS in cars, and air conditioning. It is often applied for controlling nonlinear, fast-changing processes, where quick decisions are more important than exact ones [40]. Fuzzy control, just as neural networks, can cope with uncertainty of information. It is also attractive because of the following:

- It allows for linguistic variables (like "drive fast," where "fast" is a fuzzy term).
- It is easier to understand by humans. The fuzzy rule "if X is Big and Y is Small then Z is Medium" is much easier to follow than the crisp one "if X is between 32.22 and 43.32 and Y is less than 5.2 then Z is 19.2."

To evolve controllers whose behavior is explainable, fuzzy control has been developed in Framsticks (cf. Sect. 5.5.6). It has been applied in a number of artificial life evolutionary experiments; here, we outline a variant of the popular "inverted pendulum" problem [8, 7]. This experiment tested the efficiency of evolution in optimizing fuzzy control systems, verified if the evolved systems can really explain behaviors in a human-friendly way, and compared evolved fuzzy and neural control.

The base of the pendulum was composed of three joints (J_0, J_1, J_2) equipped with two actuators (bottom and top) working in two planes; see Fig. 5.25. The top part of the pendulum was composed of four perpendicular sticks, each having a single equilibrium sensor (G_0, G_1, G_2, G_3). The sensors provide signal values from the $[-1, 1]$ range depending on the spatial orientation of the joint in which they are located. Sensory information is further processed by the control system that controls actuators; this constitutes a loop of relations between the agent and the environment known from Section 5.6.5.

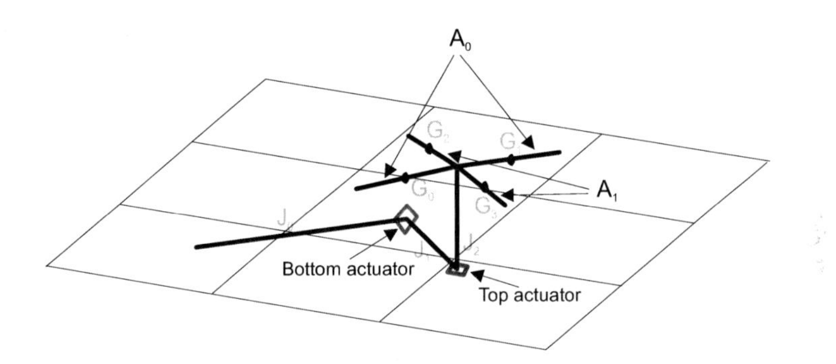

Fig. 5.25 The pendulum body structure (shown in a bent position).

The optimization task was to evolve a control system capable of keeping the head of the inverted pendulum from falling down for as long as possible. Evolved fuzzy systems were compared to evolved neural networks in the same experiment (cf. Fig. 5.26), and their performance was similar. Since evolved fuzzy systems were quite complex, they were simplified by performing an additional, short optimization phase with "add fuzzy set" and "add fuzzy rule" operators disabled. Modifications and deletions were allowed. Thus, the complexity of the control system was radically decreased without deteriorating its fitness.

Control systems considered in this experiment have four inputs and two outputs. Input signals s_0, s_1, s_2, and s_3 come from four equilibrium sensors. Based on their values, the fuzzy system sends two outputs signals for actuators: bend_bottom and bend_top. Linguistic variables for inputs (upright, leveled, and upside_down) and outputs characterizing bending directions (right, none, left) need to be defined to present the fuzzy system in a human-readable form. After they are introduced, the best evolved fuzzy system consisting of five rules can be rendered as follows:

1. **if** (s_2=leveled and s_0=leveled) **then** (bend_bottom=left and bend_top=left)
2. **if** (s_3=leveled and s_1=upside_down) **then** (bend_top=left)
3. **if** (s_1=upright) **then** (bend_bottom=left and bend_top=left)
4. **if** (s_3=upside_down) **then** (bend_bottom=right and bend_top=left)

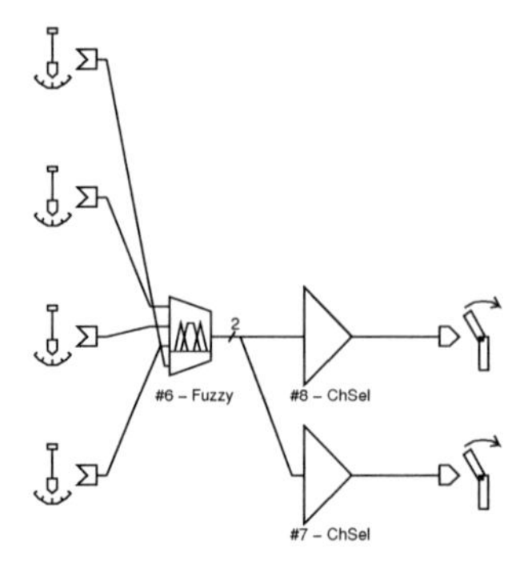

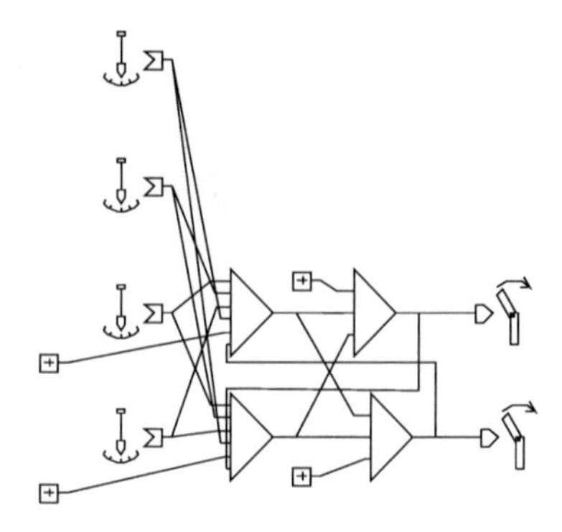

Fig. 5.26 Fuzzy and neural pendulum control. Left: a fuzzy system that provides two values in a single output (further separated by the Channel Selector neurons). Right: a sample neural network with two recurrent connections.

5. **if** (s_1=upside_down) **then** (bend_bottom=left and bend_top=none)

Although evolution of both neural and fuzzy controllers yielded similar pendulum behaviors in this experiment, careful analysis of the evolved fuzzy knowledge confirmed an additional, explanatory value of the fuzzy controller. The evolved fuzzy rules referenced to the pendulum structure and behavior are easily understandable by a human. This approach employed in evolving artificial life agents allows one to present evolved control rules in a human-readable form.

5.6.7 Other Experiments

The open architecture of Framsticks lets users define diverse genetic representations, experimental setups, interaction and communication patterns, and environments (see Sects. 5.4 and 5.4.2). Possible ideas include cooperative or competitive coevolution of species, predator–prey relationships, and multiple gene pools and populations. Sample experiment ideas related to biology include introducing geographical constraints and investigating differences in clusters obtained after a period of time, or studying two or more populations of highly different sizes. The latter, under geographical constraints, can be used to simulate and understand speciation.

A number of interesting experiments regarding evolutionary and neuro-computational bases of the emergence of complex cognition forms and a discussion about semantics of evolved neural networks, perception, and memory are presented in [26].

In the Virtual life lab at Utrecht University, Framsticks has been used to investigate evolutionary origins and emergence of behavior patterns. In

Fig. 5.27 A snapshot of the predator–prey simulation. The two dark creatures are predators, hunting down the lighter prey. Prey have evolved to smartly flee from their predators.

contrast to the standard evolutionary computation approach – with selection criteria imposed by the experimenter outside of the evolving system ("exogenous" or artificial selection) – in these studies selection emerges from within the system ("endogenous" or natural selection). Experiments are designed to include the problems of survival and reproduction in which creatures are born, survive (by eating food), reproduce (by colliding with potential mates), and die (if their energy level is insufficient). This approach enables the investigation of environmental conditions under which certain behaviors offer reproductive advantages, as in the following experiments:

- Coevolutionary processes in **predator–prey** systems are considered to result in arms races that promote *complexification*. For such complexification to emerge, the system must exhibit (semi)stable population dynamics. The primary goal of this experiment is to establish the conditions in which the simulated ecosystem is stable. Food, prey, and predator creatures are modeled in a small food chain and allowed to consume each other and reproduce (see Fig. 5.27). The resulting population dynamics are analyzed using an extended Lotka–Volterra model. When the relations between the parameters in the biological model and the simulation are established, stable conditions can be predicted which enables studies in long-term coevolutionary complexification [2].
- **Semiosis** is the establishment of connections between a sign and the signified via a situated interpretant. The segregation of the sign and the signified from the environment is not given a priori to agents: A sign becomes a part of an agent's subjective environment only if it offers benefits in terms of survival and/or reproduction. In this experiment, the evolutionary emergence of the relation between signs and signified is studied: Through natural selection, agents that have established behavioral relations between the signs and the signified are promoted. A sample of such a relation is movement toward the sign (*chemotaxis*). In varieties of this

experiment, agents can leave trails of signs (e.g., pheromones) in the environment or evolve the ability to signal to each other by using *symbols*.
- Usually, speciation occurs through geographical isolation, which disables gene flow and promotes genetic drift. Such speciation is called allopatric. This experiment reproduces another kind of speciation: **sympatric speciation** which happens in populations that live in the same geographical area.

5.7 Education with Entertainment

Simulating evolution of three-dimensional agents is not a trivial task. On the other hand, 3D creatures are very attractive and appealing to both young and older users who spend much time enjoying the simulation. Many users wish to design their own creatures, simulate them, improve, and evaluate, but designing creatures is not very obvious when it takes place on the genetic level. To simplify this process, the Framsticks graphical editor (FRED) was developed: It helps in building creatures just as CAD programs support designing 3D models. The user-friendly graphical interface (shown in Fig. 5.28), drag-and-drop operations, and instant preview allow users to develop structures of their imagination. Designing neural networks lets users understand basic principles of control systems, their architecture, and their behavior. The editor can also browse and download existing genotypes from the Internet database.

Framsticks can be used to illustrate basic notions and phenomena like genes and genetics, mutation, evolution, user-driven evolution and artificial selection, walking and swimming, artificial life simulation, and virtual world interactions. However, the simulator has numerous options and parameters that make it complicated for the first-time users to handle. Thus, a predefined set of parameters and program behaviors was created primarily for the purposes of demonstration.

Since predefined simulation parameters shift users' focus from creating and modifying to observing what is happening in the virtual world, the demonstration program was named the *Framsticks Theater*. It is an easy-to-use application that includes a number of "shows," and new shows can be added by advanced users or developers using scripts (Sect. 5.4). The shows already included have their script source files available. Each show available in the Theater has a few major options to adjust (like the number of running creatures, length, and difficulty for the "Race" show) . The list of shows (see also Fig. 5.29) includes the following:

- Biomorph – illustrates user-driven (interactive, aesthetic) evolution. Users iteratively choose and double-click a creature to create eight mutated offspring.

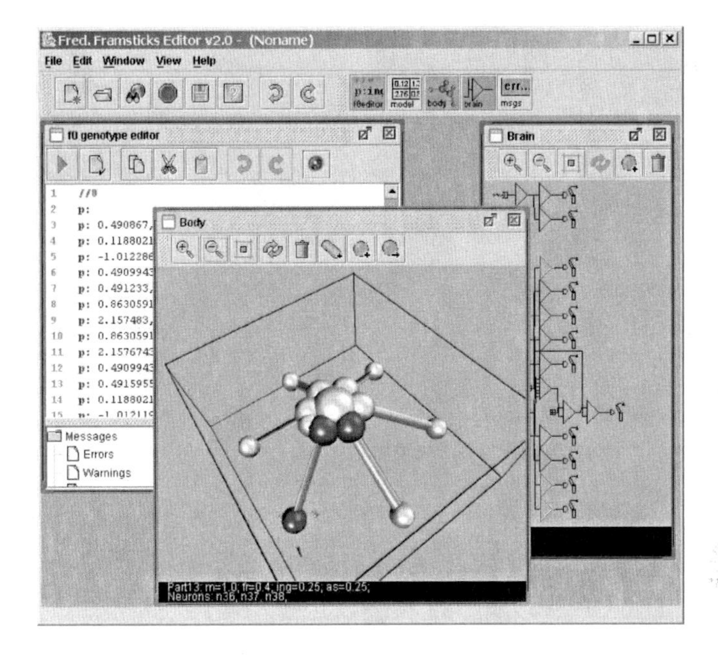

Fig. 5.28 FRED: the graphical editor of the Framsticks creatures.

- Dance – effectors of all simulated creatures are forced to work synchronously.
- Evolution – shows evolutionary optimization with predefined fitness criteria, 50 genotypes in the gene pool, and tournament selection.
- Mixed world – no evolution takes place, creatures are just simulated in a mixed land-and-water environment.
- Mutation – presents a chain of subsequent mutants.
- Presentation – shows various walking and swimming methods of creatures evolved or constructed by the Framsticks users.
- Race – creatures compete in a terrain race running to the finish line.
- Reproduction – illustrates spontaneous evolution. Each creature with a sufficient energy level produces an offspring that appears near its parent. Food is created at a constant rate and placed randomly.
- Touch of life – creatures pass life from one to another by touching.
- Framsbots – the aim of this simple game is to run away from hostile creatures and make all of them hit one another.

The Framsticks Theater can be run on stand-alone workstations as a show (artistic installations, shops, fairs), as well as for education (e.g., in biology, evolution, optimization, simulation, robotics), illustration, attractive graphical background for music, advertisement, entertainment, or screen-saving mode.

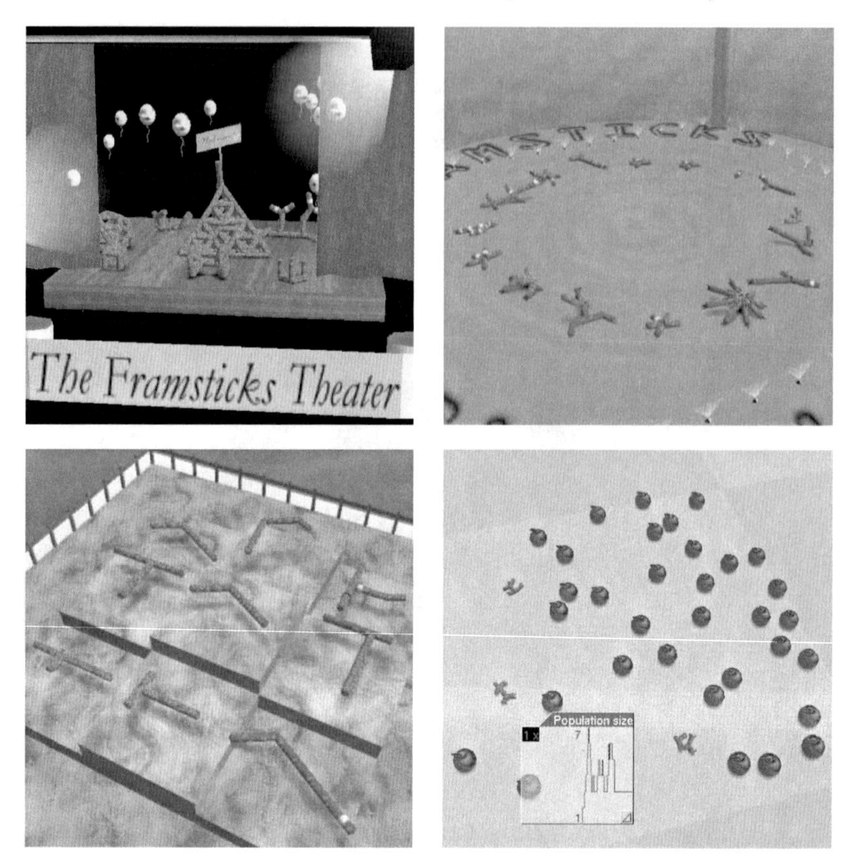

Fig. 5.29 Four Framsticks Theater shows: introduction, dance, biomorph, and reproduction.

5.8 Summary

This chapter presented Framsticks, a tool for modeling, simulation, optimization, and evolution of three-dimensional creatures. Sections 5.5, 5.6, and 5.7 demonstrated selected features and applications in education, research, and entertainment. Framsticks is developed with a vision of combining these three aspects, to make research and education attractive, playing for fun – educationally involving, and education – a demonstration and introduction to research. The software is used by cognitive scientists, biologists, roboticists, computer and other scientists and also by students and non-professionals of various ages.

Although the Framsticks system is versatile and complex, it can be simplified when some features are not needed. For example, control systems can be neglected if only static structures are of interest; genetic encoding may only allow for two-dimensional structures if 3D is not required; simulation or evolution can be restricted to a specific type of a neuron; local optimization techniques can be used if the problem at hand does not require evolutionary algorithms.

Complexity is useless when it cannot be understood or applied. This is why Framsticks software tries to present information in a human-friendly and clear way, encouraging development of automated analysis tools and helping to understand the phenomena of life and nature.

Acknowledgements This work has been supported by the Ministry of Science and Higher Education, grant No. N N519 3505 33.

References

1. Adamatzky, A.: Software review: Framsticks. Kybernetes: The International Journal of Systems & Cybernetics **29**(9/10), 1344–1351 (2000)
2. de Back, W., Wiering, M., de Jong, E.: Red Queen dynamics in a predator-prey ecosystem. Proceedings of the 8th Annual Conference on Genetic and Evolutionary Computation pp. 381–382 (2006)
3. Bentley, P.: Evolutionary Design by Computers. Morgan Kaufmann (1999)
4. Funes, P., Pollack, J.B.: Evolutionary body building: adaptive physical designs for robots. Artificial Life **4**(4), 337–357 (1998)
5. Goldberg, D.E.: Genetic Algorithms in Search, Optimization and Machine Learning. Addison-Wesley, Reading, MA (1989)
6. Gruau, F., Whitley, D., Pyeatt, L.: A comparison between cellular encoding and direct encoding for genetic neural networks. In: J.R. Koza, D.E. Goldberg, D.B. Fogel, R.R. Riolo (eds.) Proceedings of the First Annual Conference, Genetic Programming 1996, pp. 81–89. MIT Press, Cambridge, MA (1996)
7. Hapke, M., Komosinski, M.: Evolutionary design of interpretable fuzzy controllers. Foundations of Computing and Decision Sciences **33**(4), 351–367 (2008)
8. Hapke, M., Komosinski, M., Waclawski, D.: Application of evolutionarily optimized fuzzy controllers for virtual robots. In: Proceedings of the 7th Joint Conference on Information Sciences, pp. 1605–1608. Association for Intelligent Machinery, North Carolina, USA (2003)
9. Hornby, G., Pollack, J.: The advantages of generative grammatical encodings for physical design. Proceedings of the 2001 Congress on Evolutionary Computation CEC2001 pp. 600–607 (2001)
10. Ijspeert, A.J.: A 3-D biomechanical model of the salamander. In: J.C. Heudin (ed.) Proceedings of 2nd International Conference on Virtual Worlds (VW2000), Lecture Notes in Artificial Intelligence No. 1834, pp. 225–234. Springer-Verlag, Berlin (2000)
11. Jaskowski, W., Komosinski, M.: The numerical measure of symmetry for 3D stick creatures. Artificial Life Journal **14**(4), 425–443 (2008)
12. Jelonek, J., Komosinski, M.: Biologically-inspired visual-motor coordination model in a navigation problem. In: B. Gabrys, R. Howlett, L. Jain (eds.) Knowledge-Based Intelligent Information and Engineering Systems. Lecture Notes in Computer Science 4253, pp. 341–348. Springer-Verlag, Berlin (2006)
13. Jones, T., Forrest, S.: Fitness distance correlation as a measure of problem difficulty for genetic algorithms. In: L.J. Eshelman (ed.) Proceedings of the 6th International Conference on Genetic Algorithms, pp. 184–192. Morgan Kaufmann (1995)
14. Komosinski, M.: The world of Framsticks: Simulation, evolution, interaction. In: J.C. Heudin (ed.) Virtual Worlds. Lecture Notes in Artificial Intelligence No. 1834, pp. 214–224. Springer-Verlag, Berlin (2000)

15. Komosinski, M., Koczyk, G., Kubiak, M.: On estimating similarity of artificial and real organisms. Theory in Biosciences **120**(3-4), 271–286 (2001)
16. Komosinski, M., Polak, J.: Evolving free-form stick ski jumpers and their neural control systems. In: Proceedings of the National Conference on Evolutionary Computation and Global Optimization. Poland (2009)
17. Komosinski, M., Rotaru-Varga, A.: From directed to open-ended evolution in a complex simulation model. In: M.A. Bedau, J.S. McCaskill, N.H. Packard, S. Rasmussen (eds.) Artificial Life VII, pp. 293–299. MIT Press (2000)
18. Komosinski, M., Rotaru-Varga, A.: Comparison of different genotype encodings for simulated 3D agents. Artificial Life Journal **7**(4), 395–418 (2001)
19. Komosinski, M., Ulatowski, S.: Framsticks web site, `http://www.framsticks.com`
20. Komosinski, M., Ulatowski, S.: The Framsticks Tutorial. `http://www.framsticks.com/common/tutorial`
21. Komosinski, M., Ulatowski, S.: Genetic mappings in artificial genomes. Theory in Biosciences **123**(2), 125–137 (2004)
22. Lipson, H., Pollack, J.B.: Automatic design and manufacture of robotic lifeforms. Nature **406**(6799), 974–978 (2000)
23. Livio, M.: The Equation That Couldn't Be Solved. How Mathematical Genius Discovered the Language of Symmetry. Simon & Schuster, New York (2005)
24. Lund, H.H., Hallam, J., Lee, W.P.: Evolving robot morphology. In: Proceedings of IEEE 4th International Conference on Evolutionary Computation. IEEE Press, Piscataway, NJ (1997)
25. Mamdani, E.H.: Advances in the linguistic synthesis of fuzzy controllers. International Journal of Man-Machine Studies **8**(6), 669–678 (1976)
26. Mandik, P.: Synthetic neuroethology. Metaphilosophy, Special Issue on Cyberphilosophy: The Intersection of Philosophy and Computing **33**(1/2) (2002)
27. Merz, P.: Advanced fitness landscape analysis and the performance of memetic algorithms. Evolutionary Computation **12**(3), 303–325 (2004). URL `http://dx.doi.org/10.1162/1063656041774956`
28. Michalewicz, Z.: Genetic Algorithms + Data Structures = Evolution Programs. Springer-Verlag, New York (1996)
29. Moller, A.: Fluctuating asymmetry in male sexual ornaments may reliably reveal male quality. Animal Behaviour **40**, 1185–1187 (1990)
30. Moller, A.: Female swallow preference for symmetrical male sexual ornaments. Nature **357**, 238–240 (1992)
31. Reynolds, C.W.: Flocks, herds and schools: A distributed behavioral model. In: SIGGRAPH '87: Proceedings of the 14th Annual Conference on Computer Graphics and Interactive Techniques, pp. 25–34. ACM Press, New York (1987). DOI http://doi.acm.org/10.1145/37401.37406
32. Rhodes, G., Proffitt, F., Grady, J., Sumich, A.: Facial symmetry and the perception of beauty. Psychonomic Bulletin and Review **5**, 659–669 (1998)
33. Ross, T.J.: Fuzzy Logic with Engineering Applications. Wiley, New York (2004)
34. Sims, K.: Evolving 3D morphology and behavior by competition. In: R.A. Brooks, P. Maes (eds.) Proceedings of the 4th International Conference on Artificial Life, pp. 28–39. MIT Press, Boston, MA (1994)
35. Smith, R.: Open Dynamics Engine. `http://www.ode.org/` (2007)
36. Taylor, T., Massey, C.: Recent developments in the evolution of morphologies and controllers for physically simulated creatures. Artificial Life **7**(1), 77–88 (2001)
37. Ulatowski, S.: Framsticks GDK (Genotype Development Kit), `http://www.framsticks.com/dev/gdk/main`
38. Williams, T., Kelley, C.: gnuplot: a plotting utility. `http://www.gnuplot.info/`
39. Wolpert, D.H., Macready, W.G.: No free lunch theorems for optimization. IEEE Transactions on Evolutionary Computation **1**, 67–82 (1997)
40. Yager, R., Filev, D.: Foundations of Fuzzy Control. Wiley, New York (1994)
41. Zaidel, D., Aarde, S., Baig, K.: Appearance of symmetry, beauty, and health in human faces. Brain and Cognition **57(3)**, 261–263 (2005)

Part II
Lattice Worlds

Chapter 6
StarLogo TNG: Making Agent-Based Modeling Accessible and Appealing to Novices

Eric Klopfer, Hal Scheintaub, Wendy Huang, and Daniel Wendel

Computational approaches to science are radically altering the nature of scientific investigatiogn. Yet these computer programs and simulations are sparsely used in science education, and when they are used, they are typically "canned" simulations which are black boxes to students. StarLogo The Next Generation (TNG) was developed to make programming of simulations more accessible for students and teachers. StarLogo TNG builds on the StarLogo tradition of agent-based modeling for students and teachers, with the added features of a graphical programming environment and a three-dimensional (3D) world. The graphical programming environment reduces the learning curve of programming, especially syntax. The 3D graphics make for a more immersive and engaging experience for students, including making it easy to design and program their own video games. Another change to StarLogo TNG is a fundamental restructuring of the virtual machine to make it more transparent. As a result of these changes, classroom use of TNG is expanding to new areas. This chapter is concluded with a description of field tests conducted in middle and high school science classes.

6.1 Computational Modeling

New skills and knowledge will be required to successfully navigate the complex, interconnected, and rapidly changing world in which we live and work. Older paradigms alone are not sufficient to handle these challenges. The field of complexity science is providing the insights scientists and engineers need to push our thinking in new directions [16, 28], providing the tools to move beyond the reductionist point of view and look at whole systems [31, 4, 10]. This approach has been driven by advances in computer technology that have facilitated the development of advanced multi-agent simulations. This powerful new modeling approach has been called the third way of doing science, with traditional experimentation and observation/description being the other two [16]. The increasing importance of computational modeling has changed the knowledge base required to practice science. Scientists in all fields now rely on a variety of computer skills, from programming computer simulations, to analysis of large-scale datasets, to participation in the globally networked scientific communities.

While this explosion in technology and the increased importance of modern paradigms has radically altered the landscape of science and engineering practice, these changes have not yet impacted the way high school students learn and perform science. We believe that computational science and simulation can and should modernize secondary school science courses by making these tools, practices, and habits of mind an integral part of the classroom experience. Although the mere use of computer simulations may aid in understanding both the scientific concepts and the underlying systems principles, a deeper understanding is likely to be fostered through the act of creating models. These principles center on the idea that we must create a culture in which students can navigate novel problem spaces and collaboratively gather and apply data to solutions. Resnick [20] suggested that an equally important goal is for educational programs to foster creativity and imagination. In other words, students should be acquiring habits of mind that will not only enable them to address today's problems and solutions but will also allow them to venture into previously unimagined territories. To achieve this, they should have a facility with tools to create manifestations of their ideas so that they can test them and convey their ideas to others.

6.1.1 Simulations and Modeling in Schools

Science has long relied on the use of scientific models based on ordinary differential equations (ODEs) that can be solved (in simple cases) without computers. These models describe how aggregate quantities change in a system, where a variable in the model might be the size of a population or the proportion of individuals infected by a disease. The mathematics required for these models is advanced, but several commonly used modeling programs like Model-It 1.1 [29], Stella [24], and MatLab [15] have graphical interfaces that make them easy to construct. These programs have become very popular in the classroom as well as the laboratory. The user places a block on the screen for each quantity and draws arrows between the blocks to represent changes in those quantities. While this interface does not remove the need for the user to learn math, it lowers the barrier for entry. However, the abstraction required to model these systems at an aggregate level is a difficult process for many people, which often limits the utility of this modeling approach [33].

With the increase of accessibility to the Internet in schools, teachers are turning to the World Wide Web for visual and interactive tools in the form of Java applets or Flash applications to help students better understand important scientific processes. However, most of these on-line applications are simulation models whose rules and assumptions are opaque to students.

Additionally, many of the systems that are studied in the classroom are more amenable to simulation using agent-based modeling. Rather than tracking aggregate properties like population size, agent-based models track individual organisms, each of which can have its own traits. For instance, to simulate how birds flock, one might make a number of birds and have each bird modify its flight behavior based on its position relative to the other birds. A simple rule for the individual's behavior (stay a certain distance away from the nearest neighbor) might lead to a complex aggregate behavior (flocking) without the aggregate behavior being explicitly specified anywhere in the model. These emergent phenomena, where complex macro-behaviors arise from the interactions of simple micro-behaviors, are prevalent in many systems and are often difficult to understand without special tools.

6.1.2 Computer Programming

Computer programming is an ideal medium for students to express creativity, develop new ideas, learn how to communicate ideas, and collaborate with others. The merits of learning through computer programming have been discussed over the years [8, 11, 12, 18]. As a design activity, computer programming comes along with the many assets associated with "constructionist" [9] learning. Constructionism extends constructivist theories, stating that learning is at its best when people are building artifacts. These benefits of learning through building include [23]:

1. Engaging students as active participants in the activity
2. Encouraging creative problem-solving
3. Making personal connections to the materials
4. Creating interdisciplinary links
5. Promoting a sense of audience
6. Providing a context for reflection and discussion

Other learning outcomes from programming are more broadly applicable. Typical programming activities center on the notion of problem-solving, which requires students to develop a suite of skills that may be transferable to other tasks. As students learn to program algorithms, they must acquire systematic reasoning skills. Modeling systems through programs teaches students how to break down systems into their components and understand them. Many problems require students to acquire and apply mathematical skills and analysis to develop their programs and interpret the output. Together these skills comprise a suite of desirable learning outcomes for students, many of which are difficult to engender through typical curriculum.

They can also form the cornerstones of true inquiry-based science [2], providing students with a set of skills that allows them to form and test hypotheses about systems in which they are personally interested.

6.1.3 Programming in Schools

Computer programming was widely introduced to schools in the early 1980s, largely fueled by inexpensive personal computers and inspired by Papert's manifesto, Mindstorms [17], which advocated that programming in Logo was a cross-cutting tool to help students learn any subject. Logo was popularly used to enable students to explore differential geometry, art, and natural languages. Microworlds Logo [13] extended this use into middle school, where students continued to learn about mathematics, science, reading, and storytelling. KidSim/Cocoa/Stagecast Creator [5] brought modeling and simulation to children through a simple graphical programming interface, where they could learn thinking skills, creativity, and computer literacy. Subsequent work with Stagecast Creator has shown some success in building students' understand of simple mathematical concepts through building games [14]. Boxer [7] is used to help students in modeling physics and mathematical reasoning while building understanding of functions and variables. Kafai [12] introduced novel paradigms for helping students learn about mathematics and social and relational skills through programming of games. Through this paradigm, older elementary school students use Logo to develop mathematical games for younger elementary students. It is of note that most of these efforts have been targeted at elementary-school-aged children. Few efforts have targeted the middle school and high school audience.

Despite many small-scale research efforts to explore programming in schools, Papert's larger vision of pervasive programming in school curricula encountered some serious problems. Programming did not fit into traditional school structures and required teachers to give up their roles as domain experts to become learners on par with the children they were teaching [1]. Computers were used to teach typing skills, and Papert's vision of programming as a vehicle for learning seemed to be dead in the water.

With the growing popularity of the Internet, many schools have revisited the idea of computers in the classroom and rediscovered their utility in the educational process. Schools now use computers to target job-oriented skills, such as word processing and spreadsheets. Some schools encourage children to create mass media, such as computer-based music, animations, or movies. For the last several years, the state of Maine has provided laptops to all of its 7th and 8th grade students (and teachers) to make computers a ubiquitous tool in every subject in school [30].

Now that computers in school are enjoying a resurgence in popularity, it is time to reconsider programming as a vehicle toward fluency in information technology and advancing science education through building simulation and complex systems models.

6.2 Original Design Criteria

In designing an appropriate modeling tool for use in the K–12 classroom we needed to consider several design criteria. One criterion is the choice of agent-based as opposed to aggregate-based modeling. This approach is not only more amenable to the kinds of models that we would like to study in the classroom, but it is also readily adopted by novice modelers.

The next criterion is to create a modeling environment that is a "transparent box." Many of the simulations that have been used in classrooms to date are selected for the purpose of exploring a specific topic such as Mendelian genetics or ideal gases. Modeling software has been shown to be particularly successful in supporting learning around sophisticated concepts often thought to be too difficult for students to grasp [26, 32]. Such software allows students to explore systems, but they are "black box" models that do not allow the students to see the underlying models. The process of creating models – as opposed to simply using models built by someone else – not only fosters model-building skills but also helps develop a greater understanding of the concepts embedded in the model [21, 25, 32]. When learners build their own models, they can decide what topic they want to study and how they want to study it. As learners' investigations proceed, they can determine the aspects of the system on which they want to focus and refine their models as their understanding of the system grows. Perhaps most importantly, building models helps learners develop a sound understanding of both how a system works and why it works that way. For example, to simulate a cart rolling down an inclined plane in the popular Interactive Physics program, a student could drag a cart and a board onto the screen and indicate the forces that act about each object. In doing so, the student assumes the existence of an unseen model that incorporates mass, friction, gravity, and so forth, calculates the acceleration of the cart, and animates the resulting motion. It would be a much different experience to allow the student to construct the underlying model herself and have that act on the objects that she created.

We also considered the level of detail that we felt would be appropriate in student-built models. All too often, students want to create extremely intricate models that exhaustively describe systems. However, it is difficult to learn from these "systems models" [27]. It is more valuable for students to design and create more generalized "idea models" that abstract away as much about a system as possible and boil it down to the most salient element. The ability to make these abstractions and generalize scientific principles is central to the idea of modeling. For example, a group of students might want to create a model of a stream behind their school, showing each species of insect, fish, and plant in the stream. Building such a model is not only an extremely large and intricate task, but the resulting model would be extremely sensitive to the vast number of parameters. Instead, the students should be encouraged to build a model of a more generalized system that includes perhaps one animal and one plant species. The right software should support the building of such "idea models."

Based on these criteria, we originally created a computer modeling environment named StarLogo, which we describe in the next section.

6.3 StarLogo

StarLogo is a programmable modeling environment that embraces the constructionist paradigm of learning by building, and it has existed in numerous forms for many years. StarLogo was designed to enable people to build models of complex, dynamic systems. In StarLogo [19], one programs simple rules for individual behaviors of agents that "live" and move in a two-dimensional (2D) environment. For instance, a student might create rules for a bird, which describe how fast it should fly and when it should fly toward another bird. Because StarLogo makes use of graphical output, when the student watches many birds simultaneously following those rules, she can observe how patterns in the system, like flocking, arise out of the individual behaviors. Building models from the individual, or "bird" level, enables people to develop a good understanding of the system, or "flock"-level behaviors (Fig. 6.1 shows the graphical interface for the growth of an epidemic, another phenomenon that can be studied through simulation).

6.3.1 StarLogo Limitations

For years we worked with students and teachers in the Adventures in Modeling (AIM) program, a program which helps secondary school students and teachers develop an understanding of science and complex systems through simulation activities using our StarLogo programming environment. AIM offered computer modeling to science and math teachers as part of their professional development. AIM-trained teachers developed simulations that were integrated into their standard classroom practices. However, activities in which students program their own simulations have been limited to a small number of classrooms. Our research shows that there are several barriers to students developing their own models in science classes (instead of isolated computer classes). These barriers include the time it takes to teach programming basics and teacher and student comfort with the syntax of programming languages.

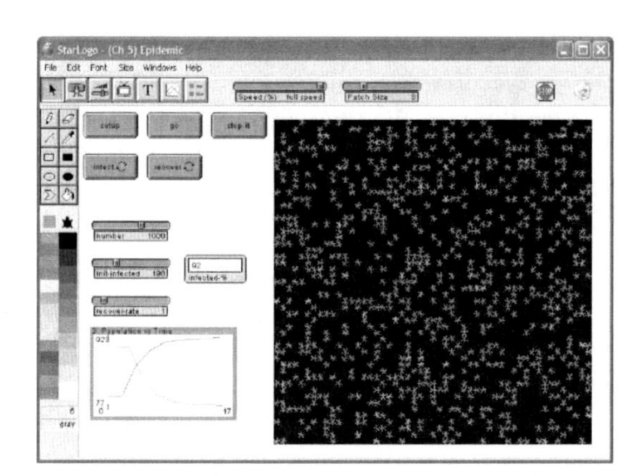

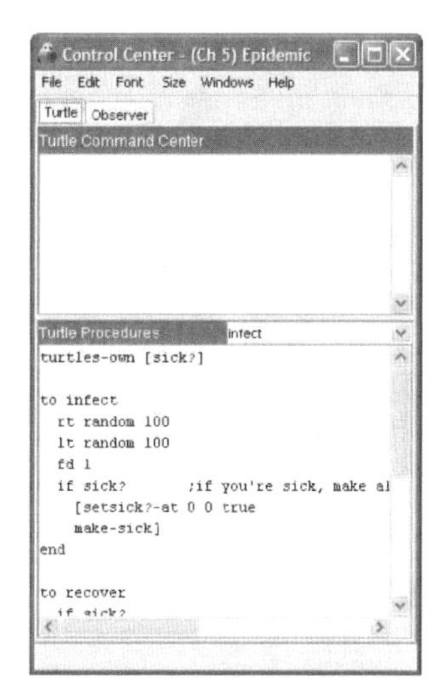

Fig. 6.1 The left-hand panel shows a StarLogo model of an epidemic in action. One can see the interface area, in which users can start and stop the model and adjust parameters, the running model where the different colors code for different states, and the graphs at the bottom tracking model data. The right-hand panel shows the code that was written to generate this model.

These barriers stand in the way of the deeper learning we have observed when students are given the opportunity to develop their own models.

6.4 StarLogo: The Next Generation

We designed StarLogo: The Next Generation (TNG) to address these problems and to promote greater engagement of students in programming. StarLogo TNG provides two significant advances over the previous version. First, the programming is now done with graphical programming blocks instead of text-based commands. This innovation adds a powerful visual component to the otherwise abstract process of programming. The programming blocks are arranged by color based on their function, which enables students to associate semantically similar programming blocks with each other. Since the blocks are puzzle-piece shaped, blocks can fit together only in syntactically sensible ways. This eliminates a significant source of program bugs that students encounter. Blocks snap together to form stacks, which visually help students organize their mental models of the program. StarLogo TNG's interface also encourages students to spatially organize these block stacks into "breed drawers," which affords an object-oriented style of programming. StarLogo TNG's second significant advance is a 3D representation of the agent world. This provides the ability to model new kinds of physical phenomena, as shown in Section 6.7, and allows students to take the perspective of an individual agent in the environment, which helps them bridge the description of individual behaviors with system-level outcomes.

StarLogo TNG retains the basic language and processes of StarLogo. Many of the favorite StarLogo complex systems simulations have been reprogrammed and included in the TNG download. Now users can see "termites" build 3D mounds in the termite simulation. In the predator-prey simulation called "Rabbits," users can choose from a variety of agent-shapes to play the role of predator and prey; and novice students can build their own African Plains predator–prey simulation in just two class periods.

6.4.1 Spaceland

As people today are inundated with high-quality 3D graphics on even the simplest of gaming platforms, students have come to expect such fidelity from their computational experiences. While the 2D "top-down" view of StarLogo is very informative for understanding the dynamics and spatial patterns of systems, it is not particularly enticing for students. Additionally, students often have difficultly making the leap between their own first-person experiences in the kinesthetic participatory simulation activities and the systems-level view of StarLogo.

Enter "Spaceland," a 3D OpenGL-based view of the world. This world can be navigated via keyboard and mouse controls and an on-screen navigator. In the aerial 3D view (Fig. 6.2), the user can zoom in on any portion of the world for a closer look and view the 3D landscape as they fly by. First-person views include the "Agent Eye" and "Over Shoulder" view (Fig. 6.2). In these views, the user sees out of the "eyes" or over the shoulder of one of the agents in the system. As the agent moves around, the user sees what the agent sees. This works particularly well in seeing how agents bump off of objects or interact with other agents. Shapes can be imported so that the agents in the system can be turtles or spheres or firefighters (Fig. 6.2). The ability to customize the characters provides a sense of personalization, concrete representation, and also of scale. Lending further to the customization capabilities are the abilities to change the "Skybox" background (the bitmap image shown in the distance) as well as the texture, color, and topography of the landscape. Since the top-down view is also still quite useful, that 2D view is seen in an inlaid "mini-map" in the lower-left corner of the screen. The 2D and 3D views can be swapped by toggling the Overhead button.

In runtime mode, users will see not just the 3D representation of the world but also the tools to edit the landscape and the runtime interface (buttons and sliders) with which a user can interact to start and stop the model or change parameters on the fly (Fig. 6.3).

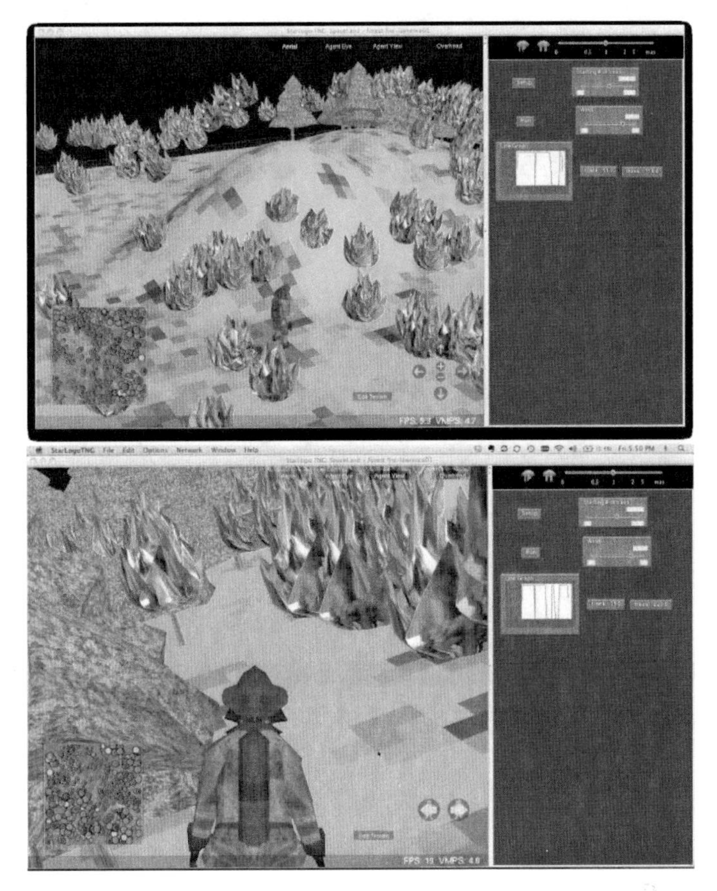

Fig. 6.2 Top-down and turtle's-eye views in TNG.

Additional 3D Benefits

Making it easier to initially engage students through a more up-to-date look, a 3D representation has additional advantages. A 3D view can make the world feel more realistic. This can actually be a limitation in modeling and simulation, in that students feel that more realistic-looking models actually represent the real world, but in the proper context, a more realistic-looking world can be used to convey useful representations of scale and context - for example, conveying the sense of a landscape or the changing of seasons. Moreover, the 3D view of the world is what people are most familiar with from their everyday experiences. Making the leap into programming already comes with a fairly high cognitive load. Providing people with as familiar an environment as possible can lighten that load and make that leap easier. Similarly, a navigable 3D landscape can provide a real sense of scale to a model that otherwise appears arbitrary and abstract to a novice modeler.

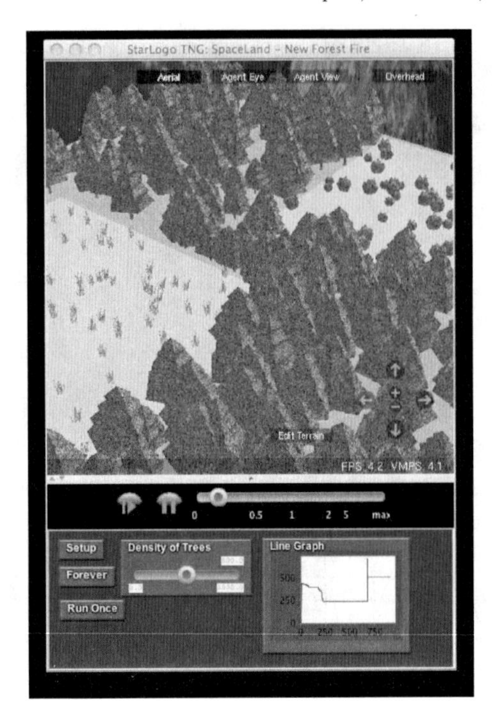

Fig. 6.3 Runtime view.

Finally, with the proliferation of 3D tools (particularly for "modding" games) we can tap into the growing library of ready-made 3D shapes and textures to customize the world.

6.4.2 StarLogoBlocks

StarLogoBlocks is a visual programming language in which pieces of code are represented by puzzle pieces on the screen. In the same way that puzzle pieces fit together to form a picture, StarLogo blocks fit together to form a program. StarLogoBlocks is inspired by LogoBlocks [3] (Fig. 6.4), a Logo-based graphical language intended to enable younger programmers to program the Programmable Brick [22] (a predecessor to the Lego Mindstorms product).

StarLogoBlocks is an instruction-flow language, where each step in the control flow of the program is represented by a block. Blocks are introduced into the programming workspace by dragging a block from a categorized palette of blocks on the left side of the interface. Users build a program by stacking a sequence of blocks from top to bottom. A stack may be topped by a procedure declaration block, thereby giving the stack a name, as well as enabling recursion. Both built-in and user-defined functions can take argument blocks, which are plugged in on the right. Data types are indicated via the block's plug and socket shapes.

The workspace is a large horizontally oriented space divided into sections: one for setup code, one for global operations, one for each breed of turtles, and one for collisions code between breeds. A philosophy we espouse in the programming environment is that a block's location in the workspace should help a user organize their program. While separating code by section appears to the user as a form of object-oriented programming, in reality the system does not enforce this. All code may be executed by any breed of turtle. Fish can swim and birds can fly, but that certainly should not prevent the user from successfully asking a fish to fly.

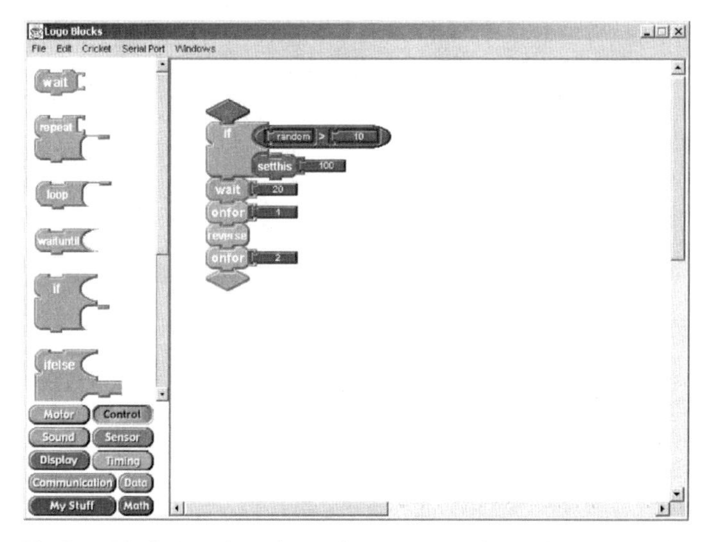

Fig. 6.4 The LogoBlocks interface. A simple program is shown that includes logical statements, random behaviors, and movement. The blocks are assembled into a complete procedure, drawing upon the palette of blocks at the left.

Graphical Affordances

The "blocks" metaphor provides several affordances to the novice programmer. In the simplest sense, it provides quick and easy access to the entire vocabulary, through the graphical categorization of commands. As programmers build models they can simply select the commands that they want from the palette and drag them out. Perhaps more importantly, the shape of the blocks is designed so that only commands that make syntactic sense will actually fit together. That means that the entire syntax is visually apparent to the programmer. Finally, it also provides a graphical overview of the control flow of the program, making apparent what happens when. For example, Fig. 6.5 shows text-based StarLogo code on the left and the equivalent graphical StarLogoBlocks code on the right for a procedure which has turtles turn around on red patches, decrease their energy, eat, and then die with a 10% chance or continue moving otherwise.

```
to run
  if pc = red
    [ rt 180 ]
  setenergy energy -- 1
  eat
  ifelse (random 100) < 10
    [ die ]
    [ move ]
end
```

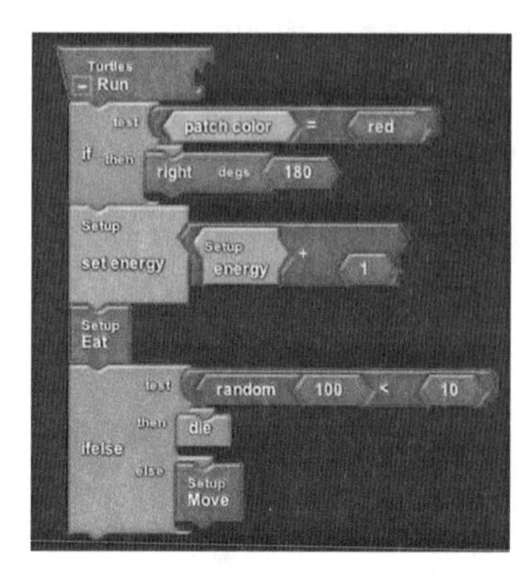

Fig. 6.5 StarLogo and StarLogoBlocks representations of the same code.

While the code in StarLogoBlocks takes up more space, it can be seen that the syntax is entirely visible to the user. In contrast, the StarLogo code has mixtures of brackets [ ], parentheses ( ), and spaces that are often confusing to users. The if-else conditional fully specifies what goes in which socket. The if predicate socket has a rounded edge, which can only fit Booleans, such as the condition shown. Similarly, functions which take integers have triangular sockets. The two other two sockets (for "then" and "else") are clearly labeled so that a programmer (or reader of a program) can clearly see what they are doing.

Blocks create more obvious indicators of the arguments taken by procedures, as compared to text-based programming. This enables us, as the software developers, to provide new kinds of control flow and Graphical User Interfaces (GUIs) for the blocks without confusing the programmer. Adding a fifth parameter to a text-based procedure results in children (and teachers) more often consulting the manual, but introducing this fifth parameter as a labeled socket makes the new facility more apparent and easier to use. Graphical programming affords changes in these dimensions because a change in the graphics can be more self-explanatory than in a text-based language.

While StarLogoBlocks and LogoBlocks share the "blocks" metaphor, StarLogoBlocks is presented with much greater challenges due to the relative complexity of the StarLogo environment. LogoBlocks programs draw from a language of dozens of commands, are typically only 10–20 lines long, have a maximum of 2 variables, have no procedure arguments or return values, and have no breeds. StarLogo programs draw from a language with hundreds of commands and can often be a hundred lines or more long. Screen real estate can become a real limit on program size. Driven by this challenge, we created a richer blocks environment with new features specifically designed to manage the complexity and size of StarLogo code.

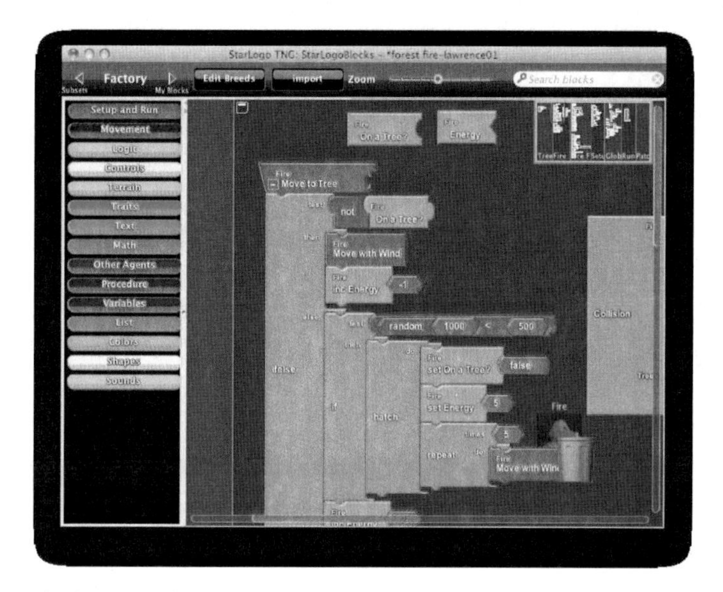

Fig. 6.6 The StarLogoBlocks interface.

One of the most important innovations is to incorporate dynamic animated responses to user actions. We use this animation to indicate what kinds of user gestures are proper and improper while the user is performing them. For example, when a user drags a large stack of blocks into an "if" statement block, the "if" block will stretch to accommodate the stack. See the "Move to Tree" procedure in Fig. 6.6. The first part of the if-else command has two commands to run (Move With Wind and inc Energy), while the else has many (starting with a nested if command inside of it). The blocks expand to fit as many commands as necessary so that the else clause could in theory have dozens of commands.

If a user tries to use a procedure parameter in a different procedure from which the parameter was declared, all of the block sockets in the incorrect stacks of blocks close up. When a user picks up a number block to insert into a list of values (which in Logo may contain values of any type), all block sockets in the list will morph from an amorphous "polymorphic" shape into the triangular shape of a number, to indicate that a number block may be placed in that socket. We plan to continue adding new kinds of animations to help prevent users from making programming errors in the system.

These animations are implemented using vector-based drawing. Vector drawing enables a second important feature: the zoomable interface. Using a zoom slider, the user can zoom the entire interface easily (see the zoom slider in the upper right of Fig. 6.6) to look closely at a procedure they are writing or to expand their view to see an overall picture of their project.

Another innovation has been "collapsible" procedures. Individual procedures can be collapsed and expanded to see or hide their contents. This allows the programmer to build many small procedures and then "roll up" the procedures that they are not using and put them on the side. Fig. 6.6 shows the "Move To Tree" procedures in its unrolled state. The individual commands are visible, but they can be hidden by clicking on the plus sign on the block. Additional advances provide for the easy creation of new procedures, including procedures with an arbitrary number of parameters and variables. Also of note is the ability to enter blocks simply by typing, a method we call "typeblocking." Typeblocking allows for quick entry of blocks, making the construction of sophisticated projects quite rapid for experts.

Together, these innovations lower the barrier of entry for programming, thus facilitating model construction/deconstruction in the context of science, math, or social science classes.

6.5 The StarLogo TNG Virtual Machine

The StarLogo TNG virtual machine (VM) is responsible for executing the commands that make up a StarLogo TNG program. It is at the core of StarLogo TNG, giving virtual life to the agents living in Spaceland and running their StarLogoBlocks programs. The VM has several advancements aimed at making StarLogo TNG programs faster, more flexible, and easier to understand than their StarLogo 2 counterparts.

6.5.1 The Starting Point: StarLogo 2

The StarLogo TNG virtual machine is based on the design of the StarLogo 2 virtual machine, which allows thousands of agents to execute programs in pseudo-parallel, meaning that execution, although serial in actuality, is switched so quickly between agents that it appears to be almost parallel. In StarLogo 2, each agent executes a handful of commands and then the VM switches to the next agent, allowing it to execute a handful of commands, and so on. Each time an agent moves, changes color, or changes any of its state, its visible representation on the screen is updated immediately. Additionally, movement commands such as forward (fd) and back (bk) are broken up into single-step pieces, effectively converting fd 5 into five copies of fd 1. This causes the execution to appear almost as if it is parallel, as each agent only takes a small step or does a small amount of execution at a time.

6.5.1.1 Synchronization

The design of StarLogo 2's VM is effective for many types of programs, but it also leads to several conceptual difficulties stemming from the problem of synchronization. In any pseudo-parallel system where two agents interact, synchronization takes place when the agents momentarily enter the same frame of time. For example, if two agents meet and each decides to take the color of the other, one of them will in actuality go first, leading to unexpected behavior. In the colors example, if a red and blue agent meet, and the red one goes first, it will take the color of the other, thereby becoming blue itself. The blue agent, then, because it executes second, will take the color of its now blue partner, producing a net result different from what is expected at first glance. Another synchronization-related difficulty is getting agents to move at the same pace. In StarLogo 2's VM, an agent with fewer commands in its program will execute the entirety of its program more quickly than one with more commands, causing that agent to appear to move more quickly even if both agents' only movement command in each iteration of the program is fd 1.

6.5.1.2 Speed Control

Another difficulty stemming from the StarLogo 2 VM is that agents cannot predictably be made to move faster or slower. As described above, adding commands to an agent's program can slow it down, but if the programmer tries to make the agent go, for example, fd 2 to make up for lost time, the command is broken down and run as two separate steps of fd 1, making fd 1 the absolute fastest speed possible. Additionally, because the extra commands slow the overall movement rate, there is no guarantee that an agent executing fd .5 as a part of its program will be moving half as fast as an agent executing fd 1, unless both agents also have an identical number of other commands. This makes any type of model where velocity needs to be explicitly controllable difficult to program in StarLogo 2.

6.5.1.3 Atomicity

A third, and perhaps the most important, drawback to the StarLogo 2 VM is that it does not allow agents to execute sequences of commands atomically, or as one command without interruption. By enforcing its pseudo-parallel, instantly visually updated execution model, the VM causes all agent commands to be reflected on the screen. For example, the command sequence left 90, forward 1, right 90 (lt 90, fd 1, rt 90 in StarLogo 2) causes the agent to turn twice, even though the final heading is the same as the initial heading. If a programmer wanted to program a sidestep, in which the agent simply slid to the left while still facing forward, he or she would be out of luck. In fact, StarLogo 2's VM requires a new built-in primitive command for every action that needs to be animated as one fluid action rather than steps.

6.5.2 StarLogo TNG: The Next Step Forward

StarLogo TNG's VM was designed with the pros and cons of StarLogo 2 in mind, and it specifically addresses the issues of synchronization, speed control, and atomicity mentioned above. The StarLogo TNG VM is similar to the StarLogo 2 VM in that it allows agents to execute commands, but its execution model is substantially different. Rather than letting each agent execute one or a handful of commands before switching to the next agent, the StarLogo TNG VM actually lets each agent run all of its commands before switching. Additionally, visible updates to the agents and the world are only displayed at the end of the execution cycle after all of the agents have finished. In this way, although agents run fully serially, the visible output is indistinguishable from that of truly parallel execution, as every agent is updated graphically at the same time.

6.5.2.1 Synchronization

Synchronization issues, although still existent in StarLogo TNG, are now explicit and more easily understandable. Whereas in StarLogo 2 one could never be sure which agent would execute a command first, in StarLogo TNG the agent with the lower ID number is guaranteed to go first, leading to behavior that can be predicted and therefore explicitly programmed. In the example with the red and blue agents, if the programmer knew the red agent would always execute first, he or she could tell the red agent to save its old color to variable before setting its new color. Then when the blue agent executes, it can look at the value in the variable rather than at the other (previously red) agent's current color. Note that such a solution only works because of the explicit nature of synchronization in StarLogo TNG. If the programmer were unsure of which agent was going to execute first, he or she would not know which agent should look at the actual color and which should look at the color stored in the variable.

6.5.2.2 Speed Control

StarLogo TNG's VM also addresses the issue of controlling movement speed. Rather than breaking up movement commands into smaller steps (which would be futile since all commands are executed before switching anyway), it executes the commands fully and relies on Spaceland to draw the changes properly. Spaceland takes the difference in location and heading between consecutive execution cycles of each agent and draws several intermediate frames of animation connecting the two, making for smoothly animated movement at the user-programmed speed rather than jerky jumps as agents change locations instantly. To the users, an agent moving forward .5 simply appears to be moving half as fast as an agent moving forward 1.

6.5.2.3 Atomicity

Probably the most powerful benefit of StarLogo TNG's revised execution model is that it allows programmers to define new atomic actions. Sidestepping, for example, is the automatic result of turning left 90, moving forward 1, and turning right 90. Because Spaceland only interpolated between the previous execution cycle and current one and because the heading ended up where it started, no heading change is animated, making for a smooth sidestep. This same pattern holds for arbitrarily complex sequences of commands, allowing the agents to perform complex measurements such as walking to every neighboring patch to see which is higher, while only appearing to move the distance between the prior execution cycle and the current one.

In fact, although extremely powerful, this atomicity can be difficult to grasp at first because it causes some commands to be executed invisibly. StarLogo TNG addresses this problem with the introduction of the yield command, which causes the agent executing it to save its current state and end its turn in the execution cycle, thereby allowing Spaceland to draw its state at that intermediate moment in its execution. Then, in the next execution cycle, the agent restores its state and continues from where it left off rather than starting from the beginning of its program again. By making the execution model more explicit, StarLogo TNG avoids confusion while giving more power and flexibility to the programmer.

6.5.3 An Addition in StarLogo TNG: Collisions

StarLogo TNG's VM also implements a new form of interaction that is easy for programmers to understand and use: collisions. The collisions paradigm allows for event-driven programming, in that users can specify commands for each type of agent to run in the event of a collision, and the VM constantly checks for collisions and tells the agents to execute the specified commands as appropriate. For example, a common collision interaction would be for one agent, the "food," to die, while the other, the "consumer," increments its energy. This pattern is also particularly useful in games where a character collects items for points or additional capabilities. An additional benefit of introducing collisions as an event-driven paradigm is that it allows the VM to calculate collisions once, for all agents, rather than having agents ask for surrounding agents, as they do in StarLogo 2.

All of the features in StarLogo TNG, including the virtual machine, were made with the goals of providing the user with more power and flexibility while simultaneously lowering the barrier to entry for novices. By starting with a proven capable VM from StarLogo 2 and improving it with the additions and adjustments described above, all based on feedback from users and careful thought, StarLogo TNG moves closer to those goals. By doing so, it opens up new possibilities and new domains for StarLogo TNG, including games and physics simulations, as discussed in Section 6.7.

6.6 StarLogo TNG Example Model

To illustrate the features of StarLogo TNG, this section describes how to build a simple model of an epidemic – a model in which red agents represent sick individuals and green agents represents healthy individuals. When a red individual touches a green one, the green one will get sick. Then we add a recovery variable that can be adjusted by the user and graph the changing number of sick and healthy agents.

6.6.1 Setup

First, we will create some agents and put them on the screen. In the Setup section of the canvas, we put together the following blocks:

The Setup block runs once to set the initial conditions of the simulation. In this case, we want to delete all agents from the previous run, create 500 turtle agents and set their color to green (to represent 500 healthy individuals), and scatter them randomly in Spaceland. Note that turtles are the default agent, but both the name and shape can be changed in the Breed Editor:

To test the Setup block, look at the Runtime space connected to Spaceland. You should see a button named Setup. Click it and you will see the agents created and dispersed as seen in Fig. 6.9.

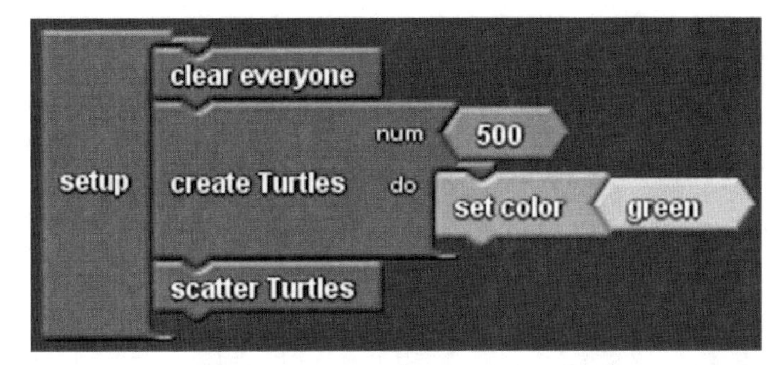

Fig. 6.7 Setup block for epidemic model.

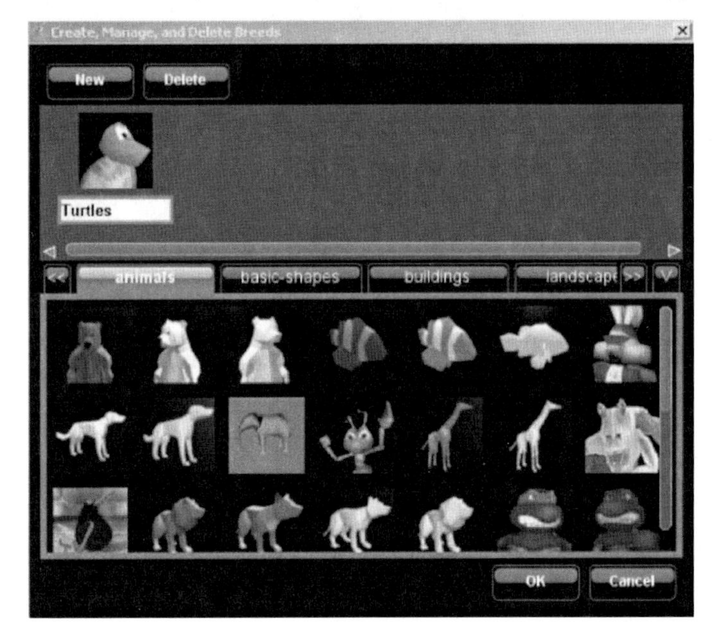

Fig. 6.8 Breed Editor.

6.6.2 Seeding the Infection

To infect some of the turtles at the start, we put together some blocks in a Run Once block and put them in the Setup area. We edit the Run Once name to say Infect to identify the purpose of this Run block (Fig. 6.10).

Notice that the Infect code is connected to a hook that is labeled "Turtles." This means that every turtle agent runs this If block once. The If block test condition randomly selects a number between 1 and 100 and compares it to 10. If the randomly selected number is less than 10, the turtle agent executes the then section of the block, which tells it to set its color to red. Since random 100 means that every number between 1 and 100 is equally likely to

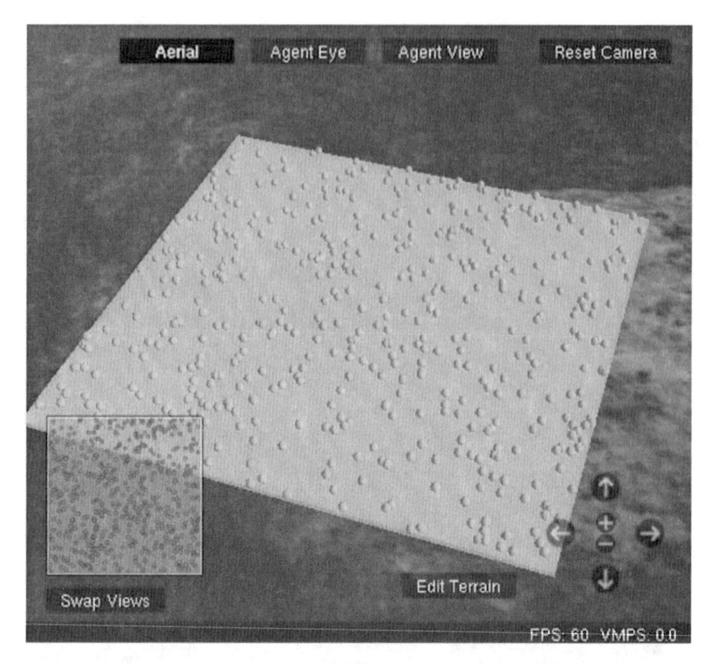

Fig. 6.9 Five hundred healthy agents created and scattered in Spaceland.

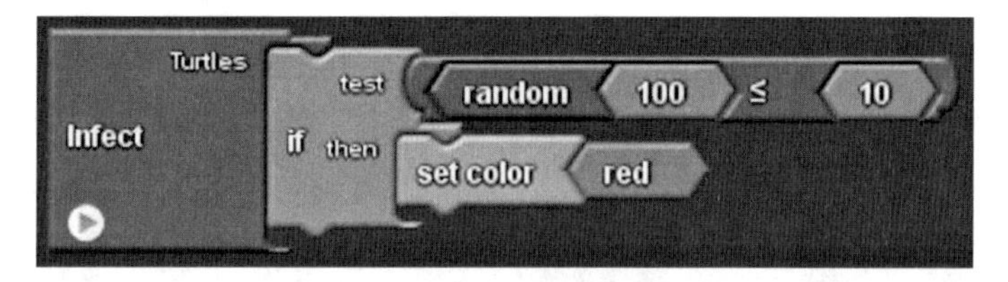

Fig. 6.10 Infect Run Once code to seed the infection at the start of the simulation.

be selected, there's a 10% chance that the number selected will be less than 10 and thus, approximately 10% of the 500 turtles will be infected.

We can test the Infect code by clicking on it in the Runtime window. The results are seen in Fig. 6.11.

6.6.3 Motion

To get the agents moving, we use a Forever block, which works as a loop. As seen in Fig. 6.12, we put the Forever block in the Runtime section of the programming canvas and attach some movement blocks to it to make the turtle agents move forward, with some random left and right "wiggle."

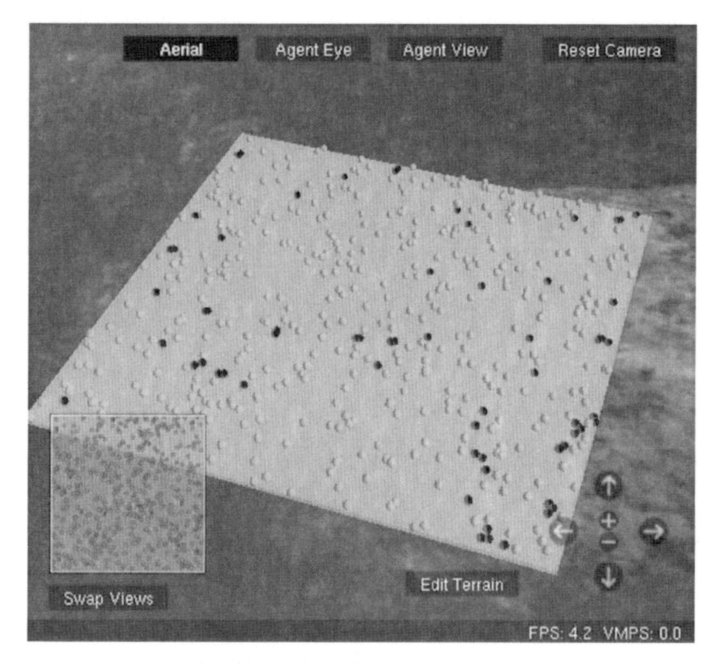

Fig. 6.11 Result of the Infect code as shown in Spaceland.

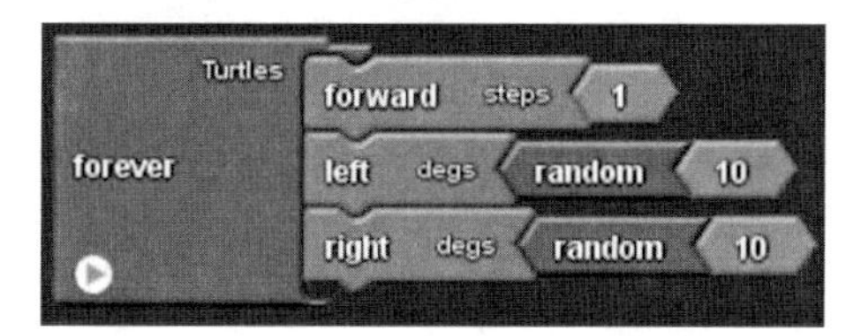

Fig. 6.12 Forever (loop) code to make the agents move.

The forever block also appears in the Runtime window and clicking on it will produce a green highlight to show that the forever block's code is running continuously in a loop until it is clicked again to stop (Fig. 6.13).

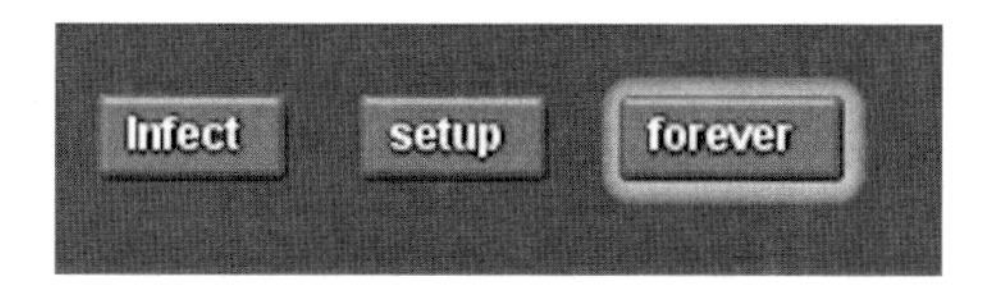

Fig. 6.13 Forever block turned on in Runtime window.

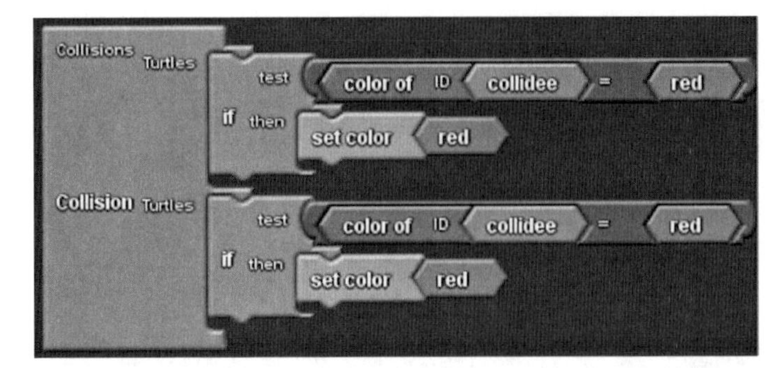

Fig. 6.14 Collision code between two turtle agents to spread the infection.

6.6.4 Infection

We program a Collision block to get the agents to infect each other when they "touch" when moving around in Spaceland. The Collisions blocks are found in the My Blocks palette, where breed-specific blocks are located. We drag the Collisions block between Turtles and Turtles to the Collisions section of the canvas (Fig. 6.14) and put in some blocks that tests if the color of the turtle being collided with is red. If it is, then the turtle agent turns its own color to red, showing that it has been infected by the other red (and sick) turtle. We use the same test for both turtles in the collision to make sure that we cover both cases where either of the turtles in the collision can be red.

Now when we run the model, we can see the infection spreading as more and more agents turn red from contact with other red agents (as seen in Fig. 6.15).

6.6.5 Recovery

To make our model more complex, we can add a recovery element, which is represented by a percent change of recovery (change color to green). Thus, during each iteration of the Forever loop, all turtle agents will have some chance of recovering, according to a percent chance variable that the user can set with a slider. A high number means a higher chance of recovering and a low number means a lower chance of recovering.

First, we create a global variable called Recovery by dragging out a "shared number" block from the Variables drawer to the Everyone page of the canvas and renaming it Recovery as seen in Fig. 6.16. Because we want the user to set the variable, we also drag out a slider block and snap it in front of the shared number block.

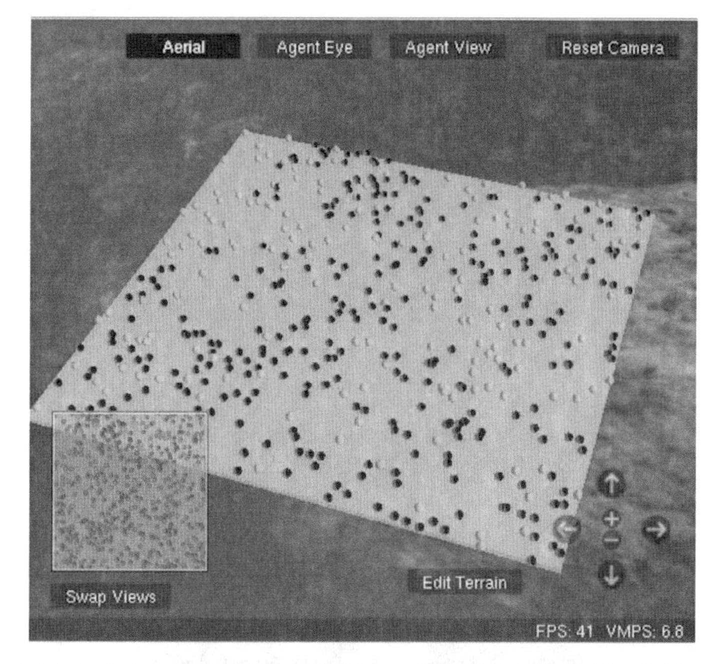

Fig. 6.15 Most of the agents infected after some time as a result of the Collision code.

Fig. 6.16 Slider block attached to variable declaration.

In the Runtime window, the slider appears as in Fig. 6.17.

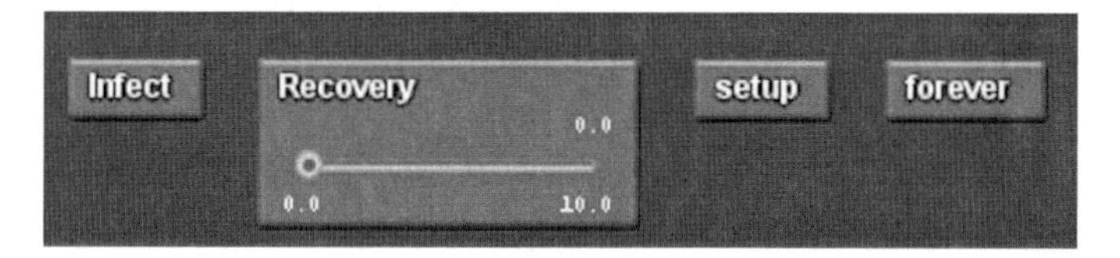

Fig. 6.17 The slider appears in the Runtime window so that the user can adjust the value of the Recovery variable.

The minimum and maximum values of the slider can be adjusted by clicking on the numbers below the slider and typing in new values.

To give the turtles a chance to recover, we write a procedure called Recover. To define a procedure, we grab a Procedure block and put it in the Turtles section of the canvas. Rename the procedure Recover and put in the code as shown in Fig. 6.18, noting that it uses the recovery variable.

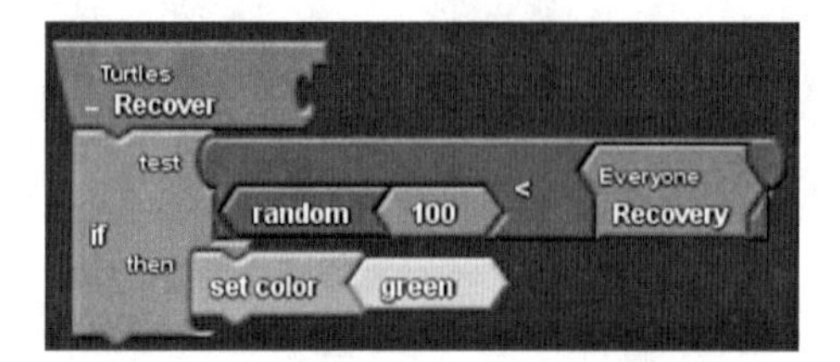

Fig. 6.18 Recover procedure for epidemic model.

Now we have the turtle agents call the Recover procedure in the Forever block so that the turtle agents run the Recover code after they move during each iteration (Fig. 6.19).

Fig. 6.19 Turtle agents call the Recover procedure as part of Forever code.

6.6.6 Graphs

It is useful to track and compare the numbers of healthy and sick turtles over time so the code in Fig. 6.20 shows how to set up a double-line graph.

Fig. 6.20 Code to set up a line braph that tracks the number of green and red agents.

The line graph appears in the Runtime window and can be double-clicked to enlarge in a separate window with other options (Fig. 6.21).

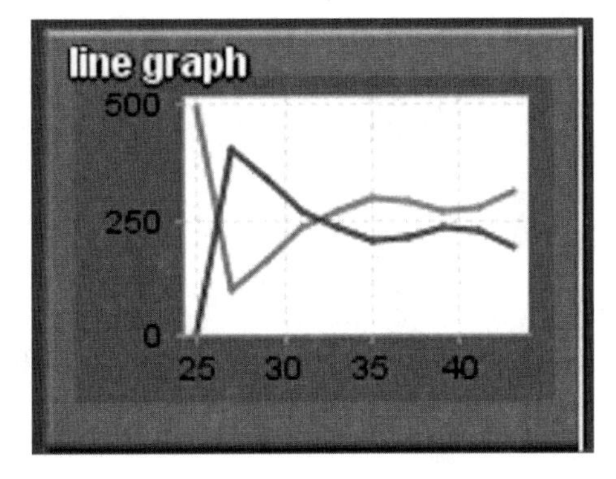

Fig. 6.21 How the line graph appears in the Runtime window.

These are just some of the basic features of programming a simulation model in StarLogo TNG: Setup, Forever, Breed Editor, Movement, Collisions, Slider, Variables, Procedures, and Line Graph.

6.7 Field Testing

StarLogo TNG has been downloaded by more than 10,000 users to date and it is being used in a variety of schools, clubs, and learning environments. We have used StarLogo TNG for introducing programming to students, teaching complex systems, and providing in-depth insights into a variety of scientific concepts. One of the most recent forays that StarLogo TNG has enabled is its use in physics classes. The following subsection is a description of this use.

6.7.1 Learning Physics Through Programming

diSessa [6] showed that fundamental physics concepts could be made accessible to students as early as the sixth grade by using simple programming activities. We believed that at the high school level, programming could help students build a deep understanding of traditionally difficult physics concepts, so we introduced StarLogo TNG programming basics through a series of physics-based activities to three high school physics classes at a private school in the Boston metropolitan area. A video-game-building activity helped establish a relationship between students and TNG that enabled them to use TNG effectively as a learning tool. Students programmed realistic motion in their games, including velocity, acceleration, friction, gravity, and other aspects of Newtonian physics. The algorithmic thinking of the 2D motion activities provided a new and useful entry point for learning physics concepts that along with labs and problem-solving formed a stable, three-legged platform for learning that aided classroom dynamics and appealed to a diversity of learning styles. While students in this study were not randomly assigned, three other comparable physics classes at the same school were used for comparison. Data were collected in the form of written assessments, lab reports, surveys, and interviews.

We found that TNG programming and simulations were useful in the physics classroom, because TNG added a new dimension to the ways students think and learn physics. For example, a physics teacher usually defines velocity as displacement/time, $v = d/t$. Then she describes the motion of an object by stating that it has a velocity of, say, 1 meter per second. With TNG, students program motion by building the code (Fig. 6.22): set x [x +1]. The 1 is called stepsize – the distance an agent moves in each time-step. Stepsize is equivalent to the concept of velocity. The code is abstract, but less so than the equation. When executing the code, students see the agent stepping 1 unit each time-step. TNG has graphing capabilities, so they can see how their agent, executing the code they built, generates a linear position versus time graph.

TNG proved to be more than a programming environment; it was a tool with which one could think and ask questions. If some students in the class are setting x [x+1] while others are setting y [y+1], there is a good chance someone in the class will ask, "What if I set x [x+1] and set y [y +1] together?" If the physics teacher creates an environment where some students are programming motion in the x direction and others in the y direction, there is a very good chance that someone in the room will "discover" that if x and y direction codes are executed simultaneously, the agent will move diagonally. TNG made the discovery of the vector addition of motions more likely.

The concept of simultaneous but independent change that is central to understanding 2D motion is difficult and frustrating to teach and learn. We hypothesized that by adding the programming of 2D motion in a virtual world, in addition to the usual mathematical analysis of that motion in the physical world, we could enhance the understanding of this difficult concept. Below are some programming activities that both introduced and supported the new concepts students would need to develop a deep understanding of 2D motion.

Students began the unit by building separate simple programming procedures that changed agent attributes like color, size, or location. Keys could be pressed individually or simultaneously, producing many humorous combinations of change.

Students recognized that the value "1" represented the distance covered in a given iteration. This number, designated "stepsize," was linked to the concept of velocity – the distance covered in a given second. With stepsize as a variable (Fig. 6.23), students could change "velocity" by using a slider that sets that variable and moves their agent in any desired direction.

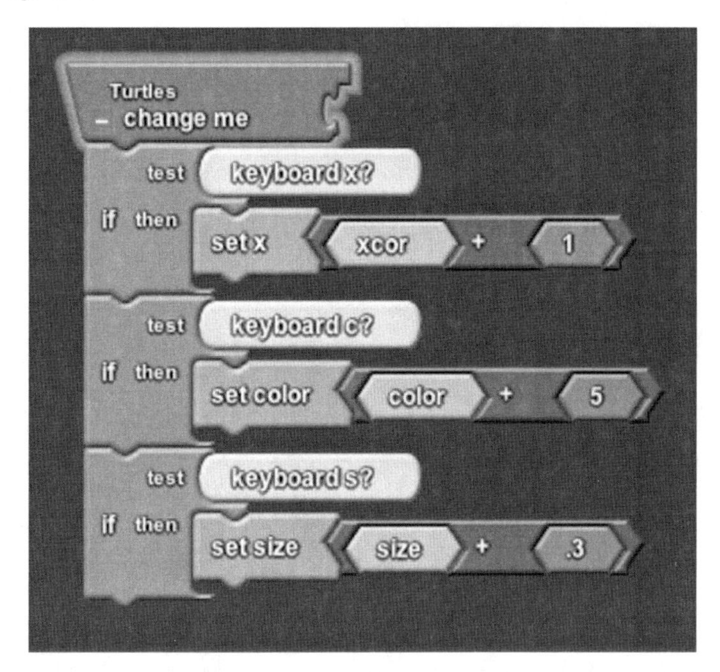

Fig. 6.22 Simultaneous and Independent change – code of keyboard controls that change agent attributes.

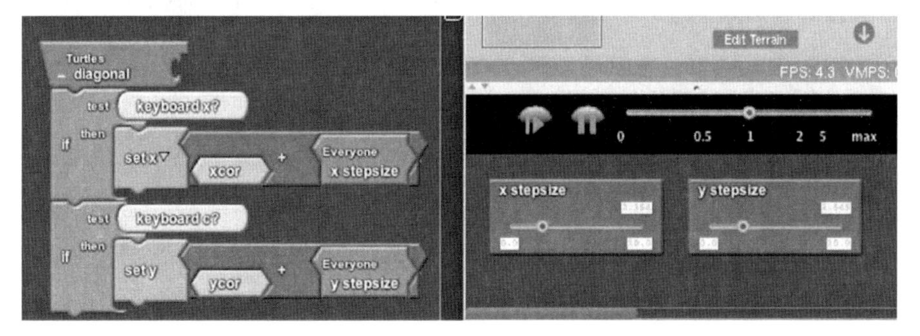

Fig. 6.23 Stepsize and Variables – code of keyboard controls for movement where the steps taken can be adjusted as slider variables.

The teaching of the advanced programming concept of variables and the addition of vertical and horizontal motions occurred naturally, linked by the students' desire to create a variety of motions for their agents. Because students were comfortable with TNG as a learning tool, the new ideas were introduced and the learning occurred in an environment where students had competence, comfort, and confidence.

The culmination of the kinematics section was a unit on projectile motion. To get realistic vertical motion, students added a procedure for the negatively directed acceleration of freefall in the z direction to independent motions on the ground. To give students experience with this hard-to-picture situation, a game was used. Students built a jumping game from instructions and the help of two students in the class who had used jumping in their video game. To get an agent to jump over an obstacle, students had to separately build yet simultaneously execute a constant-velocity, move-forward procedure and a vertical-jump procedure that employed acceleration. The goal of the game was to get a raccoon to jump over a wall and hit a target (Fig. 6.24).

Fig. 6.24 Jumping game screenshot – students adjust variables to get the racoon to jump over the wall and land on the target on the other side.

Here, solid, experiential learning is achieved with the game-like simulation. "To get the raccoon to hit the donut, I had to...I changed the horizontal stepsize smaller, and increased the raccoon's jump start speed (*initial vertical velocity*). I did this because a large horizontal stepsize made the agent exceed past the donut."

The acquisition of new skills and confidence through programming is demonstrated in this anecdote. In a junior physics class, students built swimmer-in-river simulations as part of a unit on vectors. The model has a swimmer with velocity, s, swimming at a given heading across a river of given width with current velocity, r. After building and playing with the simulation, students were assigned 2D motion problems for homework. During a class discussion about the answer to a contested problem, one student went, unprompted, to the computer and opened his swimmer model. He plugged the variables of the problem into the code of the model and ran the model. As the swimmer reached the opposite shore, he exclaimed, "I told you I was right!" When asked in an interview why he chose to use the computer rather than mathematical analysis to prove his point, he said. "This way you could ***see*** I was right." StarLogo simulations were a comforting, concrete, and useful refuge for some students in the sometimes troubling abstract world of physics equations. One student commented in an interview after the projectile motion test, "When I do TNG everything makes sense and I can do it. Once ... the questions are all on paper, it really messes me up and I have problems doing it." In one of the physics class that used StarLogo TNG, while 75% of the surveyed students agreed with the statement that the StarLogo unit was more difficult than other units, 100% of the students felt the unit was more rewarding.

6.7.2 From Models to Games

We developed and piloted StarLogo TNG curricula at two middle schools in a low-income town in Massachusetts. At one school, students learned programming by creating a first-person perspective treasure hunt game, in which the player controlled an agent using the keyboard to "collect" treasure and avoid hazards. Students had little to none prior programming experience. What is most striking from this field test is students' motivation to learn advanced programming concepts because they want to add a "cool" feature in their games, like shooting a projectile or creating a pattern in Spaceland's terrain. Also, when one student discovers how to use a particular block, such as "say" or "play sound," the knowledge quickly spread throughout the class and are incorporated into the students' own games. These two elements – a "need-to-know" desire to learn and informal peer learning – are features of classrooms where students are excited and motivated to persevere through challenging activities.

At another middle school pilot site, students made observations of and conducted experiments by manipulating variables of simulation models of forest fires and virus epidemics. The interesting twist is that they then turned these simulations into games by adding breeds and keyboard controls. The first activity took about 1 hour and the second activity took between 1 and 2 hours. So within 3 hours, students were able to both learn to use a simulation and program a basic game, with the guidance of a teacher and supporting worksheets.

When interviewed about their experiences, students shared the following:

- "You changed my mind. I thought that I was never going to be a programmer. I thought that the computer was just for work, and to look [up] stuff, and research on the Internet... it's more than just work and stuff, you can

do whatever you want on the computer. It was a process 'cause to tell you the truth. I thought it was going to be boring. I usually play video games, not make them. But it was fun."
- "It's like science, but better. It's not all textbook. You get to do what you're learning. We were doing things that we're going to remember because with Starlogo, I'm going to keep on keep on using it, using it, using it and I'm going to keep on getting better... I'm not going to forget it, like if I was reading a book or something."
- "It makes me think about how to control things. I have to think. I have to use calculations."

6.8 Conclusion

The testing and focus groups that we have used so far have suggested that the blocks design will promote the use of StarLogo as a programming tool in the classroom. Teachers can feel more secure using a tool that removes anxiety about getting the syntax right and makes learning programming easier. While it is still early to tell, it seems that it may also provide a more female-friendly programming environment than other tools.

The 3D world is "definitely cool," in the words of many of the students. While teachers have been more lukewarm in their reception of this aspect, they recognized that it will be more attractive to students. It is clear that both 2D and 3D representations will have a place in the learning environment, and the ability to customize the world and make it look like something recognizable (although not realistic – which is important in conveying that these are simply models) is a big plus for students as they approach modeling for the first time. It also allows students to invest themselves personally in their products, which is empowering and motivating.

StarLogo TNG will be in development for some time. Some of the next steps involve adding features like network connectivity and joystick control, in order to add to the attraction of those wishing to build games. However, we also will be building libraries of blocks for specific academic domains (e.g., ecology, mechanics, etc.) that will allow students to quickly get started building science games using some higher-level commands, which can, in turn, be "opened up" and modified as their skills progress.

Acknowledgements The research in this chapter was supported in part by a National Science Foundations ITEST Grant (Award 0322573).

References

1. Agalianos, A., Noss, R., Whitty, G.: Logo in mainstream schools: the struggle over the soul of an educational innovation. British Journal of Sociology of Education 22(4), 479–500 (2001)
2. American Association for the Advancement of Science: Project 2061: Benchmarks for Science Literacy. Oxford University Press, Oxford (1993)
3. Begel, A.: LogoBlocks: A graphical programming language for interacting with the world. Massachusetts Institute of Technology, Department of Electrical Engineering and Computer Science, Cambridge, MA (1996)
4. Casti, J.L.: Complexification: Explaining a paradoxical world through the science of surprise. Abacus, London (1994)
5. Cypher, A., Smith, D.C.: Kidsim: End user programming of simulations. In: I.R. Katz, R.L. Mack, L. Marks, M.B. Rosson, J. Nielsen (eds.) Proceedings of ACM CHI'95 Conference on Human Factors in Computing Systems', Vol. 1 of Papers: Programming by Example, pp. 27–34, ACM, Denver, CO (1994)
6. diSessa, A.: Changing Minds: Computers, Learning, and Literacy. MIT Press, Cambridge, MA (2000)
7. diSessa, A., Abelson, H.: Boxer: A reconstructible computational medium. Communications of the ACM 29(9), 859–868 (1986)
8. Harel, I.: Children as software designers: A constructionist approach for learning mathematics. The Journal of Mathematical Behavior 9(1), 3–93 (1990)
9. Harel, I., Papert, S. (eds): Constructionism, Ablex Publishing, Norwood, NJ (1991)
10. Holland, J.H.: Emergence: From chaos to order. Helix, Reading, MA (1998)
11. Kafai, Y.B.: Minds in Play: Computer Game Design as a Context for Children's Learning. Lawrence Erlbaum Associates, Hillsdale, NJ (1995)
12. Kafai, Y.B.: Software by kids for kids. Communications of the ACM 39(4), 38–39 (1996)
13. Logo Computer Systems Inc. Microworlds Logo, Highgate Springs, VT (2004)
14. Lucena, A.T. (n.d.): Children's understanding of place value: The design and analysis of a computer game.
15. Mathworks: Matlab, Natick, MA (1994)
16. Nature: 2020 Vision. Nature 440, 398–419, New York, NY (2006)
17. Papert, S.: Mindstorms: Children, Computers, and Powerful Ideas. Perseus Books, Cambridge, MA (1980)
18. Papert, S.: The Children's Machine: Rethinking School in the Age of the Computer. Basic Books, New York, NY (1993).
19. Resnick, M.: Turtles, Termites, and Traffic Jams: Explorations in Massively Parallel Microworlds (Complex Adaptive Systems). MIT Press, Cambridge, MA (1994)
20. Resnick, M.: Rethinking learning in the digital age. In: G. Kirkman (ed) The Global Information Technology Report: Readiness for the Networked World. Oxford University Press, Oxford (2002)
21. Resnick, M., Bruckman, A., Martin. D.: Pianos not stereos: Creating computational construction kits. Interactions 3(5), 40–50, New York, NY (1996)
22. Resnick, M., Martin, F., Sargent, R., Silverman, B.: Programmable bricks: Toys to think with. IBM Systems Journal 35(3–4), 443–452, Yorktown Heights, NY (1996)
23. Resnick, M., Rusk, N., Cooke, S.: The Computer Clubhouse. In: D. Schon, B. Sanyal, W. Mitchell (eds): High Technology and Low-Income Communities. pp. 266–286. MIT Press, Cambridge, MA (1998)
24. Roberts, N., Anderson, D., Deal, E., Garet, M., Shaffer, W.: Introduction to Computer Simulation: A System Dynamics Modeling Approach. Addison-Wesley, Reading, MA (1983)
25. Roschelle, J.: CSCL: Theory and Practice of an Emerging Paradigm. Lawrence Erlbaum Associates, Hillsdale, NJ (1996)

26. Roschelle, J., Kaput, J.: Educational software architecture and systemic impact: The promise of component software. Journal of Education Computing Research 14(3), 217–228, Baywood, NY (1996)
27. Roughgarden, J., Bergman, A., Shafir, S., Taylor, C.: Adaptive computation in ecology and evolution: A guide for future research. In: R. K. Belew and M. Mitchell (eds.). Adaptive Individuals in Evolving Populations: Models and Algorithms, Santa Fe Institute Studies in the Science of Complexity Vol. 16, pp. 25–30. Addison-Wesley, Reading, MA (1996)
28. Sander, T.I., McCabe, J.A.: The Use of Complexity Science. A Report to the U.S. Department of Education. USGPO, Washington, DC (2003)
29. Soloway, E., Pryor, A., Krajik, J., Jackson, S., Stratford, S.J., Wisnudel, M., Klein, J.T.: Scienceware model-it: Technology to support authentic science inquiry. T.H.E. Journal, 25(3), 54–56, Irvine, CA (1997)
30. State of Maine, D.o.E.: Maine Learning Technology Initiative, Augsuta, ME (2004)
31. Waldrop, M.M.: Complexity: The Emerging Science at the Edge of Order and Chaos. Simon and Schuster, New York, NY (1992)
32. White, B.: Thinkertools: Causal models, conceptual change, and science education. Cognition and Instruction 10, 1100, Philadpelphia, PA (1993)
33. Wilensky, U.: Statistical mechanics for secondary school: The GasLab modeling toolkit. International Journal of Computers for Mathematical Learning 8(1), 1–41 (special issue on agent-based modeling), New York, NY (2003)

Chapter 7
From Artificial Life to In Silico Medicine: NetLogo as a Means of Translational Knowledge Representation in Biomedical Research

Gary An and Uri Wilensky

Biomedical research today stands at a crossroads. There is a widening gulf between the extent of knowledge regarding basic mechanistic processes and the ability to integrate that information into explanatory hypotheses of system-level behavior. Techniques from the Artificial Life community can aid in bridging this gulf by providing means for visualizing and instantiating mechanistic hypotheses. This will allow the development of in silico laboratories where conceptual models can be examined, checked, and modified. NetLogo is a "low threshold, high ceiling" software toolkit that has been used to develop agent-based models (ABMs) in a multiplicity of domains and provides a good platform for the computational instantiation of biomedical knowledge. This chapter presents a brief overview of NetLogo and describes a series of ABMs of acute inflammation at multiple levels of biological organization.

7.1 Introduction

7.1.1 A Different Type of "Artificial Life"

Artificial life, as the name suggests, consists of reproducing the processes of life in a man-made, often computational, setting. A great deal of Artificial Life research has focused on examining the core properties of life by stripping away those aspects that may be present in our world resulting from historical accidents in order to develop general theories of life [43]. Software tools that have arisen to enable this type of investigation have the capabilities to reproduce those presumptive central characteristics of biological systems. However, the development of these tools has also led to a parallel but related area of investigation, one that focuses on the use of Artificial Life-derived methods to recreate and represent existing biological systems. This capability has enabled the creation of a new laboratory environment, an "in silico" environment that can serve as a vital adjunct to traditional in vitro (in a test tube) and in vivo (in an experimental animal) experimental methods. In this fashion, Artificial Life-derived techniques and the use of ABMs, in particular, can serve a vital integrating role to address some of the major hurdles facing biomedical research. In this chapter we present examples of the use of the NetLogo integrated agent-based modeling environment [79] as a laboratory environment for biomedical research. NetLogo is a widely used general-purpose agent-based modeling environment. It is designed with the principle of "low threshold, high ceiling" [67] – that is, easy learnability for novices yet high performance for researchers. As such, NetLogo is in widespread use in both research and educational contexts. This dual-audience flexibility makes NetLogo ideal for use in interdisciplinary research contexts where its generality and ease of use enable all the research team members to participate in the modeling activity and to communicate via the NetLogo model. As such, NetLogo is well suited to the context of biomedical research, where, in general, most biomedical researchers are not cross-trained in mathematical modeling or computer simulation and often do not have the spare intellectual capital to invest in acquiring the expertise required to master general-purpose computer programming. The "low threshold, high ceiling" property of NetLogo has the following potential benefits as a biomedical research tool:

1. The rapidity with which non-computer-programming biomedical researchers can produce a tangible in silico representation of their existing mental models provides vital early positive feedback to encourage continued pursuit of a new methodology.
2. Enabling the researchers themselves to do the actual coding and modeling provides a basic practical literacy with respect to the application of simulation and computational tools to biomedical problems and will facilitate future communications with computer scientists/applied mathematicians as these applications develop and may evolve to other platforms.
3. The actual exercise of creating a NetLogo model enables formalization of existing mental models, can stimulate beneficial introspection on the part of the biomedical researcher regarding his/her underlying assumptions, and help hone in on further directions in a particular research plan.
4. The ease of "reading" and widespread use of NetLogo enable researchers to communicate and disseminate their work clearly and broadly.

The examples of NetLogo models presented in this chapter take advantage of all these benefits, with the overall goal of demonstrating dynamic knowledge representation of a particular multi-scale biomedical system: acute inflammation.

7.1.2 Modern Medicine and Limits of Reductionism: The Translational Challenge

Over the last 50–75 years biomedical research has made huge strides. The advent of molecular biology, arising from the discovery of DNA, opened a new mechanistic frontier for the examination and analysis of biological systems. Based on the principle of reductionism, the concept that finer and finer-grained analysis of components and mechanisms would provide underlying core principles and understanding, biomedical research generated huge volumes of data and hypotheses regarding the basic processes associated with health and disease. However, starting in the 1970s, it was becoming evident that biological behavior and the health/disease dynamic were very much more than the sum of their parts. The efficacy of biomedical research to provide advances in the areas of infectious disease, public health, and surgical technique was not being reproduced in addressing "systems" diseases such as cancer, critical illness, autoimmunity, diabetes, and acquired immune deficiency syndrome (AIDS). This simmering crisis was crystallized in the United States by a publication by the United States Food and Drug Administration (USFDA) of a monograph titled *Innovation or Stagnation: Challenge and Opportunity on the Critical Path to New Medical Products* [27]. The central problem addressed was the steadily decreasing efficiency of the output of health-care research dollars between 1993 and 2003 in terms of release of effective medical therapeutics. This represented a widening translational gulf between the bench and the bedside.

What is the basis of the translational divide? To a great degree the translational challenge arises from a combination of the multi-scale, multi-hierarchical nature of biological systems and the existing research structures that have evolved to study them. The hierarchical structure of biological systems is well recognized: gene to protein/enzyme to cell to tissue to organ to organism. The existence of these levels (which can be thought of as representing successive levels of emergent phenomena) presents significant barriers to the inference of cause-and-effect mechanistic knowledge at one level to behavior at a higher one. This epistemological barrier between scales of biological organization is one of the hallmarks of complex systems and why these systems need to be studied in an integrated fashion [94]. Conversely, the organization of the biomedical research community that has evolved to study these different scales has been based on a reductionist paradigm. The treatment of each of these levels as a separate focus of investigation has led to a compartmentalized structure and organization of the biomedical research community. As pointed out in the USFDA Critical Path statement referenced above, the consequences of structure are seen primarily in attempts to develop effective therapies for diseases resulting from disorders of internal regulatory processes. Examples of such diseases are cancer, autoimmune disorders, and sepsis, all of which demonstrate complex, non-linear behavior and are insufficiently characterized when their components are studied in isolation. The investigation of such processes therefore presents a significant challenge that must be met by the development of translational methodologies that need to function as bridges both vertically from the bench to the bedside and link horizontally across multiple researchers focused on different diseases.

Thus, there is a growing recognition within the biomedical research community of the limits of reductionism and the need to apply a systems- level approach to attempt to reintegrate the mechanistic knowledge being generated [68, 71, 59]. In particular, it was recognized that such attempts to apply complex systems analysis to biomedical problems should not strive to supercede reductionist methods, but rather provide a synthetic adjunct to ongoing research endeavors [68, 8]. The development of in silico biomedical research is the response to this recognition.

Accomplishing this goal requires dealing with the "nine hundred pound gorilla" in the room of biomedical research: a seemingly unsolvable paradox between the volume of information and the completeness of information. On one hand, the sheer volume of biomedical knowledge that has been (and continues to be generated) is overwhelming. It is extremely challenging for a single researcher to have even semi-comprehensive knowledge of the state of even a small fraction of this information; expanding the scope of this knowledge to approach the integrative goal needed for translational interpretation is functionally impossible. However, even if it were possible for a researcher to know everything that was published regarding and connected to a particular disease process or biological function, that knowledge would still be incomplete, as the multiplicity of components and their interactions necessarily means that there likely will always be additional information yet to be identified. Therefore, the common charges against the use of computer simulation and mathematical modeling focus on these two paradoxical aspects: (1) there is too much information to include and (2) something is left out.

The solution to this paradox lies in maintaining appropriate expectations and placing the development of synthetic translational tools in the appropriate context: Mathematical modeling and computer simulation are not intended to be a panacea to the challenge of integrating biomedical knowledge, but rather as improvements on methods currently employed. The synthetic process of integrating experimental results into hypotheses and conceptual models relies, as it is currently executed, heavily upon intuition. As such, it is poorly formalized and thus difficult to probe and parse when things do not turn out as expected. What is sought, therefore, are methods of formalizing the process of knowledge representation, particularly in terms of the dynamic instantiation of knowledge as a means of hypothesis visualization and evaluation. The goal is to produce probeable synthetic constructs that can be tested by both their creator and other researchers and unambiguously communicated to the community as a whole. In this context, the use of computational models should be considered a means of "conceptual model verification" [90], in which mental or conceptual models that are generated by researchers from their understanding of the literature and used to guide their research are brought to life such that the behavioral consequences of the underlying beliefs can be evaluated. It is in addressing the synthetic aspect of science that the lessons of Artificial Life, and agent-based modeling in particular, come to the fore.

7.1.3 Agent-Based Dynamic Knowledge Representation and In Silico Laboratories

One of the major lessons from Artificial Life research is that biologically complex behaviors can be generated and substantial insights can be obtained, with relatively simple, qualitative models. The fact is that biological systems are robust, existing in a wide range of conditions while retaining a great degree of stability with respect to form and function. When Artificial Life is used to examine core biological processes (such as evolution, swarm behavior, and morphogenesis), the search for the minimal rule-set that can recognizably reproduce the desired behavior is seen as a proxy for identifying the most general and therefore most basic principles. If, however, the goal is to narrow the focus of investigation and increase the resolution of the behaviors under study, then the progressive addition of more and more details provides a mechanism for approaching a more realistic model. While one must always be careful to conflate a plausible solution with an exact solution, the correlative relationship between a set of generative mechanisms and observed recognizable behaviors has a strong tradition within both the Artificial Life and bioscience community as a basis of inference and hypothesis formation. In particular, this has been formalized as Pattern Oriented Analysis as a means of developing and interpreting ABMs, one of the primary computational techniques used in the Artificial Life community [31].

The development of ABMs is a computational modeling methodology that employs computational objects, and is rule based, discrete event and discrete time. Agent-based modeling focuses on the rules and interactions between the individual components ("agents") of a system, generating populations of those components and simulating their interactions in a virtual world to create an in silico experimental model. There are several characteristics of an ABM that sets it apart from other rule-based modeling systems:

- ABMs are *spatial*. As the field of agent-based modeling was much influenced by work in two-dimensional (2D) cellular automata many ABMs are grid based. This legacy enables spatial representation of structural relationships within a system. Non-mathematicians can model fairly complex topologies with relative ease, leading to more intuitive knowledge translation into a model. The spatial nature of an ABM also supports modeling agents with limited knowledge (i.e., input constrained by locality rules that determine its immediate environment). The emphasis on behavior driven by local interactions also matches closely with the mechanisms of stimulus and response observed in biology.

- ABMs utilize *parallelism*. In an ABM, each agent class has multiple instances within the model, forming a population of agents that interact in (an usually emulated) parallel processing environment. Thus, heterogeneous individual agent behavior within a population of agents results in systemic dynamics that yield observable output that mirrors the behavior at the higher-hierarchical level. A classic example of this is how relatively simple interaction rules among birds can lead to sophisticated flocking patterns [58, 74].
- ABMs incorporate *stochasticity*. Many systems, particularly biological ones, include behaviors that appear to be random. "Appear to" is an important distinction, since what may appear to be random is actually deterministically chaotic. However, from a practical point of view, despite the fact that a particular system may follow deterministic rules, at the observational level it is impossible to actually define the initial conditions with enough fidelity to apply formal deterministic mathematics. Thus, capturing the sensitivity of a system to initial conditions is obscured by limitations on the resolution in characterizing the state of the system. Agent-based modeling addresses this issue via the generation of populations of agents, and a subsequent distribution of agent-behaviors. It is possible to establish probabilities of a particular behavior for an agent population as a whole (i.e., an experimentally determined response distribution of a particular cell type to a particular stimulus). This allows the generation of a probability function for the behavior of a single agent-type, which is, in turn, incorporated into the agent-type's rules. For instance, a rule may look like "In the presence of Stimulus A there is a 80% chance Receptor B will be activated." When the entire ABM consisting of a population of agents is instantiated, each individual agent follows a particular trajectory of behavior as its behavior rules' probabilities are resolved with each step of the ABM's run. The behavior of the overall system is generated by the aggregated behavioral trajectories of the individual agents, each model run producing one instance of system behavior. Performing multiple runs of the ABM thus generates a "population" of behavioral outputs to produce system behavior spaces consistent with epidemiological biological observation.
- ABMs reproduce *emergent properties*. Due to the parallelism, intrinsic stochasticity, and locally-based agent rules, a central hallmark of agent-based modeling is the fact that they generate systemic dynamics that often could not have been reasonably inferred from examination of the rules of the agents, resulting in so-called emergent behavior. To return to the example of the bird flock, superficial observation would seem to suggest the need for an overall leader to generate flock behavior, and therefore rules would seem to need to include a means of determining rules for flock-wide

command and control communication. This, however, is not nature's way: Birds function on a series of locally constrained interaction rules and the flocking behavior emerges from the aggregate of these interactions [58]. The capacity to generate emergent behavior is a vital advantage of using an ABM for conceptual model verification, as it is often the paradoxical, non-intuitive nature of emergent behavior that "breaks" a conceptual model. The structure of ABMs facilitates the development of aggregated multi-scale models [8, 100]. They have an intrinsically modular structure via the classification of agents based on similar rules of behavior. ABM rules are often expressed as conditional statements ("if–then" statements), making agent-based modeling suited to expressing the hypotheses that are generated from basic scientific research. Individual agents encapsulate mechanistic knowledge in the form of a set of rules concerning a particular component. The importance of this encapsulation in agent-based modeling (as opposed to the compressed representation of knowledge with a mathematical formula, such as a biochemical rate law) is the placement of the mechanistic knowledge within a compartmentalized object. Instantiating agents and their governing rules in a virtual world creates an in silico experimental environment, or a virtual laboratory [6, 7, 65]. In doing so, agent-based modeling goes beyond the mere instantiation of this knowledge as a single case by concurrently generating multiple instances of a particular encapsulation/object. Because of this property, an ABM is an expansion of mere rule-based and object-oriented methods. Multiple individual instances have differing initial conditions by virtue of existing in a heterogeneous environment. Because stochastic components are embedded in their rule systems (a well-recognized property of biological objects [45]), individual agents have differing behavioral trajectories as the ABM is executed. This results in population-level dynamics derived from the generation of these multiple trajectories, population dynamics that, when viewed in aggregate, form the nested, multi-scalar/hierarchical organization of biological systems [90, 91, 92].

In this environment, researchers can instantiate thought experiments in an in silico environment, to test the veracity and validity of their conceptual models by comparing the simulated experiments against more traditional in vitro and in vivo experimental results. ABMs have been used to study biomedical processes such as sepsis [6, 7, 9], cancer [100], inflammatory cell trafficking [11, 64], and wound-healing [49, 73].

One vital aspect of the use of agent-based modeling is to perform the trans-hierarchical function desired in an integrative modeling framework. This moves toward the goal of communicable dynamic knowledge representation, easing the way for biomedical researchers to translate their conceptual models into executable form. While the era of multi-disciplinarily trained researchers is dawning, it is most likely that for the foreseeable future the majority of biomedical researchers will not be extremely facile in the use of computational tools and methods. Agent-based modeling, by its object-oriented nature, maps well to the current expression of biomedical knowledge. It is generally more intuitive for non-mathematicians/computer scientists to use. Nonetheless, it is a daunting software engineering task to provide a user-friendly ABM development environment for computer/math novices while including sufficient, comprehensive capabilities to capture the full richness of ABM utilization. Fortunately, the introduction and ongoing development of NetLogo provides just such a platform.

7.2 Facilitating Dynamic Knowledge Representation: The NetLogo Toolkit

7.2.1 Description and Origins of NetLogo

NetLogo is a general-purpose agent-based modeling environment for modeling complex systems. It includes the core NetLogo modeling language as well as an integrated modeling environment with a graphical user interface and many modeling tools. NetLogo was designed by Wilensky in the mid-1990s. Having collaborated with Resnick on StarLogo in the late 1980s and early 1990s [56, 55, 93, 57, 94], Wilensky set out to remedy limitations of StarLogo and create the next generation of the Logo-based agent-based modeling environment. Like StarLogo, NetLogo borrowed from the Logo programming language [25, 52] both core syntactic language elements and the central object of a "turtle," the default name for a mobile agent in NetLogo. Basic NetLogo also includes stationary agents called "patches" and connective agents called "links." New classes of agents can be defined using the "breeds" construct, so users can populate their models with wolves and sheep, atoms and molecules, or buyers and sellers. Constructing a NetLogo model involves choosing agents to represent aspects of the to-be-modeled system and giving them rules of behavior and interaction which they execute at every tick of a clock. One can then run the model and observe the emergent behavior of the system. Patches are typically used to model aspects of the environment, such as earth, grass, or atmosphere, that are computationally active, though stationary. So, at each time tick, grass may grow and be available for sheep agents to eat, earth may absorb percolating oil, and atmosphere patches may harbor carbon dioxide molecules [77, 76, 66]. In most natural science models, mobile agents affect each other through local interaction, thus the spatiality of the patches becomes the vehicle for agent interaction. Agents interact with their spatial neighbors. In many social science models, spatiality is not the dominant form of interaction. Agent communication can happen at a distance and is governed by social connection between (human) agents. The "links" agents typically model these social connections. In such models, agents interact with their "link neighbors," the agents with whom they are linked regardless of their spatial locations.

NetLogo comes with an extensive library of sample models (over 300 as of release 4.0.2 of NetLogo in December 2007) from a wide variety of content domains [80], including natural sciences such as physics, chemistry, and biology, and social sciences such as sociology, psychology, and linguistics, and engineering domains such as materials science, industrial engineering, and computer science, and professional disciplines such as medicine, law, and business. NetLogo is freely available from [86] and comes with extensive documentation, code examples, and tutorials. For a good introduction to the methodology of agent-based modeling, including exploring constructing, testing and verifying models, see the textbook by Wilensky and Rand [90].

7.2.2 Design Philosophy Behind NetLogo

A core design principle of NetLogo, also originating in the early Logo community, is the principle of "low threshold, high ceiling" (sometimes rendered as "low threshold, no ceiling") [52, 67]. What this principle means, on the one hand, is that the language should be simple enough for a novice without programming background to be able to learn and productively use within minutes of starting. On the other hand, the language should be powerful and robust enough to be used by researchers in their scientific work. These twin design objectives are traditionally seen as conflicting: One either designs an educational tool for novices and students or one designs a professional tool suitable for scientists in their research. However, Wilensky started out with the hypothesis that these goals could be reconciled in one environment and, together with his colleagues at the Center for Connected Learning and Computer-Based Modeling (CCL), has been continuously developing and refining NetLogo so as to be more useful to each of these communities.

Core aspects of NetLogo design addressed to novices are natural-language-inspired syntax, graphical world present at start-up, easy-to-build drag and drop interfaces, and extensive help, documentation, sample models, and code examples. Core design features addressed to power users include an extensions application programming interface (API) through which users can write their own NetLogo libraries (in Java), support for replicability of model runs, programmatic control of agent execution order, and support for probability distributions of agent variables and for a variety of network layouts.

Although the criterion of low threshold is typically interpreted as being targeted to novices and high ceiling being targeted to researchers and advanced users, the design rationale for NetLogo hypothesized that both goals are actually important to both communities [82, 67]. On the high-ceiling side, the reasoning is that novices benefit from being able to smoothly move from elementary modeling to more advanced modeling without having to change modeling languages or platforms. Moreover, and this is important for educational adoption, schools are much more easily persuaded to adopt software that is used in universities and in research contexts – one need not convince them that the software helps the students learn better, one can just point them to the scientific and commercial uses and they become interested in giving their students the same tool as they will use later on in life. On the low-threshold side, we have argued that low threshold is very important for researchers as well. Many researchers do not consider themselves programmers, and therefore, once they have created a conceptual model, they delegate the task of implementing the model to a programmer. This practice can lead to mismatches in understanding between the two parties and, consequently, to incorrect models [89]. Moreover, if the researcher can build the model himself/herself, then the researcher can much more rapidly and fluidly explore the design space of the model and make better progress in his/her research. Finally, the easy readability of a low-threshold language enables researchers to read and verify each others' models, leading to greater cumulativity in the scientific community. From the widespread adoption of NetLogo in both scientific research communities and in educational contexts, we conclude that the hypothesized compatibility of the two design goals has been successfully implemented. NetLogo has a large and active community numbering in the tens of thousands, with over 160,000 downloads in 2008 (and millions of web page hits). Its discussion lists are very active and include researchers, educators, students, businesspeople and policy-makers. NetLogo is in use in scientific research across a wide range of content domains and there is a rapidly increasing inclusion of NetLogo models in scientific research publications (e.g., [4, 37, 18, 42, 44, 17, 32, 38, 40, 64, 62, 29, 69]). This list is not intended to be comprehensive, as it is constantly being added to; interested parties should visit the NetLogo homepage and look in the "Community Models" section for an updated list of projects [85].

In a recent comparison of research-oriented agent-based modeling platforms [54], the authors admitted that, at first, they did not plan to include NetLogo in the comparison set, as, given its educational intent, they assumed it would be too limited for research purposes. However, after examining it more closely, they found that it was eminently suitable for research: "we found we could implement all our test models in NetLogo, with far less effort than for other platforms," that "NetLogo is by far the most professional platform in its appearance and documentation," and that it "is quite likely the most widely used platform for ABMs."

NetLogo is also in use in many thousands of educational contexts, especially in middle schools, high schools, and universities as well as museums and other informal learning contexts. Wilensky and colleagues have designed a number of effective model-based curricula for use in both precollegiate and undergraduate contexts. Topics covered include probability and statistics, kinetic molecular theory, reaction chemistry, electricity and magnetism, ecology, evolution, robotics, materials science, and microeconomics [12, 87, 2, 13, 1, 14, 88, 95, 61]. The CCL and colleagues have conducted hundreds of NetLogo workshops for teachers, researchers, students, and would-be modelers of all stripes.

7.2.3 NetLogo Features

NetLogo is a full-featured agent-based modeling environment and contains many sophisticated capabilities suitable for modeling Artificial Life. NetLogo is in continuous active development. Since version 1.0 in 1999, NetLogo has advanced considerably on both low-threshold and high-ceiling dimensions. Notable features include the HubNet [98] architecture that supports participatory simulations [57, 80, 19, 97, 41], where users can assume the role of agents alongside virtual agents, facilitating social science experimentation on a large scale [39, 30]. NetLogo also contains (1) NetLogoLab[15] that connects NetLogo to external hardware-based actuators and sensors and enables the grounding of models in local real-world data, (2) BehaviorSpace [96], a model analysis package that enables automated parameter sweeping, experimental analysis, and parameter space visualization, and (3) System Dynamics Modeler [83], a stocks and flows simulator that can be used alone or integrated with multi-agent simulations to create hybrid models that combine ABM and System Dynamics features [72]. Furthermore, NetLogo 4.0.2 was released in December of 2007 and contains many new and improved features. For instance, a major component of NetLogo is its models library, which contains hundreds of pre-built models, each with detailed explanations and extensive curricular activities, ready to be used as seed models. In the latest version of NetLogo, there are new models of biological evolution, mathematics, physics, neural networks, evolution of social cooperation, linguistics, psychology, management, and geographic information software (GIS). Some major new components that have been recently added include the integration of language primitives that facilitate the building, analysis, and examination of network models, enhancements to analysis and visualization of multiple model runs, and enhanced facilities for user-authored extensions of the core language. Users have taken advantage of the latter to build many such extensions. Of particular note to researchers and Artificial Life modelers are two new extensions: an extension that enables NetLogo to import GIS data files into a model and to generally play well with existing GIS software and a second extension that provides a library of genetic algorithm primitives for use in evolutionary computation.

The features and design philosophy of NetLogo offer several benefits to its use in the biomedical community. In keeping with its legacy as an educational tool, we note that there is great potential for use of NetLogo in medical school education, and we have made plans to integrate NetLogo into medical education here at Northwestern University. As alluded to above in Section 7.1, its "low threshold, high ceiling" characteristics make NetLogo well suited to implementation as a computational adjunct to current biomedical research. Evidence of its utility can be demonstrated in the range of research areas where NetLogo has been applied: intracellular signaling [10], acute inflammation [6, 7, 9, 72], inflammatory cell trafficking [64, 11], wound healing [49], and morphogenesis [46]. This chapter will focus on demonstrating how NetLogo can be used to effect multi-scale knowledge integration through the use of agent-based modeling, as well as demonstrating some of the capabilities of the NetLogo toolkit, particularly with respect to three-dimensional (3D) topologies.

7.3 NetLogo Models of Acute Inflammation

This chapter presents a series of NetLogo ABMs representing multiple scales of organization and integration. These models all involve aspects of the acute inflammatory response (AIR). The inflammatory response represents the body's initial reaction to injury and infection and is a ubiquitous process found in all tissues. In addition to dramatic expression in the face of severe infection or injury (as seen in sepsis and massive trauma), inflammation also provides a key link between damage and repair, as the healing process relies upon signals initiated during the AIR. Furthermore, it is increasingly recognized that inflammatory processes are essential to the maintenance of normal homeostasis, as the body exists in an ever-changing environment and is involved in such ubiquitous conditions as atherosclerosis, obesity, and aging. Therefore, acute inflammation is a prototypical example of a multi-scale biocomplex system and thus suited to examination and characterization using agent-based modeling.

The series of NetLogo ABMs have been developed at multiple levels of resolution, extending from intracellular signaling leading up to simulated organ function and organ–organ interactions. The specific model reference system for these ABMs is the clinical manifestation of multi-scale disordered acute inflammation, termed systemic inflammatory response syndrome (SIRS), multiple organ failure (MOF), and/or sepsis. These clinical entities form a continuum of disseminated disordered inflammation in response to severe levels of injury and/or infection and represent one of the greatest clinical challenges in the current health-care environment.

7.3.1 NetLogo ABM of Intracellular Processes: The Response to Inflammatory Signals

Perhaps the greatest translational challenge for biomedical research exists in the step between intracellular mechanism and the behavior of cellular populations. The extensive characterization of intracellular processes forms the bulk of ongoing biomedical research, and the ability to integrate this information is the subject to the burgeoning field of Systems Biology. While there are notable exceptions (see [16, 53] for examples of ABMs used to characterize intracellular processes), in general, the systems biology community has used more traditional methods of mathematical modeling based on differential equations and stochastic algorithms. This method has been successful in modeling kinetic processes in well-mixed systems, classically manifested as biochemical reactions. However, at a basic level, biochemical processes involve discrete events between molecules and particles. If the goal is to characterize these processes in the context of their influence on cellular behavior, it is possible to accept a level of abstraction that eliminates the details of the molecule–molecule interaction by labeling them generically as interaction events. The emphasis is shifted to representing the interaction–event. Accomplishing this requires two realizations (1) Molecules do not have volition to direct their movement to find their reaction partner and (2) cells are not bags of molecules (i.e., cells are not well mixed systems), and because molecules have no volition, spatial and environmental conditions within the cell must somehow direct signaling pathways by increasing the likelihood that participants in steps of a signal cascade will actually contact each other. Incorporating a spatial component to the characterization of signaling pathways, such as relating enzymes to the internal cytoskeletal architecture or simulating the effects of molecular crowding (that suggest that biochemical rate constants are dependent on an intracellular context [59]), can accommodate this. A modeling system can utilize abstraction at the level of the signaling event, without details as to what happens at a molecular structural level during the event. Interaction rates can be qualitatively scaled, as similar classes of interactions act within the same general time frame, and the emphasis on physical proximity renders fine parsing of these rates unnecessary. Rather, the focus is on characterizing conditions that lead to interaction events: molecular movement across space, likelihood of interaction events occurring, and the ordering of signaling enzymes. This leads to a particle view of signal transduction, where interactions within a reaction cascade follow a spatial architecture that is defined by the sequence of the signaling pathway. "Particles" are used to represent signaling events, and viewing the trajectory of the particle through the various reaction spaces can simulate transduction through a signal cascade. This modeling architecture is termed Spatially Configured Stochastic Reaction Chambers (SCSRC) [10]. Of note, the SCSRC utilizes and expands upon one particular type of basic NetLogo model, the GasLab models in the Models Library representing gas behavior via particle dynamics [78, 80, 81]. The transition from discrete particle behavior to the global system behavior expressible via the Ideal Gas Law equations provides an analogy for the underlying precept of the SCSRC: modeling biochemical reactions, which are usually expressed as rate equations, from a discrete, particle-based standpoint. This is an example of how

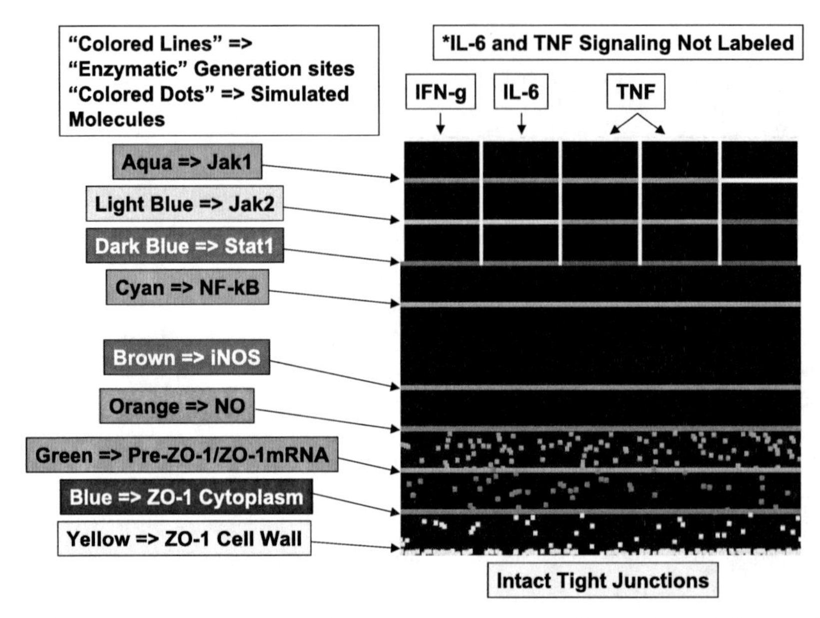

Fig. 7.1 Architecture of NetLogo ABM of intracellular signaling and metabolism: A SC-SRC model of a gut epithelial cell.

NetLogo's extensive model library can serve as a source of seed models to aid in the development of specific biomedical models.

In the SCSRC each simulation space represents a single cell. The space is a grid of 2D squares. The agent level is at the particle level, where each agent represents an abstracted molecule within the signal transduction cascade. The space is subdivided into a series of smaller rectangular compartments. The particular example presented here represents the pro-inflammatory signaling events in a gut epithelial cell and the effect those events have on tight-junction protein metabolism, an important component of epithelial barrier function (Fig. 7.1). The design features of the SCSRC are derived from the two central premises listed above: (1) Molecules do not behave volitionally and (2) spatial and structural factors influence the sequence of molecular interactions and signal transduction. The movement rules for agent-molecules in the SCSRC follow a semi–Brownian random-walk while they are simulating the molecular events of signal transduction. The subdivided compartments are reaction spaces that represent the close proximity of spatially located signaling enzymes. This proximity simulates either the arrangement of enzymes along cytoskeletal elements or the effects of molecular crowding in the cytoplasm. The apposed borders of the reaction spaces represent sequential enzymes of the signal cascade. The entry of a molecule-agent into the reaction space through one of these "enzyme" borders represents the chemical reaction event catalyzed by the enzyme, producing the signal molecule and introducing it into the reaction space. The molecule-agent moves in a random fashion eventually encountering the opposite border representing the next enzyme in the signaling pathway, and is transformed as it passes through that border into the next reaction chamber. As mentioned above, the specifics of the chemical reactions are abstracted into a state transition for the molecule-agent. Primarily, the state transition is merely a change in the labeling of that molecule-agent (to keep track of the signal as it propagates), but occasionally it results in altering the way it interacts with subsequent enzyme-borders. For instance, inhibitory activity is simulated by agent–border interactions that lead to the affected areas of the enzyme border being unable to execute the signaling state transition for subsequent encounters with up stream molecule-agents. The number of molecule-agents of a particular type represents the strength of a signal. The spatial configuration of the chambers of the SCSRC is defined by the sequence of a signaling/synthetic pathway. Each enzymatic step is represented by the horizontal "bars" abutting a reaction chamber. Specific qualitative types of enzymatic reactions, such as signal amplification, inhibition, or activation, can be specified in the encounter rules between the agent-molecules and the enzyme-borders. The reference model for the molecular processes in this SCSRC model is a well-described human cultured enterocyte model (Caco–2) and its responses to inflammatory mediators, including nitric oxide (NO) and a pro-inflammatory cytokine mix ("cytomix") that includes tumor necrosis factor (TNF), interleukin–1 (IL–1), and interferon-gamma (IFN–gamma) [33, 36, 35, 60]. Figure 7.1 demonstrates the structural architecture of these signaling pathways. This is a screenshot of the SCSRC model of a single gut epithelial cell, representing the pro–inflammatory signaling and tight-junction protein components. The inflammatory signaling complexes are at the top of the model and tight-junction protein pathway chambers are at the bottom. "Dots" present in each chamber represent molecules being synthesized and transported by the various enzyme complexes. In this figure, the only "Dots" present are in the tight-junction proteins present in the baseline cellular state.The response of the model to inflammatory stimuli can be seen in the screenshots displayed in Fig. 7.2.

7.3.2 NetLogo ABM of In Vitro Gut Epithelial Barrier Experiments

This subsection describes the translation of the mechanisms and behaviors of the SCSRC into a cell-as-agent-level ABM as seen in an in vitro experimental model of gut epithelial barrier function (epithelial barrier agent-based model = EBABM) [9]. As opposed to the agents being molecules and signaling enzymes in the SCSRC, each agent in the EBABM represents a single cell. The output of the SCSRC has been translated into a series of algebraic relationships for agent state variables corresponding to the molecular agent classes in the SCSRC. The EBABM utilizes a model architecture derived from another

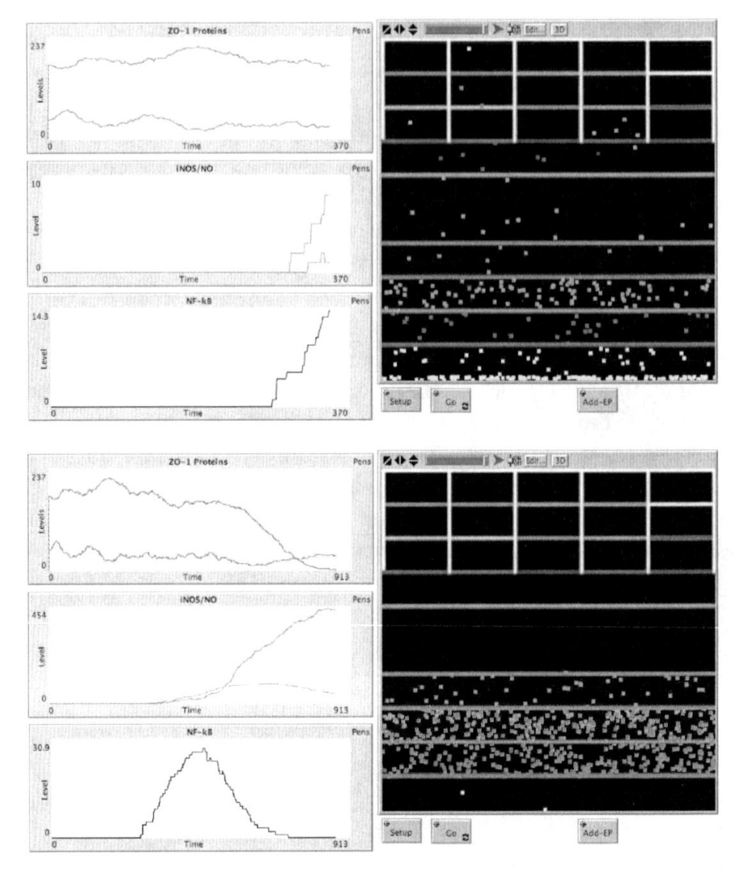

Fig. 7.2 Screenshots of gut epithelial cell SCSRC in response to inflammatory stimuli. The top panel is a screenshot of the SCSRC just after the addition of the inflammatory signal. "Dots" representing signaling molecules can be seen in each chamber. The graphs demonstrate the rise in levels of "NF–kB," and the activation of "iNOS" and "NO." The bottom panel is a screenshot of the SCSRC near the end of a run. The signaling molecules have disappeared, "iNOS" is starting to decrease while residual "NO" is still present.

paradigmatic type of NetLogo model, the "patch-centric" cellular automata-like 2D grid models [75, 76, 66].

The topology of the EBABM is a 2D square grid. The grid has 21×21 cells, in each of which there is an epithelial cell agent ("epi-cell"). The size of this grid was chosen as a representative portion of a total cell culture surface for reasons of computational efficiency; the processes being modeled by the EBABM are proportional to the cell surface area and the model could be, if desired, scaled up to any size. There are also two additional simulation "spaces" – one layer representing the apical extracellular space (from which the diffusate originates) and another layer representing the basal extracellu-

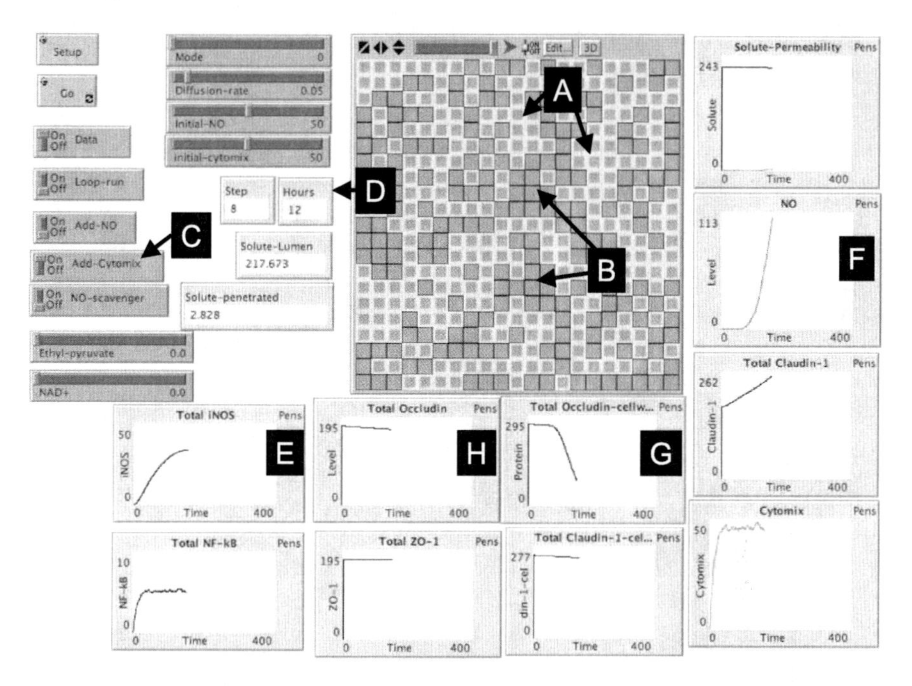

Fig. 7.3 Screenshot of the graphical user interface of the EBABM. Control buttons are on the left; Graphical Output of the simulation is in the center. In the Graphical Output, Caco–2 agents are seen as squares, those with intact tight junctions with light borders (letter A), those with failed tight junctions are bordered in black (letter B). This particular run is with the addition of cytomix (letter C) seen after 12 hours of incubation (letter D). The heterogeneous pattern of tight-junction failure can be seen in the Graphical Output. Graphs of variables corresponding to levels of mediators (letters E and F) and tight-junction proteins (letters G and H) are at the bottom and right. This figure is reprinted under the terms of the Creative Commons license from [9].

lar space (into which the diffusate flows if there is permeability failure). A screenshot of the EBABM during an experimental run can be seen in Fig. 7.3.

Each epi-cell has eight immediate neighbors, and at each contact point there is a simulated tight junction (TJ). The integrity of the TJ requires both apposed epi-cells to have adequate production and localization of TJ proteins.

The importance of the EBABM lies in the translational function it plays as a transitioning step between the intracellular behavior represented and modeled using the SCSRC, and the aggregated cell-type models that are described in the following subsections. The EBABM is thus a critical validation step in the modular construction of a multi–scale ABM architecture.

7.3.3 NetLogo ABM of Organ Function: Gut Ischemia and Inflammation

The next level of ABM development is intended to simulate organs as a synthesis of two distinct hypotheses of disseminated inflammation and organ failure: viewing disordered systemic inflammation as either a disease of the endothelium [5] or a disease of epithelial barrier function [26]. The endothelial surface is the primary communication and interaction surface between the body's tissues and the blood, which carries inflammatory cells and mediators. Endothelial activation is a necessary aspect of the initiation and propagation of inflammation, particularly in the expansion of local inflammation to systemic inflammation [5]. On the other hand, end-organ dysfunction related to inflammation can be seen as primarily manifest in a failure of epithelial barrier function. Pulmonary, enteric, hepatic, and renal organ systems all display epithelial barrier dysfunction that has consequences at the macro-organ level (impaired gas exchange in the lung, loss of immunological competence in the gut, decreased synthetic function in the liver, and impaired clearance and resorptive capacity in the kidney) [26]. The organ-level ABM reconciles these two hypotheses by integrating the epithelial barrier component, used to represent the consequence of individual organ failure, and the endothelial/inflammatory cell component that provides the binding interaction space that generates, communicates, and propagates the inflammatory response [9]. The primary cell classes in this architecture are endothelial cells, blood-borne inflammatory cells (with their attendant subtypes), and epithelial cells. Therefore the structure of the organ ABM involves the 3D linkage of the cellular surface ABMs already developed representing these two systems, the EBABM representing epithelial function and a previously published endothelial/inflammatory cell ABM [6, 7]. The result is a bilayer organ model (see Fig. 7.4).

Both the original endothelial/inflammatory cell ABM and the EBABM were developed as 2D models. In order to create the bilayer topology of the organ ABM, it was necessary to convert both of these models to the 3D version of NetLogo [84], with each model represented as a layer of agents projected in the XY plane. The two layers were then juxtaposed: the endothelial layer below and the epithelial layer above along the Z axis. The simulated blood vessel luminal space occupied another XY plane one place inferior to the endothelial surface along the Z axis. Inflammatory cells move only in this plane. The organ luminal space occupied the XY plane at one place superior to the epithelial axis along the Z axis. This space contains the "diffusate" that leaks into the gut in cases of epithelial tight-junction failure.

In vivo models that examine the inflammatory behavior of the gut either look at a local effect from direct occlusion of gut arterial flow [70, 63] or as a result of some systemic insult, be it hemorrhagic shock, endotoxin administration [50], or burn injury [47, 24]. These studies suggest that the primary

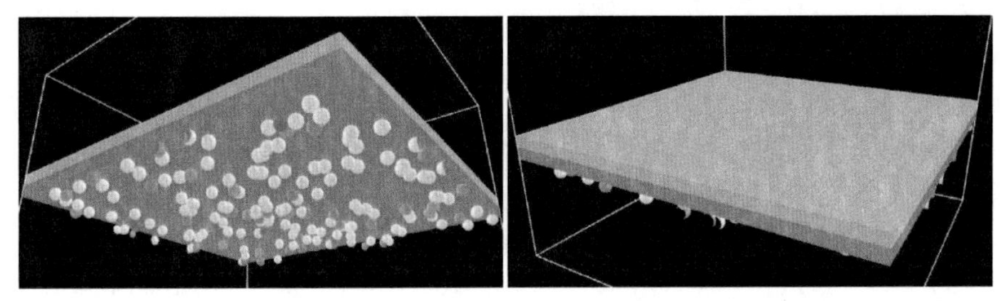

Fig. 7.4 Screenshots of bilayer configuration of the gut ABM, following the structure for hollow organs described in the text. The left panel is the view of bilayer from endothelial surface. The lower surface consists endothelial cell agents, with spherical inflammatory cell agents seen just below. Inflammatory cell agents move in the plane immediately below the endothelial surface. The right panel is the view of bilayer from epithelial surface. Each cube represents an epithelial cell agent, governed by rules transferred from the EBABM. Impairment of TJ protein metabolism is shown by darkening of the shade of the epithelial cell agent, with the epithelial cell agents eventually turning black and changing their shape to a "cone" when TJs have failed. This figure is reprinted under the terms of the Creative Commons license from [9].

process that initiates inflammation in the gut is ischemia and reperfusion and the subsequent effects on the endothelial surfaces within the gut. The measurable outputs of the reference models exist at different scales. At the cellular level, tight-junction integrity and epithelial barrier function is one measured endpoint [99, 35], however, the organ as a whole also has an output: the nature of the mesenteric lymph. Multiple studies suggest that ischemia to the gut (and subsequent inflammation) leads to the excretion of an as-of-yet unidentified substance in the mesenteric lymph that has pro-inflammatory qualities. Some characteristics of the substance can be identified from the literature: It is an acellular, aqueous substance [21], is greater than 100 kDa in size [3], does not correspond to any currently recognized cytokine, and is bound or inactivated by albumin [51]. The time course of the production of the substance is identified to some degree [22, 20] but it is unclear if it arises from a late production of inflamed cells or is a product of cellular degeneration or apoptosis (programmed cellular death) or is a transudated bacterial product from the intestinal lumen. The uncertainty with respect to an identified mediator provides a good example of how the ABM architecture deals with incomplete knowledge. Based on the characteristics defined above, we make an ***hypothesis*** regarding this substance with respect to its origin, but we acknowledge that this is, to a great degree, a "best guess." Doing so establishes a knowledge bifurcation point, allowing the development of potential experiments and/or data that would nullify the particular hypotheses. A specific example will be demonstrated next.

The nature of the initial perturbation was altered to match that seen in the reference experiments (i.e., tissue ischemia). With the premise that the inflammatory response was generated at the endothelial surface, the initial perturbation was modeled focusing at the endothelial layer, with the response of the epithelial component being subsequently driven by the output of the endothelial–inflammatory cell interactions. Rather than having a localized insult with either infectious agents (simulating infection) or sterile endothelial damage (simulating tissue trauma), as was the case in the base endothelial–inflammatory cell ABM, gut ischemia was modeled as a percentage of the total endothelial surface rendered ischemic. The degree (or percentage affected) of the initial ischemia was controlled with a slider in the NetLogo interface. Therefore, "Percentage Gut Ischemia" (= %Isch) represents the independent variable as initial perturbation for this model. To address the issue of modeling the production of the post-ischemic, pro-inflammatory lymph, attention is focused on linking the knowledge that has been acquired regarding the characteristics of the substance and relating this information to the components of the organ ABM. The known characteristics listed above are used to exclude potential candidate substances/actors from consideration. Specifically, this group comprises any of the cellular agents and any of the included cytokines. Therefore, the search is limited to the following:

1. An as-of-yet unidentified compound linked to cellular damage. An example of such a compound would be high-mobility box protein 1 (HMGB–1), which to date has not been looked for in the post-ischemic mesenteric lymph. In the organ ABM, this variable is termed "cell-damage-byproduct," and it is calculated as a function of total endothelial damage with a set decay rate consistent with that of other bio-active compounds associated with inflammation.
2. A luminal compound that diffuses in response to TJ barrier failure. This would correspond to potential byproducts of gut bacterial metabolism, or bacterial toxins, or other soluble aspects of the gut luminal environment that would leak into the gut tissue by virtue of the loss of barrier function. This variable is represented by "gut-leak," which is equal to the "solute" (from the EBABM) that penetrates the failed barrier.
3. A downstream metabolite of compounds generated by the inflammatory process. These would most likely be compounds generated by superoxide and nitric oxide (NO) reactions. For purposes of these simulations, levels of NO will be used as a proxy for this possible candidate.

Therefore, the goal of the organ ABM simulation runs will be to examine the time course levels of these three values and identify which one (if any) matches the reported time course effects of the post-ischemic mesenteric lymph. The first step was to determine the greatest non-lethal level for "Percentage Ischemia" (%Isch). It should be noted again that the name of this variable is descriptive for how it is implemented in the ABM, and not intended to match quantitatively, per se, with measured ischemia in vivo. Rather, %Isch is representative of the initial conditions for the simulation that will produce a pattern of simulation behavior that matches that of the in vivo system [31]. A parameter sweep of this value was performed, using a previously described method [6] with the goal of identifying the greatest non-lethal level of %Isch, which was %Isch = 35. The output from the organ ABM with %Isch = 35 evaluated the time courses for three global output variables: "cell-damage-byproduct," "gut-leak" and NO. The pro-inflammatory properties of the post-ischemic mesenteric lymph were noted to increase the most at 3 hours and 6 hours and remain out to 24 hours [22, 20]. Examining the time courses of these three global output variables the candidate compound that most closely approximates the pattern identified in the literature was the "cell-damage-byproduct."

As discussed above, this possible source of the unknown compound in the post-ischemic mesenteric lymph is based on the recognition of certain late pro-inflammatory mediators produced by activated and damaged cells, HMGB–1 being the most studied as a possible key mediator in the pathogenesis of sepsis [48]. To date, there have been no studies examining the production or presence of HMGB–1 in the post-ischemic mesenteric lymph. However, based on the information generated by the organ ABM, and placed in the context of the knowledge framework concerning the characteristics of pro–inflammatory mesenteric lymph, we will make a hypothesis that some later byproduct of damaged gut tissue, rather than a diffused material or direct metabolite of first-pass inflammatory mediators, is the responsible compound in the post-ischemic mesenteric lymph. It is recognized that this is "guided speculation;" however, it also demonstrates how the construction and use of models in the ABM architecture is an evolving process that parallels the development and refinement of conceptual models. As will be seen in the next subsection, the next scale of biological organization to be addressed in the ABM architecture involves the extension and integration of this hypothesis.

7.3.4 NetLogo ABM of Multi-organ Interactions: The Gut–Pulmonary Axis of Inflammation

The ultimate goal of all of these modeling endeavors is to create some facsimile of clinical conditions, with the hope of developing a platform that represents the complexity seen clinically. In patients, organs do not exist in isolation; their mutually complementary functions interact to sustain the organism as a whole. Disease states can lead to a breakdown of these interactions, causing a cascade effect, as single-organ dysfunction can lead to multi-system failure. Sepsis and MOF are characterized by a progressive breakdown in these interactions, leading to recognizable patterns of linked organ failure [28]. Therefore the next scale of biological organization represented in the multi-scale ABM architecture is that of organ–organ interaction [9]. The gut–pulmonary axis of MOF [47, 22, 23] is used as the initial example of organ-to-organ crosstalk. This relationship is relatively well defined pathophysiologically (although not completely, as indicated by the uncertainty of the identity of the pro-inflammatory compound in the post-ischemic mesenteric lymph) and represents an example of multi–organ effects of disseminated disordered inflammation. Disordered acute inflammation of the lung is termed Acute Respiratory Distress Syndrome (ARDS) and is manifested primarily by impaired endothelial and epithelial barrier function, leading to pulmonary edema. This leads to impaired oxygenation of arterial blood, requiring support of the patient with mechanical ventilation. While the comprehensive pathogenesis of ARDS involves additional subsequent issues related, to a great degree, to the consequences of mechanical ventilation (specifically the effects of barotrauma and shear forces on the airways, and the persistent propagation of inflammation that results), for purposes of this initial demonstration only the initiating events associated with the development of ARDS will be modeled. Those events concern the production and release into the mesenteric lymph by ischemic gut (resulting from shock) of various pro-inflammatory mediators and their effects both on circulating inflammatory cells and the pulmonary endothelium as they circulate back to the lung via the mesenteric lymph (as discussed above). At this point, the hypothesis regarding the nature of the pro-inflammatory mediator in the mesenteric lymph is extended to the assumption that, for modeling purposes, the levels of "cell-damage-byproduct" will be the proxy for the unidentified compound that is produced in the ischemic gut and circulated to the lung, leading to inflammation of the pulmonary endothelium.

Drawing upon the endothelial–epithelial bilayer configuration for a hollow organ, a pulmonary ABM was developed utilizing the same endothelial–inflammatory cell component as the gut ABM and using rules for pulmonary epithelial cells with respect to tight-junction metabolism and epithelial barrier function [34]. The functional consequence of the intact pulmonary epithelial barrier is effective oxygenation of arterial blood (expressed at the endothelial lumen) via diffusion from the alveolar epithelial surface. Pulmonary barrier failure manifests as increased egress of fluid from the endothelial lumen into the alveolar space, affecting the transfer of alveolar oxygen to the endothelial surface. Thus, systemic oxygenation may be altered with the consequence that progressive pulmonary dysfunction would feed back to the system as whole. This leads to impaired oxygenation into the endothelial lumen, which is summed across the surface of the model to produce a measure of systemic arterial oxygen content.

The topology of the linked gut and pulmonary ABMs consists of two parallel bilayer planes, each bilayer representing one of the organ ABMs (Fig. 7.5).

The Z-axis orientation of both bilayers is the same: to allow conservation of the agent rules for equivalent agent classes (i.e., endothelial–epithelial–lumen relationships are consistent). The simulated blood flow continues to be modeled by movement in the XY plane immediately inferior to the endothelial surface. Blood flow between organs is simulated by adding a "perfusion" variable. For purposes of the model, large-caliber blood vessels and the heart are treated as biologically inert with respect to inflammation. Similarly, the

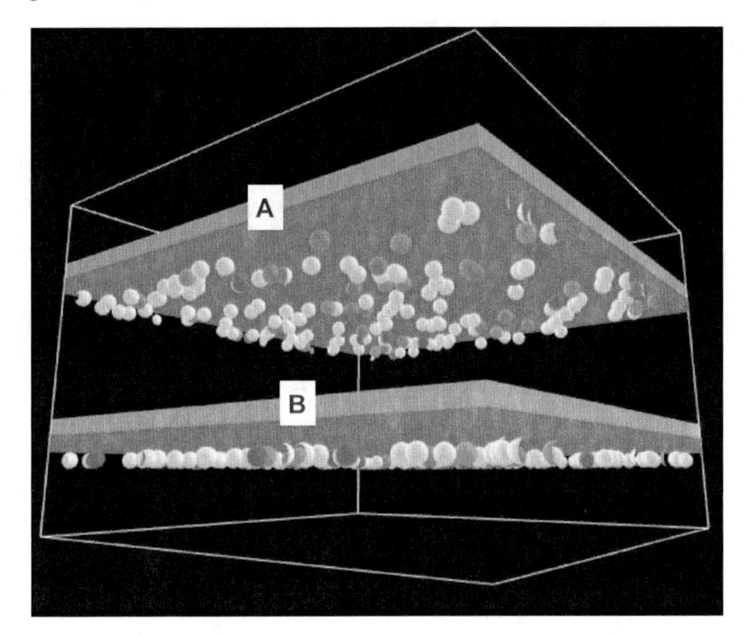

Fig. 7.5 Screenshot of multi-bilayer gut–lung axis ABM. The letter A labels the pulmonary bilayer, with cubes in the upper portion of the bilayer representing pulmonary epithelial cell agents, while the lower surface represents pulmonary endothelial cell agents, and below are spherical inflammatory cell agents. The letter B labels the gut bilayer, with a similar configuration, epithelial cell agents above and endothelial cell agents below. Circulating inflammatory cell agents move between these two bilayers in the fasion described in the text. This figure is reprinted under the terms of the Creative Commons license from [9].

flow of the mesenteric lymph is transferred from the gut ABM endothelial space to the lung ABM endothelial space.

The effects of mesenteric ischemia on pulmonary barrier dysfunction were then evaluated using a parameter sweep of %Isch to identify the inflection point between lethal and sub-lethal perturbation levels. Screenshots of both these outcomes can be seen in Fig. 7.6.

This parameter sweep demonstrated that the corresponding lethality of mesenteric ischemia in the gut–lung ABM is significantly increased as compared to the gut ABM alone, dropping the sub-lethal %Isch from 35 for the gut ABM to 11 for the gut–lung ABM. This results from the addition of the lung ABM and its effect of decreasing the maximally available "oxy" to non-perturbed endothelial agents via the consequence of pulmonary epithelial barrier function ("pulm–edema"). With increasing pulmonary edema and worsening oxygen delivery, gut epithelial agents "die" owing to a decrease in the available maximal "oxy" level to below the threshold for generalized endothelial agent activation. The impaired systemic oxygenation due to a pulmonary leak arises from pulmonary epithelial barrier failure. At the sublethal

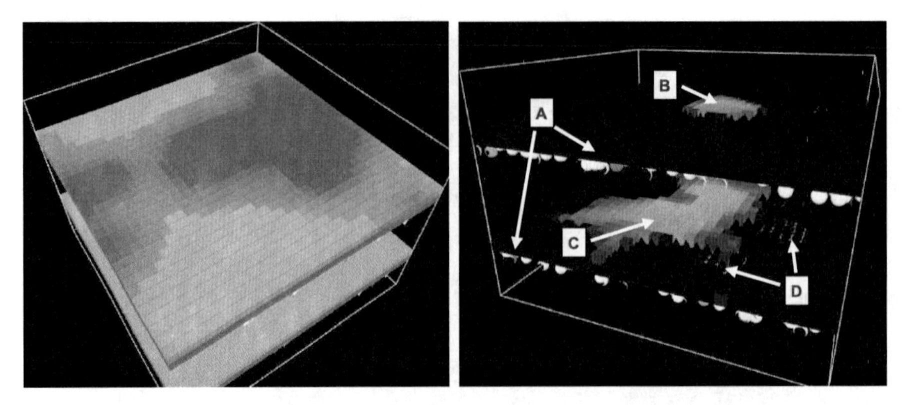

Fig. 7.6 Effect of gut ischemia on pulmonary barrier dysfunction and pulmonary edema: sublethal and lethal outcomes. The left panel is a screenshot of a representative run with a sub-lethal initial %Isch = 11 over a 72-hour run. Pulmonary epithelial cells show gradual recovery as inflammation subsides up to the screenshot time of 72 hours. The right panel demonstrates a "lethal" initial level of %Isch = 13, where impaired oxygen delivery from the lung leads to greater ischemia and cellular death in the gut. This run is terminated at 24 hours because endothelial damage is nearly complete. The letter A points to black cubes representing "dead" endothelial cell agents. The letter B points to the only remaining intact pulmonary epithelial cell agents. The letter C points to the only remaining intact gut epithelial cell agents. The letter D points to the only remaining patches of surviving endothelial agents (seen through the cones of "dead" epithelial cell agents). This figure is reprinted under the terms of the Creative Commons license from [9].

%Isch level of 11 the system is able to correct itself, with attenuation of gut ischemic damage and recovery of the pulmonary epithelisl cells, mostly occuring by 72 hours. This pattern is consistent with that seen clinically in the recovery of pulmonary edema secondary to inflammatory causes. However, the transition to lethal outcome is accomplished by only a slight increase of %Isch to 13, where the oxygen delivery consequences of pulmonary epithelial failure leads to a forward feedback loop with progressive gut bilayer ischemia. Thus, the survival space of the system would appear to be greatly limited, and it may initially suggest that this model would be unsuited to examining the range of dynamics of interest in the study of sepsis. However, it should be noted that the high lethality of mesenteric ischemia, which implies the presence of hemodynamic shock, is historically correct. Shock states, prior to the development of fluid resuscitation and respiratory support, were nearly universally fatal. This is the circumstance that is being represented with the gut–lung ABM at this point. If the goal is to simulate the clinical conditions associated with sepsis and MOF, then it is necessary to simulate the effects of organ support, to shift the survival space to the right. Doing so reproduces the fact that sepsis and MOF are diseases of the intensive care unit (ICU), arising only after the advances of resuscitative, surgical, antimicrobial, and organ-supportive care allowed the maintenance of patients in situations where they previously would have died. Therefore, sepsis and MOF can be thought of as a previously unexplored behavior space of systemic inflammation, one where the inflammatory system is functioning beyond its evolutionarily defined design parameters [6, 7].

Therefore, a very abstract means of organ support is modeled in the form of "supplementary oxygen." This function increases the amount of "oxy" that is able to be diffused through the pulmonary epithelial barrier and is the qualitative equivalent of increasing the fraction of inspired oxygen, and therefore alveolar oxygen, to increase the partial pressure of oxygen diffused in the blood. It is qualitative insomuch as there is no attempt to reproduce the dynamics of gas exchange, or the binding of hemoglobin to oxygen in the blood, or the effects of redistributed ventilation–perfusion matching in the lung as a result of hypoxia. This degree of detail is beyond the scope of this initial demonstration model; however, the qualitative behavioral effects do show that this type of support, even abstractly modeled, increases the richness of the behavior of the model as a whole and can extend the examinable behavior space of the model to situations that can approximate the effects of organ support in the ICU. The corresponding changes in outcome with this type of simulated organ support can be seen in Fig. 7.7.

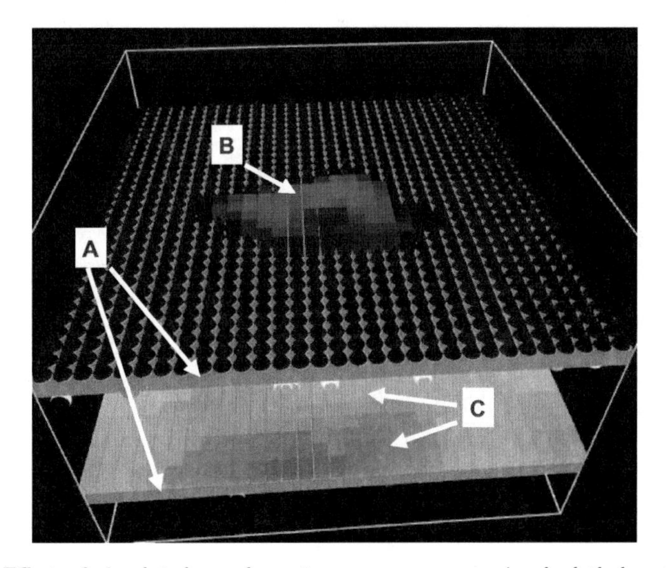

Fig. 7.7 Effect of simulated supplementary oxygen on previously lethal gut ischemia. This is a screenshot of a representative run with an initial %Isch = 15 and the addition of simulated organ support in the form of "Supplementary Oxygen" at 50%. The stabilization and initiation of recovery of pulmonary epithelial TJs at 72 hours is consistent with the clinical time course of ARDS due to an episode of shock. The letter A demonstrates the intact endothelial agent layer due to "Supplementary Oxygen" support (compare with Fig. 7.6, the panel on the right). The letter B demonstrates the recovering pulmonary epithelial cell agents. The letter C demonstrates intact and recovering gut epithelial cell agents. This figure is reprinted under the terms of the Creative Commons license from [9].

In Fig. 7.7, the initial %Isch = 15, greater than the lethal level seen in Fig. 7.6; however, the survival of the overall system is enhanced by blunting the negative consequences of impaired pulmonary function on the gut. The effect of "Supplementary Oxygen" is additive to the level of "oxy" generated by the lung ABM and distributed to the endothelial surface, effectively blunting the effect of the resulting pulmonary edema and keeping the "oxy" level above the threshold ischemic level for activation of the generalized endothelial cell agent population. As a result, the endothelial surface is maintained through the period of most intense inflammation and this allows the epithelial cells to begin recovery of their TJs. This dynamic is consistent with current managment of these types of patients and thus needs to be effectively modeled if the appropriate disease state is to be examined.

This sequence illustrates an important point in creating translational models of disease states. The tendency may be to attempt to model the pathological state being studied (i.e., creating a model of sepsis). However, it needs to be remembered that pathological states result from transitions from normal physiological behavior, and if the intent of a model is to facilitate the eventual transition from disease back to health, then the normal mechanism must be the basis of a translational model. The need to capture transitions from one state to another takes on further importance when the pathological state results, as with sepsis, from medical/clinical interventions. Therefore, the architecture of a modeling structure needs to be flexible enough to accommodate the addition and integration of these factors, and it is hoped that the presented modular structure of the ABM architecture demonstrates this capability.

7.4 Conclusion

The biomedical research community today faces a challenge that has paradoxically arisen from its own success: As greater amounts of information become available at increasingly finer levels of biological mechanism, it becomes progressively more difficult for individual researchers to survey and integrate information effectively, even within their own area of expertise. It still falls upon the individual researcher to create mental models to guide the direction of his/her individual research and, in aggregate, form the components of the evolving structure of community knowledge. However, the formal expression of mental models remains poorly defined, leading to limitations in the ability to share, critique, and evolve the knowledge represented in these conceptual models, particularly across disciplines.

These limitations can be overcome by developing methods of formal dynamic knowledge representation to enable researchers to express and communicate their mental models more effectively. By being able to "see" the consequences of a particular hypothesis–structure/mental model, the mechanistic consequences of each hypothesis can be observed and evaluated. In addition, this type of dynamic knowledge representation enables the instantiation of thought experiments, of trying out possible alternative solutions, so long as these hypotheses and assumptions are made explicit. Again, this draws upon the experience in the Artificial Life community by creating alternative worlds driven by these proposed rules. These models can aid in the scientific process by providing a transparent framework for this type of speculation, which can then be used as departure points for the planning and design of further wet lab experiments and measurements.

In short, the agent-based paradigm, with its defining characteristics of encapsulation, modularity, and parallelism, can provide an over-arching design architecture for the computational representation of biological systems. However, for this to be effective, there needs to be participation on the part of the biomedical community and its participants. Modeling and simulation toolkits such as NetLogo serve a vital role in giving novices to computer modeling an opportunity to represent and visualize their conceptual models. In particular, NetLogo provides a highly effective mixture of ease-of-use and modeling capability to make the initial foray into this arena most rewarding. It is hoped that the increasing use of this type of knowledge representation and communication will foster the further development of virtual laboratories and in silico investigations.

Acknowledgements This work was supported in part by the National Institute of Disability Rehabilitation Research (NIDRR) Grant H133E070024 and the National Science Foundation (HCC) Grant 0713619. Thanks to Seth Tisue and the CCL's NetLogo development team and also to Bill Rand for comments on drafts of this chapter.

References

1. Abrahamson, D., Wilensky, U.: Problab: A computer-supported unit in probability and statistics. In: M.J. Hoines, A.A. Fuglestad (eds.) 28th Annual Meeting of the International Group for the Psychology of Mathematics Education, vol. 1. Bergen University College, Bergen, Norway (2004)
2. Abrahamson, D., Wilensky, U.: ProbLab Curriculum (Computer Program). Center for Connected Learning and Computer Based Modeling, Northwestern University (2004). `http://ccl.northwestern.edu/curriculum/problab`
3. Adams Jr., C.A., Xu, D.Z., Lu, Q., Deitch, E.A.: Factors larger than 100 kd in post-hemorrhagic shock mesenteric lymph are toxic for endothelial cells. Surgery **129**(3), 351–363 (2001)
4. Agar, M.: An anthropological problem, a complex solution. Human Organization **63**(4), 411–418 (2004)
5. Aird, W.C.: Vascular bed-specific hemostasis: role of endothelium in sepsis pathogenesis. Crit Care Med **29**(7 Suppl), S28–S34; discussion S34–S35 (2001)
6. An, G.: Agent-based computer simulation and sirs: building a bridge between basic science and clinical trials. Shock **16**(4), 266–273 (2001)
7. An, G.: In silico experiments of existing and hypothetical cytokine-directed clinical trials using agent-based modeling. Crit Care Med **32**(10), 2050–2060 (2004)

8. An, G.: Concepts for developing a collaborative in silico model of the acute inflammatory response using agent-based modeling. J Crit Care **21**(1), 105–110; discussion 110–111 (2006)
9. An, G.: Introduction of an agent-based multi-scale modular architecture for dynamic knowledge representation of acute inflammation. Theor Biol Med Model **5**(1), 11 (2008)
10. An, G.: A model of tlr4 signaling and tolerance using a qualitative, particle-event-based method: Introduction of spatially configured stochastic reaction chambers (sc-src). Mathematical Biosciences **217**(1), 9 (2008)
11. Bailey, A.M., Thorne, B.C., Peirce, S.M.: Multi-cell agent-based simulation of the microvasculature to study the dynamics of circulating inflammatory cell trafficking. Ann Biomed Eng **35**(6), 916–936 (2007)
12. Berland, M., Wilensky, U.: Virtual robotics in a collaborative contructionist learning environment. In: The Annual Meeting of the American Educational Research Association. San Diego, CA (April 12–16, 2004)
13. Blikstein, P., Wilensky, U.: Materialsim: an agent-based simulation toolkit for materials science learning. In: International Conference on Engineering Education. Gainesville, FL (2004)
14. Blikstein, P., Wilensky, U.: MaterialSim curriculum (Computer Program). Center for Connected Learning and Computer Based Modeling, Northwestern University (2004). `http://ccl.northwestern.edu/curriculum/materialsim`
15. Blikstein, P., Wilensky, U.: NetLogoLab (Computer Program). Center for Connected Learning and Computer Based Modeling, Northwestern University (2005). `http://ccl.northwestern.edu/curriculum/NetLogoLab`
16. Broderick, G., Ru'aini, M., Chan, E., Ellison, M.J.: A life-like virtual cell membrane using discrete automata. In Silico Biol **5**(2), 163–178 (2005)
17. Bryson, J., Caulfield, T., Drugowitsch, J.: Integrating life-like action selection into cycle-based agent simulation environments. In: M. North, D. Sallach, C. Macal (eds.) Agent 2005: Generative Social Processes, Models and Mechanisms, pp. 67–82. Argonne National Laboratory and the University of Chicago, Chicago, IL (2006)
18. Centola, D., Willer, R., Macy, M.: The emperor's dilemma: A computational model of self-enforcing norms 1. American Journal of Sociology **110**(4), 1009–1040 (2005)
19. Colella, V.: Participatory simulations: Building collaborative understanding through immersive dynamic modeling. Journal of the Learning Sciences **9**(4), 471–500 (2000)
20. Davidson, M.T., Deitch, E.A., Lu, Q., Osband, A., Feketeova, E., Nemeth, Z.H., Hasko, G., Xu, D.Z.: A study of the biologic activity of trauma-hemorrhagic shock mesenteric lymph over time and the relative role of cytokines. Surgery **136**(1), 32–41 (2004)
21. Dayal, S.D., Hasko, G., Lu, Q., Xu, D.Z., Caruso, J.M., Sambol, J.T., Deitch, E.A.: Trauma/hemorrhagic shock mesenteric lymph upregulates adhesion molecule expression and il-6 production in human umbilical vein endothelial cells. Shock **17**(6), 491–495 (2002)
22. Deitch, E.A., Adams, C., Lu, Q., Xu, D.Z.: A time course study of the protective effect of mesenteric lymph duct ligation on hemorrhagic shock-induced pulmonary injury and the toxic effects of lymph from shocked rats on endothelial cell monolayer permeability. Surgery **129**(1), 39–47 (2001)
23. Deitch, E.A., Adams, C.A., Lu, Q., Xu, D.Z.: Mesenteric lymph from rats subjected to trauma-hemorrhagic shock are injurious to rat pulmonary microvascular endothelial cells as well as human umbilical vein endothelial cells. Shock **16**(4), 290–293 (2001)
24. Deitch, E.A., Shi, H.P., Lu, Q., Feketeova, E., Skurnick, J., Xu, D.Z.: Mesenteric lymph from burned rats induces endothelial cell injury and activates neutrophils. Crit Care Med **32**(2), 533–538 (2004)
25. Feurzeig, W., Papert, S., Bloom, M., Grant, R., Solomon, C.: Programming-languages as a conceptual framework for teaching mathematics. ACM SIGCUE Outlook **4**(2), 13–17 (1970)

26. Fink, M.P., Delude, R.L.: Epithelial barrier dysfunction: a unifying theme to explain the pathogenesis of multiple organ dysfunction at the cellular level. Crit Care Clin **21**(2), 177–196 (2005)
27. Food, U., Administration, D.: Innovation or stagnation: Challenge and opportunity on the critical path to new medical products (2004). http://www.fda.gov/oc/initiatives/criticalpath/whitepaper.html
28. Godin, P.J., Buchman, T.G.: Uncoupling of biological oscillators: a complementary hypothesis concerning the pathogenesis of multiple organ dysfunction syndrome. Crit Care Med **24**(7), 1107–1116 (1996)
29. Goldstone, R., Roberts, M., Gureckis, T.: Emergent processes in group behavior. Curr Direct Psych Sci **17**(1), 10–15 (2008)
30. Goldstone, R., Wilensky, U.: Promoting transfer through complex systems principles. Journal of the Learning Sciences **17**, 465–516 (2008)
31. Grimm, V., Revilla, E., Berger, U., F., J., Mooij, W., Railsback, S., Thulke, H.H., Weiner, J., Wiegand, T.: Pattern-oriented modeling of agent-based complex systems: Lessons from ecology. Science **310**, 987–991 (2005)
32. Hammond, R., Axelrod, R.: The evolution of ethnocentrism. J Conflict Resolu **50**(6), 926 (2006)
33. Han, X., Fink, M.P., Delude, R.L.: Proinflammatory cytokines cause NO*-dependent and -independent changes in expression and localization of tight junction proteins in intestinal epithelial cells. Shock **19**(3), 229–237 (2003)
34. Han, X., Fink, M.P., Uchiyama, T., Yang, R., Delude, R.L.: Increased iNOS activity is essential for pulmonary epithelial tight junction dysfunction in endotoxemic mice. Am J Physiol Lung Cell Mol Physiol **286**(2), L259–L267 (2004)
35. Han, X., Fink, M.P., Yang, R., Delude, R.L.: Increased iNOS activity is essential for intestinal epithelial tight junction dysfunction in endotoxemic mice. Shock **21**(3), 261–270 (2004)
36. Han, X., Uchiyama, T., Sappington, P.L., Yaguchi, A., Yang, R., Fink, M.P., Delude, R.L.: Nad+ ameliorates inflammation-induced epithelial barrier dysfunction in cultured enterocytes and mouse ileal mucosa. J Pharmacol Exp Ther **307**(2), 443–449 (2003)
37. Henein, C., White, T. (eds.): Agent-Based Modelling of Forces in Crowds, *Lecture Notes in Computer Science*, vol. 3415. Springer-Verlag (2004)
38. Hills, T.: Animal foraging and the evolution of goal-directed cognition. Cogni Sci **30**(1), 3–41 (2006)
39. Jacobson, M., Wilensky, U.: Complex systems in education: Scientific and educational importance and research challenges for the learning sciences. J Learni Sci **15**(1), 11–34 (2006)
40. Joyce, D., Kennison, J., Densmore, O., Guerin, S., Barr, S., Charles, E., Thompson, N.: My way or the highway: A more naturalistic model of altruism tested in an iterative prisoners' dilemma. J Arti Soc Social Sim **9**(2), 4 (2006)
41. Klopfer, E., Yoon, S., Perry, J.: Using palm technology in participatory simulations of complex systems: A new take on ubiquitous and accessible mobile computing. J Sci Ed Techno **14**(3), 285–297 (2005)
42. Koehler, M., Tivnan, B., Bloedorn, E.: Generating fraud: Agent based financial network modeling. In: Proceedings of the North American Association for Computation Social and Organization Science (NAACSOS 2005). Notre Dame, IN (2005). http://www.casos.cs.cmu.edu/events/conferences/2005/2005_proceedings/Koehler.pdf
43. Langton, C.: "What is artificial life?" (2000)
44. Laver, M.: Policy and the dynamics of political competition. Am Political Sci Rev **99**(2), 263–281 (2005)
45. Lipniacki, T., Paszek, P., Brasier, A.R., Luxon, B.A., Kimmel, M.: Stochastic regulation in early immune response. Biophys J **90**(3), 725–742 (2006)

46. Longo, D., Peirce, S.M., Skalak, T.C., Davidson, L., Marsden, M., Dzamba, B., DeSimone, D.W.: Multicellular computer simulation of morphogenesis: blastocoel roof thinning and matrix assembly in xenopus laevis. Dev Biol **271**(1), 210–222 (2004)
47. Magnotti, L.J., Xu, D.Z., Lu, Q., Deitch, E.A.: Gut-derived mesenteric lymph: A link between burn and lung injury. Arch Surg **134**(12), 1333–1340; discussion 1340–1341 (1999)
48. Mantell, L.L., Parrish, W.R., Ulloa, L.: Hmgb-1 as a therapeutic target for infectious and inflammatory disorders. Shock **25**(1), 4–11 (2006)
49. Mi, Q., Riviere, B., Clermont, G., Steed, D.L., Vodovotz, Y.: Agent-based model of inflammation and wound healing: Insights into diabetic foot ulcer pathology and the role of transforming growth factor-beta1. Wound Repair Regen **15**(5), 671–682 (2007)
50. Mishima, S., Xu, D., Lu, Q., Deitch, E.A.: The relationships among nitric oxide production, bacterial translocation, and intestinal injury after endotoxin challenge in vivo. J Trauma **44**(1), 175–182 (1998)
51. Osband, A.J., Deitch, E.A., Hauser, C.J., Lu, Q., Zaets, S., Berezina, T., Machiedo, G.W., Rajwani, K.K., Xu, D.Z.: Albumin protects against gut-induced lung injury in vitro and in vivo. Ann Surg **240**(2), 331–339 (2004)
52. Papert, S.: Mindstorms: Children, Computers, and Powerful Ideas. Basic Books, New York (1980)
53. Pogson, M., Smallwood, R., Qwarnstrom, E., Holcombe, M.: Formal agent-based modelling of intracellular chemical interactions. Biosystems **85**(1), 37–45 (2006)
54. Railsback, S., Lytinen, S., Jackson, S.: Agent-based simulation platforms: Review and development recommendations. SIMULATION **82**(9), 609 (2006)
55. Resnick, M.: Turtles, Termites and Traffic Jams: Explorations in Massively Parallel Microworlds. MIT Press, Cambridge, MA (1994)
56. Resnick, M., Wilensky, U.: Beyond the deterministic, centralized mindsets: A new thinking for new science. In: Annual Meeting of the American Educational Research Association. Atlanta, GA (1993)
57. Resnick, M., Wilensky, U.: Diving into complexity: Developing probabilistic decentralized thinking through role-playing activities. J Learn Sci **7**(2), 153–171 (1998)
58. Reynolds, C.: Flocks, herds, and schools: A distributed behavioral model in computer graphics. In: SIGGRAPH '87, pp. 25–34 (1987)
59. Ridgway, D., Broderick, G., Lopez-Campistrous, A., Ru'aini, M., Winter, P., Hamilton, M., Boulanger, P., Kovalenko, A., Ellison, M.J.: Coarse-grained molecular simulation of diffusion and reaction kinetics in a crowded virtual cytoplasm. Biophys J **94**, 3748–3759 (2008)
60. Sappington, P.L., Han, X., Yang, R., Delude, R.L., Fink, M.P.: Ethyl pyruvate ameliorates intestinal epithelial barrier dysfunction in endotoxemic mice and immunostimulated caco-2 enterocytic monolayers. J Pharmacol Exp Ther **304**(1), 464–476 (2003)
61. Sengupta, P., Wilensky, U.: N.I.E.L.S: An emergent multi-agent based modeling environment for learning physics. In: Proceedings of the 4th International Joint Conference on Autonomous Agents and Multiagent Systems (AAMAS 2005). Utrecht, The Netherlands (2005)
62. Smith, E., Conrey, F.: Agent-based modeling: A new approach for theory building in social psychology. Personality Social Psychol Rev **11**(1), 87 (2007)
63. Stallion, A., Kou, T.D., Latifi, S.Q., Miller, K.A., Dahms, B.B., Dudgeon, D.L., Levine, A.D.: Ischemia/reperfusion: A clinically relevant model of intestinal injury yielding systemic inflammation. J Pediatr Surg **40**(3), 470–477 (2005)
64. Thorne, B.C., Bailey, A.M., Benedict, K., Peirce-Cottler, S.: Modeling blood vessel growth and leukocyte extravasation in ischemic injury: An integrated agent-based and finite element analysis approach. J Crit Care **21**(4), 346 (2006)
65. Thorne, B.C., Bailey, A.M., Peirce, S.M.: Combining experiments with multi-cell agent-based modeling to study biological tissue patterning. Brief Bioinform **8**(4), 245–257 (2007)

66. Tinker, R., Wilensky, U.: NetLogo Climate Change Model (Computer Program). Center for Connected Learning and Computer-Based Modeling, Northwestern University (2007). http://ccl.northwestern.edu/netlogo/models/ClimateChange
67. Tisue, S., Wilensky, U.: Netlogo: Design and implementation of a multi-agent modeling environment. In: C.M. Macal, D. Sallach, M.J. North (eds.) Proceedings of the Agent2004 Conference on Social Dynamics: Interaction, Reflexivity and Emergence, pp. 161–184. Argonne National Laboratory and the University of Chicago, IL (2004)
68. Tjardes, T., Neugebauer, E.: Sepsis research in the next millennium: concentrate on the software rather than the hardware. Shock **17**(1), 1–8 (2002)
69. Troutman, C., Clark, B., Goldrick, M.: Social networks and intraspeaker variation during periods of language change. In: 31st Annual Penn Linguistics Colloquium, vol. 14, pp. 323–338 (2008)
70. Uchiyama, T., Delude, R.L., Fink, M.P.: Dose-dependent effects of ethyl pyruvate in mice subjected to mesenteric ischemia and reperfusion. Intensive Care Med **29**(11), 2050–2058 (2003)
71. Vodovotz, Y., Clermont, G., Hunt, C.A., Lefering, R., Bartels, J., Seydel, R., Hotchkiss, J., Ta'asan, S., Neugebauer, E., An, G.: Evidence-based modeling of critical illness: An initial consensus from the society for complexity in acute illness. J Crit Care **22**(1), 77–84 (2007)
72. Wakeland, W., Macovsky, L., An, G.: A hybrid simulation for studying the acute inflammatory response. In: Proceedings of the 2007 Spring Simulation Multiconference (Agent Directed Simulation Symposium), vol. 1, pp. 39–46 (2007)
73. Walker, D.C., Hill, G., Wood, S.M., Smallwood, R.H., Southgate, J.: Agent-based computational modeling of wounded epithelial cell monolayers. IEEE Trans Nanobioscience **3**(3), 153–63 (2004)
74. Wilensky, U.: NetLogo Flocking Model (Computer Program). Center for Connected Learning and Computer-Based Modeling, Northwestern University (1998). http://ccl.northwestern.edu/netlogo/models/Flocking
75. Wilensky, U.: NetLogo Life Model (Computer Program). Center for Connected Learning and Computer-Based Modeling, Northwestern University (1998). http://ccl.northwestern.edu/netlogo/models/Life
76. Wilensky, U.: NetLogo Percolation Model (Computer Program). Center for Connected Learning and Computer-Based Modeling, Northwestern University (1998). http://ccl.northwestern.edu/netlogo/models/Percolation
77. Wilensky, U.: NetLogo Wolf Sheep Predation model (Computer Program). Center for Connected Learning and Computer-Based Modeling, Northwestern University (1998). http://ccl.northwestern.edu/netlogo/models/WolfSheepPredation
78. Wilensky, U. (ed.): GasLab: An extensible modeling toolkit for exploring micro-and-macro-views of gases. Computer Modeling and Simulation in Science Education. Springer-Verlag, Berlin (1999)
79. Wilensky, U.: NetLogo (Computer Program). Center for Connected Learning and Computer-Based Modeling, Northwestern University (1999). http://ccl.northwestern.edu/netlogo
80. Wilensky, U.: NetLogo Models Library (Computer Program). Center for Connected Learning and Computer-Based Modeling, Northwestern University (1999). http://ccl.northwestern.edu/netlogo/models
81. Wilensky, U.: GasLab Curriculum (Computer Program). Center for Connected Learning and Computer-Based Modeling, Northwestern University (2000). http://ccl.northwestern.edu/cm/GasLab/
82. Wilensky, U.: Modeling nature's emergent patterns with multi-agent languages. In: Proceedings of Eurologo 2001. Linz, Austria (2001)
83. Wilensky, U.: Systems Dynamics Modeler (Computer Program). Center for Connected Learning and Computer Based Modeling, Northwestern University (2005). http://ccl.northwestern.edu/netlogo/docs/systemdynamics.html

84. Wilensky, U.: NetLogo 3D Preview 5 (Computer Program). Center for Connected Learning and Computer Based Modeling, Northwestern University (2007). http://ccl.northwestern.edu/netlogo/whatsnew3d.html
85. Wilensky, U.: NetLogo Community Models. Center for Connected Learning and Computer Based Modeling, Northwestern University (2008). http://ccl.northwestern.edu/netlogo/models/community/
86. Wilensky, U.: NetLogo website. Center for Connected Learning and Computer Based Modeling, Northwestern University (2008). http://ccl.northwestern.edu/netlogo
87. Wilensky, U., Levy, S., Novak, M.: Connected Chemistry Curriculum (Computer Program). Center for Connected Learning and Computer Based Modeling, Northwestern University (2004). http://ccl.northwestern.edu/curriculum/chemistry
88. Wilensky, U., Novak, M., Rand, W.: BEAGLE evolution curriculum (Computer Program). Center for Connected Learning and Computer Based Modeling, Northwestern University (2005). http://ccl.northwestern.edu/curriculum/simevolution/beagle.shtml
89. Wilensky, U., Rand, W.: Making models match: Replicating agent-based models. J Arti Soc Social Sim **10**, 42 (2007)
90. Wilensky, U., Rand, W.: An Introduction to Agent-Based Modeling: Modeling Natural, Social and Engineered Complex Systems with NetLogo. MIT Press, Cambridge, MA (2009)
91. Wilensky, U., Reisman, K.: Connectedscience: Learning biology through constructing and testing computational theories – an embodied modeling approach. In: Y. Bar-Yam (ed.) Second International Conference on Complex Systems, vol. 234, pp. 1–12. InterJournal of Complex Systems, Nashua, NH (1998)
92. Wilensky, U., Reisman, K.: Thinking like a wolf, a sheep or a firefly: Learning biology through constructing and testing computational theories. Cognition and Instruction **24**(2), 171–209 (2006)
93. Wilensky, U., Resnick, M.: New thinking for new sciences: Constructionist approaches for exploring complexity. In: Annual Conference of the American Educational Research Association. San Francisco, CA (1995)
94. Wilensky, U., Resnick, M.: Thinking in levels: A dynamic systems approach to making sense of the world. J Sci Ed Technol **8**(1), 3–19 (1999)
95. Wilensky, U., Sengupta, P.: N.I.E.L.S. curriculum (Computer Program). Center for Connected Learning and Computer Based Modeling, Northwestern University (2005). http://ccl.northwestern.edu/curriculum/niels
96. Wilensky, U., Shargel, B.: BehaviorSpace (Computer Program). Center for Connected Learning and Computer Based Modeling, Northwestern University (2002). http://ccl.northwestern.edu/netlogo/behaviorspace
97. Wilensky, U., Stroup, W.: Networked gridlock: Students enacting complex dynamic phenomena with the hubnet architecture. In: B. Fishman, S. O-Connor-Divelbiss (eds.) Proceedings of the Fourth Annual International Conference of the Learning Sciences, pp. 282–289. Ann Arbor, MI (2000)
98. Wilensky, U., Stroup, W.: Participatory simulations: Envisioning the networked classroom as a way to support systems learning for all. In: Annual Meeting of the American Research Education Association. New Orleans, LA (2002)
99. Yang, R., Gallo, D.J., Baust, J.J., Watkins, S.K., Delude, R.L., Fink, M.P.: Effect of hemorrhagic shock on gut barrier function and expression of stress-related genes in normal and gnotobiotic mice. Am J Physiol Regul Integr Comp Physiol **283**(5), R1263–R1274 (2002)
100. Zhang, L., Athale, C.A., Deisboeck, T.S.: Development of a three-dimensional multiscale agent-based tumor model: simulating gene-protein interaction profiles, cell phenotypes and multicellular patterns in brain cancer. J Theor Biol **244**(1), 96–107 (2007)

Chapter 8
Discrete Dynamics Lab: Tools for Investigating Cellular Automata and Discrete Dynamical Networks

Andrew Wuensche

DDLab is interactive graphics software for creating, visualizing, and analyzing many aspects of Cellular Automata, Random Boolean Networks, and Discrete Dynamical Networks in general and studying their behavior, both from the time-series perspective – space-time patterns, and from the state-space perspective – attractor basins. DDLab is relevant to research, applications, and education in the fields of complexity, self-organization, emergent phenomena, chaos, collision-based computing, neural networks, content addressable memory, genetic regulatory networks, dynamical encryption, generative art and music, and the study of the abstract mathematical/physical/dynamical phenomena in their own right.

8.1 Introduction

Networks of sparsely interconnected elements with discrete values and updating in parallel are central to our understanding of a wide range of natural and artificial phenomena drawn from many areas of science: from physics to biology to cognition; to social and economic organization; to parallel computation and artificial life; to complex systems of all kinds.

Abstract, idealized networks – Cellular Automata (CA), Random Boolean Networks (RBN), and Discrete Dynamical Networks in general (DDN) – provide insights into complexity in nature by providing the simplest models of self-organization and bottom-up emergence. They are also fascinating in themselves as mathematical, physical, dynamical, and computational systems with a large body of literature devoted to their study [1, 5, 6, 8, 9, 14, 15, 16].

The dynamics that play out on these signaling – "decision-making" – discrete networks are difficult if not impossible to investigate by classical mathematics; numerical methods are therefore essential.

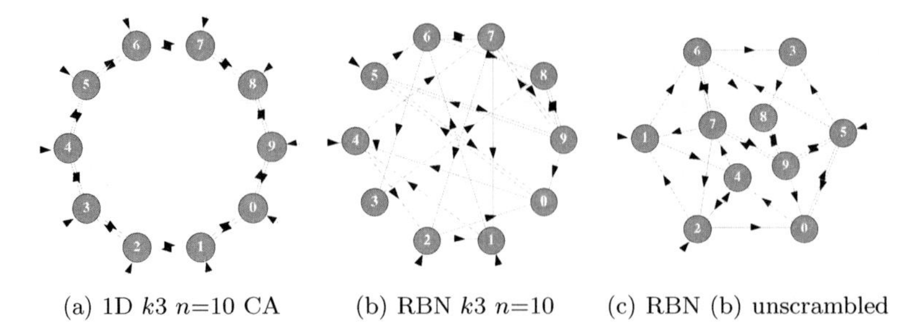

(a) 1D $k3$ n=10 CA (b) RBN $k3$ n=10 (c) RBN (b) unscrambled

Fig. 8.1 Network-graphs of (a) the simplest one-dimensional CA, n=10 cells and k=3 inputs per cell, one from each neighbor and one from itself. Periodic boundary conditions make the cells form a ring. Wolfram called these "elementary" CA [13]. (b) n=10, k=3 RBN wired at random, and (c) the same RBN unscrambled – nodes rearranged.

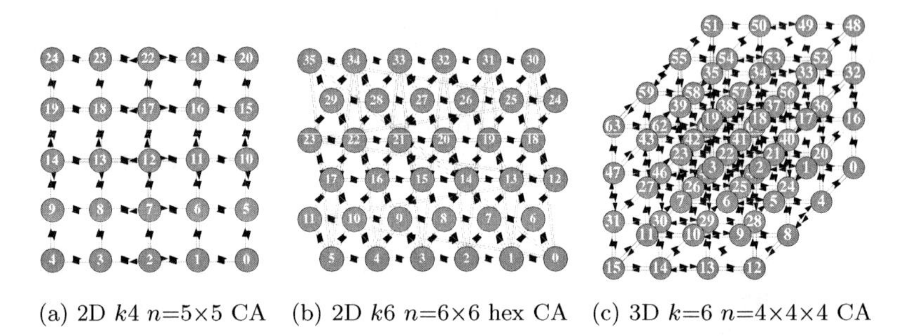

(a) 2D $k4$ n=5×5 CA (b) 2D $k6$ n=6×6 hex CA (c) 3D k=6 n=4×4×4 CA

Fig. 8.2 Network-graphs of (a) a small two-dimensional CA, (b) a 2D CA on a hexagonal lattice, (c) a 3D CA. These network-graphs can be created and manipulated in DDLab.

8.1.1 What Is DDLab?

DDLab is interactive graphics software, able to construct, visualize, and manipulate a hierarchy of discrete systems: CA, RBN, DDN (Figs. 8.1, 8.2 and 8.4), intermediate or hybrid networks, and random maps – and investigate and visualize many aspects of the dynamics *on* the networks, both from the time-series perspective – space-time patterns, and from the state-space perspective – attractor basins, with interactive graphical methods, as well as data gathering, analysis, and statistics. "Attractor basins" refers to any state transition graph: basin of attraction fields, single basins, or sub-trees.

Fig. 8.3 gives a glimpse of the main themes in DDLab and also the broad and slippery categories of the systems available:

- CA: Cellular Automata: a local neighborhood of k inputs (1D, 2D, 3D) and one rule (but possibly a mix of rules to extend the definition).

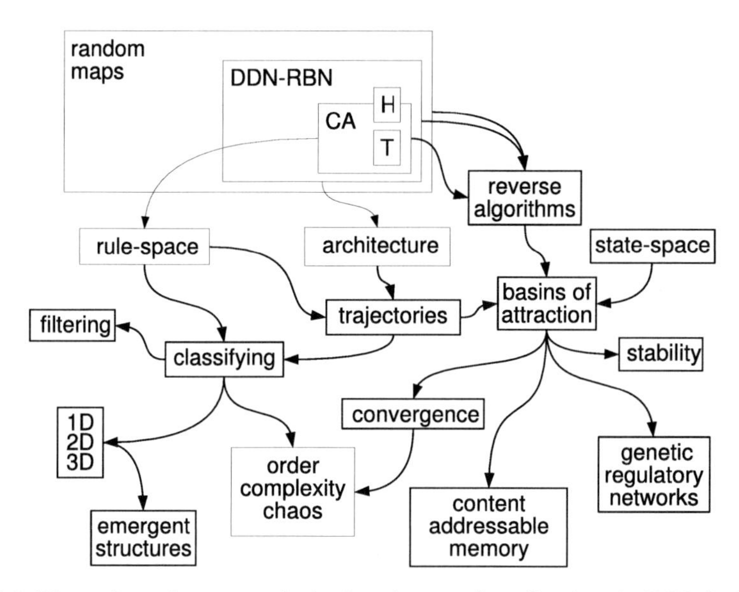

Fig. 8.3 The various themes, methods, functions, and applications in DDLab, loosely connected. *Top left*: The expanding hierarchy of networks: CA → RBN/DDN → within the super-set of random maps (directed graphs with out-degree one), imposing decreasing constraints on the dynamics. There are also multiple sub-categories; for example, totalistic rules (T), hybrids (H), and networks of networks.

- RBN: Random Boolean Networks: random wiring of k inputs (but possibly with mixed k) and a mix of rules (but possibly one rule).
- DDN: Discrete Dynamical Networks: same as RBN, but allowing a value range $v \geq 2$. Binary CA is a special case of RBN, and RBN and multi-value CA are special cases of DDN.
- Random maps: directed graphs with out-degree one, where each state in state-space is *assigned* a successor. CA, RBN, and DDN, which are usually sparsely connected ($k \ll n$) are all special cases of random maps, but if fully connected ($k = n$), rule-mix CA and the rest are equivalent to random maps.

DDLab is used widely in research and education, and has been applied to study complexity, emergence, and self-organization [8, 9, 14, 16, 23, 27], in the behavior of bio-molecular networks such as neural [18, 20] and genetic [7, 11, 4, 22] networks, in many other disparate areas. As well as scientific applications, DDLab's imagery has featured in art exhibitions [24], as the light show at raves, and has been applied for generative music [2].

There are currently versions of DDLab for Mac, Linux, Unix, Irix, Cygwin, and DOS. The source code is written in C; it may be made available on request subject to some conditions, although the intention is to make the code open source in the near future.

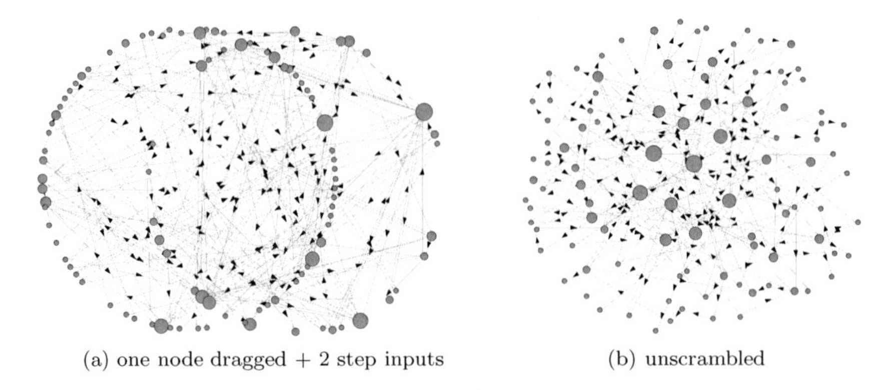

(a) one node dragged + 2 step inputs (b) unscrambled

Fig. 8.4 Network graphs showing wiring with a power-law distribution of both inputs ($k = 1$ to 10) and outputs, $n = 100$. Nodes are scaled by their number of inputs. (a) A circle layout, but with one node dragged together with its two-step inputs. (b) The same network unscrambled – nodes rearranged.

DDLab generates space-time patterns in one, two, or three dimensions and also constructs attractor basins – graphs that link network states according to their transitions, analogous to Poincaré's "phase portrait" that provided powerful insights in continuous dynamics. A key insight is that the dynamics on the networks converge, thus falling into a number of basins of attraction. This is the network's content addressable memory, its ability to hierarchically categorize all possible patterns (state-space), as a function of the precise network architecture [18, 20]. Relating this to space-time patterns in CA, high convergence implies order and low convergence implies disorder or chaos [16]. The most interesting emergent structures occur at the transition, sometimes called the "edge of chaos" [9, 23].

DDLab is generalized for multi-value logic. Up to eight values (or colors) are available, instead of just Boolean logic (two values: 0, 1) in early versions. Of course, with just two values selected, DDLab behaves like the older versions. Multi-values open up new possibilities for dynamical behavior and modeling – for example, fractional gene activation and reaction-diffusion.

Although DDLab is designed to run CA/DDN both forward (space-time patterns) and backward (attractor basins), it can be constrained to run forward-only for various types of totalistic and outer-totalistic rules, reducing memory load by cutting out all basin of attraction functions. This allows larger neighborhoods (max-$k = 25$ instead of 13). In 2D, the neighborhoods are pre-defined to make hexagonal as well as square lattices. Many interesting cellular automaton rules with "life"-like and other complex dynamics can be found in totalistic multi-value rule-space, in 3D as well as 2D [26].

DDLab is an applications program, allowing network parameters and the graphics presentation to be flexibly set, reviewed, and altered interactively, including changes on-the-fly. This chapter provides a brief of history and

Fig. 8.5 The basin of attraction field of a small random Boolean network, $k = 3$, $n = 13$. The $2^{13} = 8192$ states in state-space are organized into 15 basins, with attractor periods ranging between 1 and 7 and basin volume between 68 and 2724 [1.2 second]. The arrow points to the basin shown in more detail in Fig. 8.6.

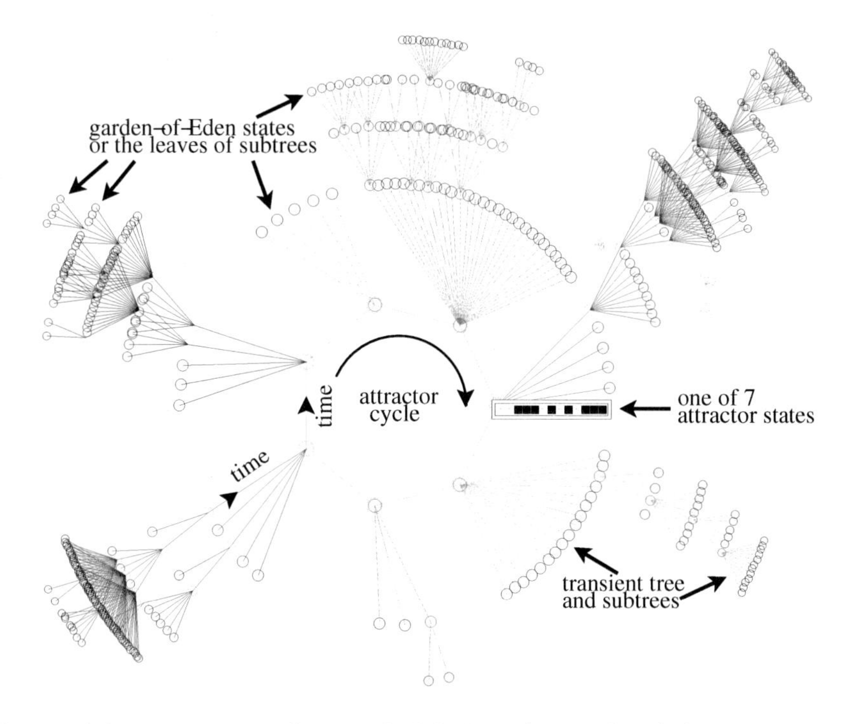

Fig. 8.6 A basin of attraction (one of 15) of the RBN ($n = 13$, $k = 3$) shown in Fig. 8.5. The basin links 604 states, of which 523 are garden-of-Eden states. The attractor period = 7, and one of the attractor states is shown in detail as a bit pattern. The direction of time is inward from garden-of-Eden states to the attractor, then clockwise. [0.56 second]

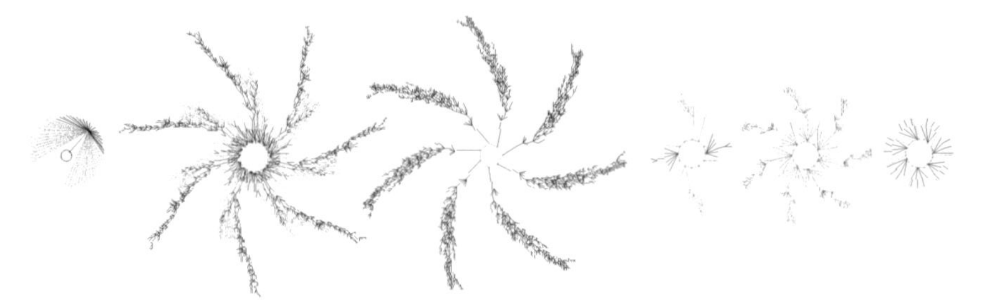

Fig. 8.7 The basin of attraction field of a binary ($v = 2$) CA, neighborhood $k = 3$, lattice size $n = 14$ [0.2 second]. There are in fact 15 basins, but equivalent basins have been suppressed, leaving just the 6 prototypes; note also the equivalent trees typical of CA [16] but not in RBN. State nodes have been omitted. This is the famous "elementary" rule 110 [13], which is computationally universal.

gives the flavor of DDLab with a range of examples. The figures presented are drawn with various DDLab functions, including vector PostScript. The operating manual [25] (a new update is in progress) describes all of the many functions and includes a "quick start" chapter. DDLab, together with many examples, publications, and all other information, is available in [25].

DDLab remains free shareware for personal, non-commercial, users. Any other users, including commercial users, companies, government agencies, and research or educational institutions, must register for a license [25].

8.1.2 A Brief History

Two friends in London in the late 1980s became interested in CA – as to why, that is another story. They were not academic scientists. The only article they had read on the subject was "Statistical Mechanics of Cellular Automata" [13], the first of Wolfram's many influential papers [14].

The question soon arose: Can CA be run backward as well as forward? Not knowing that this was supposed to be impossible, an algorithm was invented the next day working on an Apple2 – computing a state's predecessors for rule 150. Extending the algorithm for any rule took more time and effort. Basins of attraction were computed and painfully drawn by hand.

This eventually led to the publication of their book in 1992, in the Santa Fe Institute series in the sciences of complexity: *The Global Dynamics of Cellular Automata*, subtitled *An Atlas of Basin of Attraction Fields of One-Dimensional Cellular Automata* [16]. A floppy disk was attached inside the back cover, with DOS software for drawing space-time patterns and basins of attraction – the origin of what later became DDLab.

The book introduced a reverse algorithm for finding the pre-images (predecessors) of states for any finite 1D binary CA with periodic boundary

Fig. 8.8 The front covers of Wuensche and Lesser's (1992) *The Global Dynamics of Cellular Automata* [16] and Kauffman's (1993) *The Origins of Order* [6]. The attractor basins of a CA and RBN were computed and drawn with the precursor of DDLab.

conditions, making it possible to reveal the precise nature of their "basins of attraction" – represented by state transition graphs (i.e., states linked into trees rooted on attractor cycles, which could be drawn automatically). CA could henceforth be studied from a state-space perspective as well as from a time-series perspective.

Two important personalities at the Santa Fe Institute (SFI) at the time were Stuart Kauffman and Chris Langton. Stuart Kauffman was famous for his Random Boolean Network (RBN) model of genetic regulation [5, 6], where cell types are seen as attractors in the dynamics on the genetic network. A meeting was arranged in the summer of 1990 in his office at the Department of Biochemistry and Biophysics, University of Pennsylvania – "OK, you've got 2 minutes" – but he was amazed to see the hand-drawn "atlas" of basins of attraction. Professor Kauffman ordered: "take the next flight to Santa Fe."

A long relationship with SFI began, also with Chris Langton, founder of Artificial Life, which sprang from his important early CA work [8, 9]. Langton wrote the preface [10] to *The Global Dynamics of Cellular Automata*, comparing the work with Poincaré, and promoted its publication. Kauffman later wrote a generous review [7].

I continued my research with the new methods, gaining new insights [17, 18, 20]. As to the software, the next task I set for myself was to find an reverse algorithm for Kauffman's RBN; the CA reverse algorithm could cope with different rules at each cell, but for inputs that could arrive from anywhere – random wiring instead of wiring from a 1D CA neighborhood – that required a completely different algorithm. I invented this algorithm in 1992, in time to provide Kauffman with the cover image for his *The Origins of Order* [6]

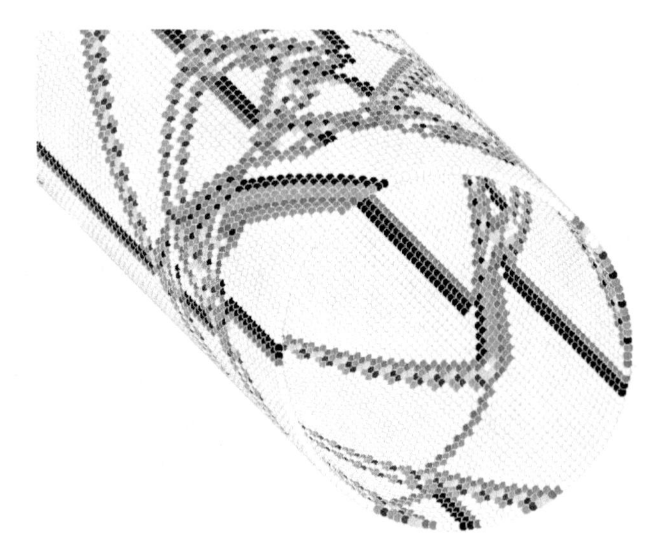

Fig. 8.9 A 1D space-time pattern shown as a ring of cells. scrolling diagonally towards the top left. The present moment is the ring at the bottom right. The space-time pattern is colored according to neighborhood, and has been filtered. 1D binary CA, k=5, hex rule e9 f6 a8 15, n=150 (as in Fig.8.16).

(Fig. 8.8) and to present my results [18] at the Artificial Life III conference in Santa Fe in 1992.

In 1992 I had also started a doctorate at COGS, University of Sussex (completed in 1997 [21]), but by 1995 I was settled in Santa Fe and working at the Santa Fe Institute – in Langton's group. They where starting the Swarm project, whereas I focused on porting the DOS version of DDLab to Unix, then to Linux, with some help from the Swarm team. I named the software "Discrete Dynamics Lab" in 1993; the first DDLab website appeared in 1996 on ALife Online and then was hosted by SFI itself in 1997.

I collaborated a great deal with Kauffman and his colleagues [4] in theoretical biology – and do still. It was at his insistence that I added a number of new RBN features to DDLab – for example, damage-spread, canalizing functions, the Derrida plot, and the attractor and skeleton frequency histograms.

8.1.3 DDLab Updates

DDlab has continued to grow and evolve, adding new ideas, features, and platforms. Multi-value logic was added in 2003 and, recently, vector PostScript for most graphic output. Below is a summary of the most significant updates with official releases on the DDLab website.

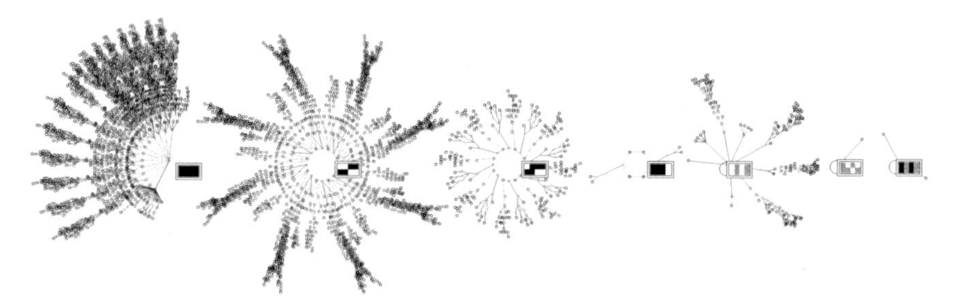

Fig. 8.10 Basin of attraction field of a three-value CA, $[v, k, n] = [3, 3, 8]$, rule 020120122021201201011011022. There are in fact 17 basins, but equivalent basins have been suppressed leaving just the 7 prototypes. One attractor state is shown for each basin [0.16 second].

- 1995: Although still in DOS, this version already had many of the important DDLab functions, described in a 63-page manual [19]: CA and RBN, 1D, 2D and 2D+time space-time patterns, cell color by neighborhood look-up or frozen, look-up frequency and entropy, sub-trees, basins of attraction and the basin of attraction field (for a range of network size), reverse algorithms for CA, RBN, and random graphs, in-degree frequency, 1D and 2D wiring graphic and spread-sheet to review and amend the network, garden-of-Eden density, mutation, state-space matrix, learning and forgetting algorithms, filing for the network – rules – wiring.
- 1997: Unix and Linux added; sequential updating, noise, hypercube network, categorizing rule-space, attractor and skeleton frequency histograms for large networks.
- 1999: 3D networks and neighborhoods, biases for random wiring, untangling wiring, effective neighborhoods, inverse problem – pruning a fully connected network, wiring – degrees of separation, setting rules and wiring for parts of the network – layers – rows – range, filtering, value frequency color code.
- 2002: Irix added; New Manual (421 pages, 209 figures), network-graph and attractor jump-graph, where nodes can be dragged, rearranged or unravelled and space-time patterns run on the graph, 1D circle wiring graphic, wiring – power-law distribution, chain-rules for encryption, Derrida coefficient, fractal return-map.
- 2003: Mac added; all aspects of DDLab generalized for multi-value logic (up to 8 values or colors), instead of just Boolean logic (0, 1). Option to constrain DDLab to run forward only for totalistic rules, allowing larger neighborhoods (up to $k = 25$), hexagonal lattices and neighborhoods, unslanting 1D space-time patterns, 2D and 3D falling-of-edge boundary conditions, Post functions.

- 2005: Cygwin added; outer-totalistic rules, reaction–diffusion dynamics, diagonal scrolling of 1D circle and 2D space-time patterns.
- 2008: A layout-graph for the basin of attraction field allowing basins to be dragged/rearranged into any position. Vector PostScript for most DDLab graphic output: space-time patterns, attractor basins, network-graph, jump-graph and layout-graph; partial-order updating; many other significant improvements. In the pipeline is the updated manual for the latest version of DDLab.

8.1.4 Basins of Attraction

In continuous dynamical systems, all possible time-series make up the vector field, represented by the system's phase portrait. This is the the field of flow imposed on phase-space by the system's dynamical rule. A set of attractors, be they fixed point, limit cycles, or chaotic, attract various regions of the vector field, the basins of attraction (paraphrasing Langton [9]).

Analogous concepts and terminology apply to discrete networks; although an important difference is that transients merge outside the attractor, within trajectories, far from equilibrium [20] – in continuous systems they cannot.

8.1.5 Times for Attractor Basins

The elapsed times to generate attractor basins are displayed in DDLab. To give an idea of these times, they are shown in the relevant figure captions in square brackets, i.e., [30.1 seconds] (1.66GHz Laptop running Linux/Ubuntu).

8.1.6 Forward and Backward in Time

DDLab looks at the dynamics on networks in two ways:

- from a time-series perspective, *local* dynamics, running forward
- from a state-space perspective, *global* dynamics, running backward

Running forward in time generates the network's space-time patterns from a given initial state. Many alternative graphical representations of space-time patterns and methods for gathering and analyzing data are available to illustrate different aspects of local dynamics, including "filtering" to show up emergent structures more clearly, as in Fig. 8.16.

Running backward in time generates multiple predecessors rather than a trajectory of unique successors. This procedure reconstructs the branching sub-tree of ancestor patterns rooted on a particular state. States without predecessors are disclosed – the so-called garden-of-Eden states, the leaves of the sub-trees.

Subtrees, basins of attraction (with a topology of trees rooted on attractor cycles), or the entire basin of attraction field can be displayed as directed graphs in real time, with many presentation options and methods for gathering/analyzing data. The attractor basins of "random maps" may be generated, with or without some bias in the mapping.

Attractor basins represent the network's "memory" by their hierarchical categorization of state-space; each basin is categorized by its attractor and each sub-tree by its root [18, 20]. Learning/forgetting algorithms allow attaching/detaching sets of states as predecessors of a given state by automatically mutating rules or changing connections.

Fig. 8.11 provides a summary of the idea of how state-space is organized into attractors and basins of attraction in discrete systems, where there is one future but many pasts.

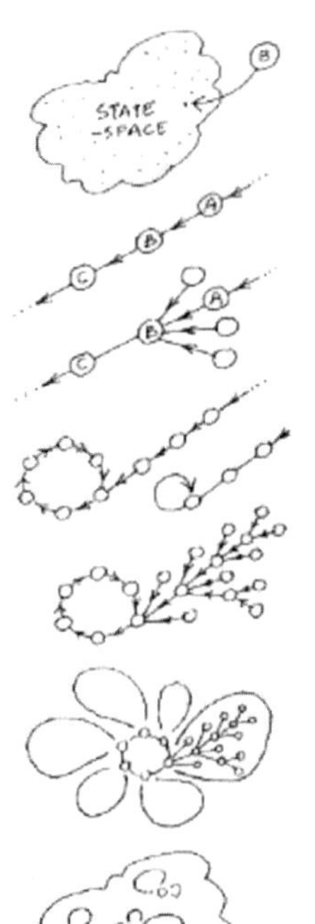

For a binary network size n, a state B might be 1010...0110 – there are 2^n states in *State-space* (v^n for multi-value)

A *trajectory* in state-space, $A \rightarrow B \rightarrow C$

B may have other *pre-images* besides A (the *in-degree*), which can be directly computed with reverse algorithms. States with zero pre-images (leaves) are known as *garden-of-Eden* states.

The trajectory must arrive at an *attractor*, a cycle of states with period one or more.

Find the pre-images of an attractor state (excluding the one on the attractor) – then the pre-images of the pre-images until all leaf states have been reached – the graph of linked states is the *transient tree* – part of the transient tree is a sub-tree defined by its root.

Construct each transient tree (if any) from each attractor state. The complete graph is the *basin of attraction.*

Find every attractor and construct its basin of attraction – this is the *basin of attraction field* – all states in state-space linked by the dynamics – each discrete dynamical network imposes a specific basin of attraction field on state-space.

Fig. 8.11 State-space and basins of attraction – the idea: one future, many pasts.

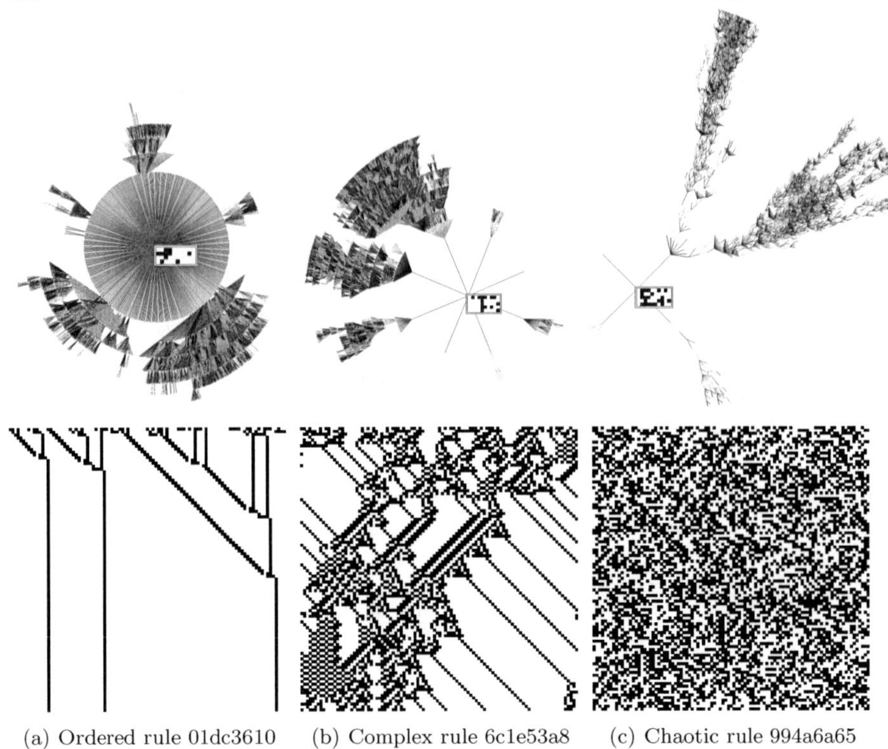

(a) Ordered rule 01dc3610 (b) Complex rule 6c1e53a8 (c) Chaotic rule 994a6a65

Fig. 8.12 Ordered, complex, and chaotic dynamics of 1D binary CA are illustrated by the space-time patterns and sub-trees of three typical $k = 5$ rules (shown in hex). The bottom row shows the space-time patterns from the same random initial state. The bit strings ($n = 100$) of successive time-steps (represented by white and black squares) are shown horizontally one below the other; time proceeds down. Above each space-time pattern is a typical sub-tree for the corresponding rule. The 1D sub-tree roots are shown in 2D: $n = 40$ (4×10) for the ordered rule and $n = 50$ (5×10) for the complex and chaotic rules. The root states were reached by first iterating the system forward by a few steps from a random initial state, then tracing the sub-tree backward. The ordered sub-tree is complete; the complex and chaotic sub-trees were stopped after 12 and 50 backward time-steps. Note that the convergence in the sub-trees, their branchiness or typical in-degree, relates to order-chaos in space-time patterns, where order has high, chaos low, convergence. [(a) 30.1 second, (b) 42.3 second, (c) 5.8 second].

8.2 DDLab's Interface and Initial Choices

DDLab's graphical user interface allows setting, viewing, and amending network parameters and provides presentation, output, and analysis options, by responding to prompts or accepting defaults. Navigating the interface takes a bit of practice, but it is designed to be both powerful and flexible, bypassing the more specialized options unless required – with CA the simplest to set up.

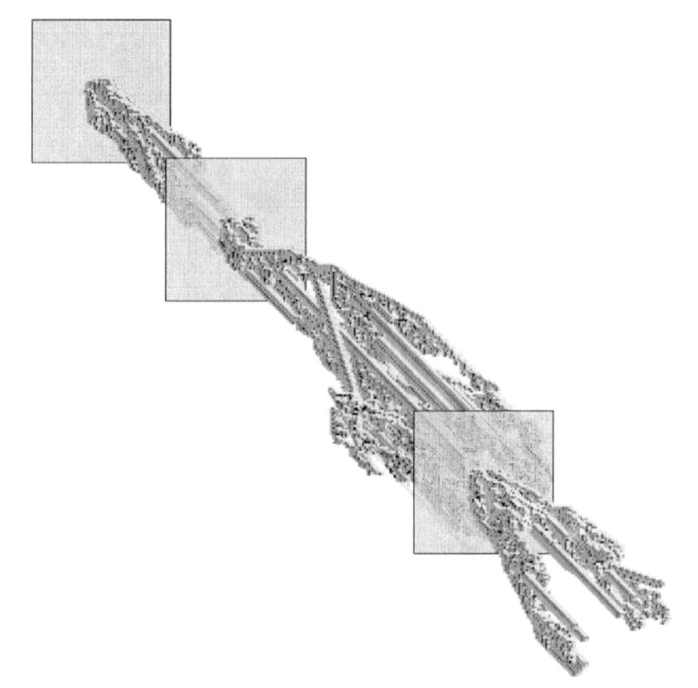

Fig. 8.13 Space-time patterns of the Game-of-Life scrolling diagonally upward ($k = 9$, $n = 66 \times 66$). States are stacked in front of each other in an isometric projection, with new seeds set at intervals and alternate time-steps skipped.

The prompts are fairly self-explanatory. They present themselves in a main sequence for the most common 1D CA parameters and also in a number of context-dependent pop-up prompt windows for 2D, 3D, random wiring, various special settings, and during an output run if interrupted, or when the run stops. A flashing cursor prompts for input. To navigate the interface:

- Enter **return** for a working default, which gives the next prompt. Repeat **return** to move forward through the prompt sequence. Eventually something interesting will happen even if no actual entries are made. All prompts have a default and some retain previous entries. An alternative to **return** is a click of the left mouse button.
- Enter **q** to backtrack to the preceding prompt. Repeat **q** to backtrack up the prompt sequence. **q** will also revise an entry and interrupt a running process being generated, such as space-time patterns or attractor basins. An alternative to **q** is a click of the right mouse button.
- Any time a top-center notice is showing with **accept defaults-d**, enter **d** to skip a series of specialist prompts.
- Enter appropriate input from the keyboard in response to a prompt, followed by **return** to move to the next prompt. Revise with **q** or **backspace**.
- To exit DDLab enter **Ctrl-q** at any prompt, followed by **q** (except in DOD – press **q** to backtrack until exit).

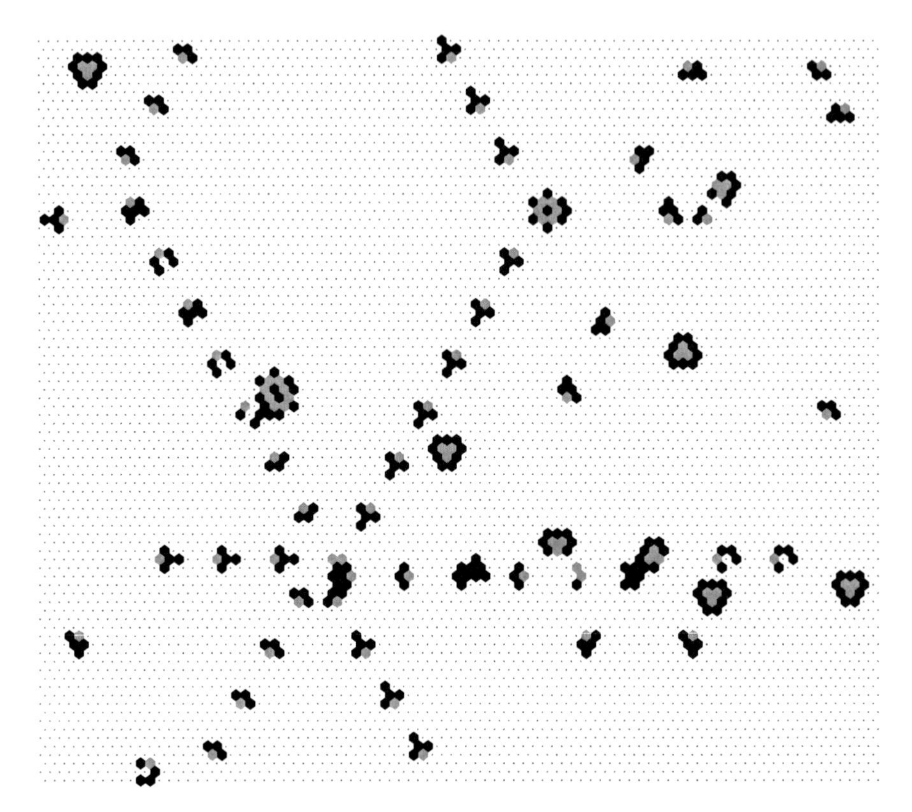

Fig. 8.14 A 2D CA space-time pattern on a hexagonal lattice. $n = 88 \times 88$, $[v, k] = [3, 7]$. The k-totalistic rule 0002001200212202212002221220222221210 gives rise to a zoo of interacting mobile and static emergent structures [27]: spiral glider-guns, mobile glider-guns, and self-reproduction by glider collisions.

8.2.1 TFO, SEED, and FIELD Modes

Some initial choices in the prompt sequence set the stage for all subsequent DDLab operations. TFO-mode constrains DDLab to run totalistic rules forward-only. This reduces memory load by cutting out full look-up tables and all attractor functions allowing larger neighborhoods as shown in Table 8.2.

If DDLab is not constrained in TFO-mode, there is a further choice: FIELD-mode to show the whole basin of attraction field, or SEED-mode to show something that requires an initial state (a space-time pattern, a single basin of attraction, or a sub-tree). At the conclusion of an output run FIELD-mode and SEED-mode can be toggled without having to backtrack.

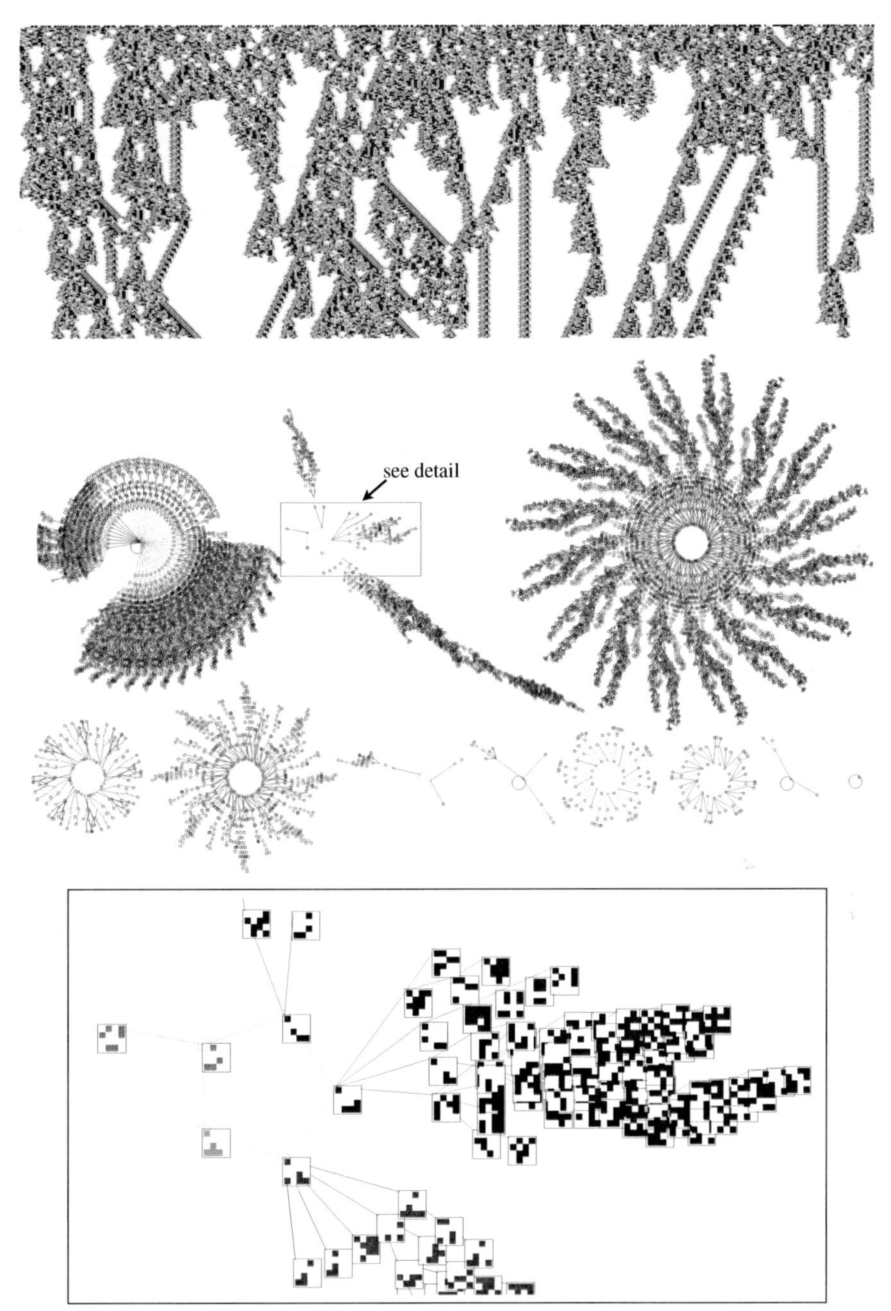

Fig. 8.15 *Top*: The space-time pattern of a 1D binary CA where interacting gliders emerge [23]. $n = 700$, $k = 7$, 250 times-steps from a random initial state. *Center*: The basin of attraction field for the same rule, $n = 16$. The 2^{16} states in state-space are connected into 89 basins of attraction, but only the 11 nonequivalent basins are shown, with symmetries characteristic of CA [6.9 seconds]. *Bottom*: A detail of the second basin in the basin of attraction field, where states are shown as 4×4 bit patterns.

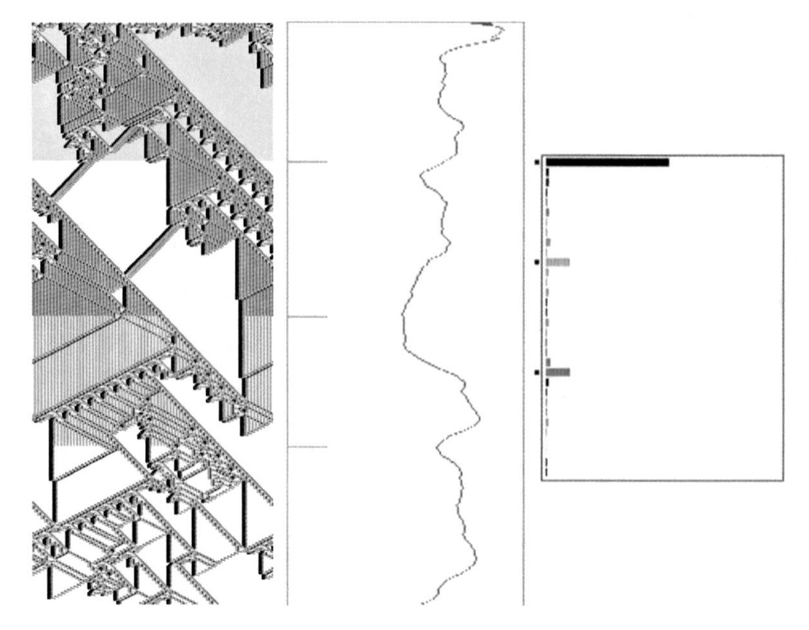

Fig. 8.16 A space-time pattern of a complex 1D CA, $[v, k] = [2, 5]$, $n = 150$, hex rule e9 f6 a8 15 (as in Fig. 8.9); 350 time-steps, and some analysis shown by default. *Left*: The space-time pattern colored according to neighborhood look-up, and progressively "filtered" on-the-fly three times, suppressing the background domain to show up interacting gliders more clearly. *Center*: The input-entropy plot of – *right*: the look-up frequency histogram, relative to a moving window of 10 time-steps. The filtered neighborhoods are indicated with a dot. The entropy variance is a sign of emergent structure.

8.3 Network Parameters

The parameters to define the network are presented in roughly the following sequence, but this depends on the mode chosen in Sect. 8.2.1: TFO, SEED, and FIELD modes and subsequent choices. To make corrections, backtrack with **q**. Depending on what is selected, some of these parameters may not apply or there may be extra options in pop-up windows.

$$\text{value-range } v \rightarrow \text{system size } n \rightarrow \text{neighborhood size } k \text{ or } k\text{-mix}$$
$$\rightarrow \text{dimension 1D, 2D, 3D} \rightarrow \text{wiring scheme}$$
$$\rightarrow \text{rule or rule-scheme} \rightarrow \text{initial state}$$

Wiring, rules, and the initial state can also be amended during an output run or at its conclusion, including changes on-the-fly.

These parameters, described in the sub-sections below, may appear somewhat intimidating with their many alternatives and diversions, but DDLab provides defaults to all prompts and bypasses the more specialized prompts if not needed.

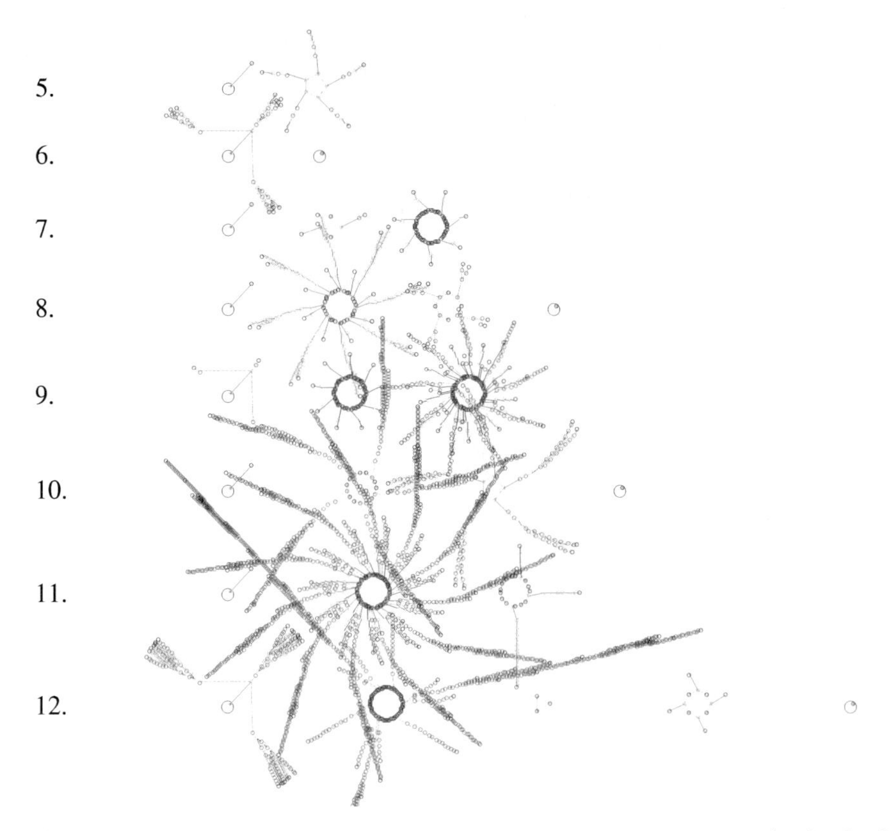

Fig. 8.17 Basin of attraction fields for a range of network size $n = 5 - 12$, $[v, k] = [2, 3]$, rule 30 [0.59 second].

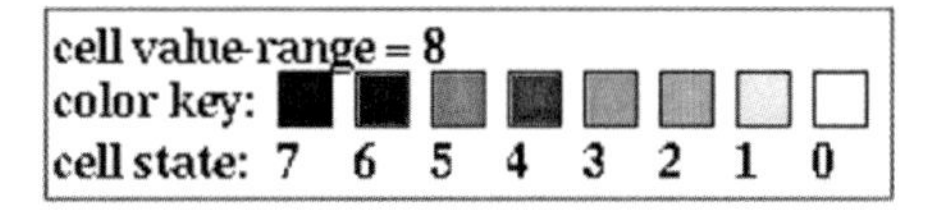

Fig. 8.18 The cell value color key window that appears when the value-range is selected – here for $v = 8$. The values themselves are indexed from 7 to 0.

8.3.1 Value Range, *v*

The value-range v is the number of possible cell-states, or colors, or letters in the cell's "alphabet." v can be set from 2 to 8 only at this early prompt. If v=2, DDLab behaves as in the old binary version. The value-range is indexed from 0 to 7, and is assigned a color code.

As v is increased, max-k will decrease (Table 8.1), and max-n in FIELD-mode will also decrease (Table 8.2).

Fig. 8.19 Snapshot of a 2D CA on a hexagonal lattice. $n = 40 \times 40$, where spirals have emerged. $[v,k] = [3,6]$. The k-totalistic rule depends on just the frequency of the $v = 3$ colors (2,1,0) in the $k = 6$ neighborhood, as shown below:

$$
\left.\begin{array}{r}
\text{black: } 2 \rightarrow 6\ 5\ 5\ 4\ 4\ 4\ 3\ 3\ 3\ 3\ 2\ 2\ 2\ 2\ 2\ 1\ 1\ 1\ 1\ 1\ 1\ 0\ 0\ 0\ 0\ 0\ 0\ 0 \\
\text{red: } 1 \rightarrow 0\ 1\ 0\ 2\ 1\ 0\ 3\ 2\ 1\ 0\ 4\ 3\ 2\ 1\ 0\ 5\ 4\ 3\ 2\ 1\ 0\ 6\ 5\ 4\ 3\ 2\ 1\ 0 \\
\text{white: } 0 \rightarrow 0\ 0\ 1\ 0\ 1\ 2\ 0\ 1\ 2\ 3\ 0\ 1\ 2\ 3\ 4\ 0\ 1\ 2\ 3\ 4\ 5\ 0\ 1\ 2\ 3\ 4\ 5\ 6
\end{array}\right\}\ \text{frequencies}
$$

$$
\begin{array}{l}
\downarrow\ \downarrow \\
0\ 0\ 2\ 2\ 0\ 0\ 0\ 2\ 2\ 0\ 0\ 2\ 2\ 0\ 0\ 1\ 1\ 2\ 2\ 2\ 0\ 0\ 0\ 2\ 1\ 1\ 1\ 0 \leftarrow \text{rule-table}
\end{array}
$$

8.3.2 Network Size, n

In FIELD-mode (for basin of attraction fields) the maximum network size (max-n) decreases with increasing value-range v as set out in Table 8.1. Note that, in practice, smaller sizes than max-n are recommended because of expected time and memory constraints.

In TFO-mode, and SEED-mode (for space-time patterns, single basins, or sub-trees), the network size is limited to max-n=65,025. A 2D square network is limited to (i,j)=255×255 and a 3D cubic network to $(i,j,h) = 40\times40\times40$, but within the limits of $n = 65025$ there can be any combination of (i,j) and (i,j,h).

v	2	3	4	5	6	7	8
max-n	31	20	15	13	12	11	10

Table 8.1 For a basin of attraction field (FIELD-mode): the maximum network size, max-n, for different value-ranges v.

Max-n works fine for space-time patterns, but for single basins and sub-trees, in practice much smaller sizes are appropriate, except when generating sub-trees for maximally chaotic CA chain-rules (Sect. 8.8).

The network size n for 1D is set early on in the prompt sequence, but this is superseded if a 2D (i, j) or 3D (i, j, h) network is selected in a subsequent prompt window.

Unconstrained FIELD/SEED-mode			Constrained TFO-mode		
v	max-k	table size S	v	max-k	table size S
2	13	8162	2	25	26
3	9	19683	3	25	351
4	7	16484	4	25	3276
5	6	15629	5	25	23551
6	5	16807	6	17	26334
7	5	16807	7	13	27132
8	4	4096	8	11	31824

Table 8.2 The maximum neighborhood size, max-k, for value-ranges v. *Left*: For FIELD-mode or SEED-mode, with a full (unconstrained) rule-table. *Right*: For TFO-mode, for running totalistic rules forward-only, where the totalistic rule-tables are shorter, so max-k can be larger. In both cases the size S of the rule-tables for v and max-k is shown.

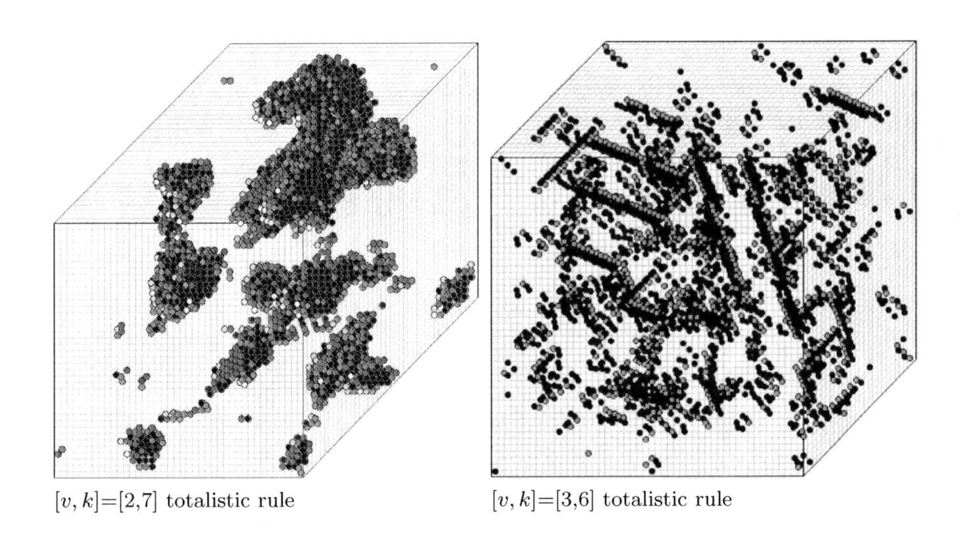

Fig. 8.20 Snapshots of 3D CA from random initial states. The projection is axonometric seen from below, as if looking up at the inside of a cage. *Left*: $n = 40 \times 40 \times 40$. The binary totalistic rule 11101000, $k = 7$, but with a bias of 1s of 45%. Cells are shown colored according to neighborhood look-up for a clearer picture, instead of by value: (0, 1). *Right*: $n = 50 \times 26 \times 50$, $[v, k] = [3, 6]$. The k-totalistic rule 0200001020100200002200120110 allows the emergence of gliders and other complex structures.

8.3.3 Neighborhood Size, k

The neighborhood size, k, is the number of input wires to each cell. If k is not homogeneous, the k-mix needs to be defined.

k can be set from 1 to max-k, but an effective $k = 0$ is also possible. Permissible max-k decreases with increasing value-range v because look-up tables may become too large. A full look-up table v^k is larger than a k-totalistic look-up table $(v+k-1)!/(k!(v-1)!)$, so max-k depends on whether DDLab was constrained in TFO-mode or unconstrained in SEED-mode or FIELD-mode. Max-k, for increasing value-range v, for both cases is given in Table 8.2, which also show the size of the corresponding rule-tables S.

The k-mix may be set and modified in a variety of ways, including defining the proportions of different k's to be allocated at random in the network or a scale-free distribution as in Fig. 8.4. A k-mix may be saved/loaded, but it is also implicit in the wiring scheme.

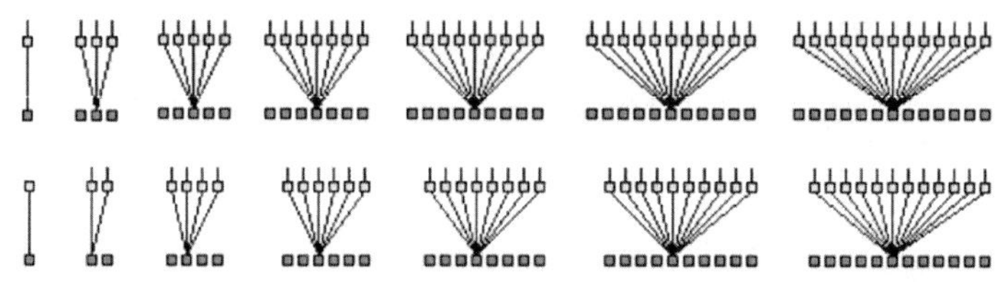

1D neighborhoods: for even k the extra asymmetric cell is on the right.

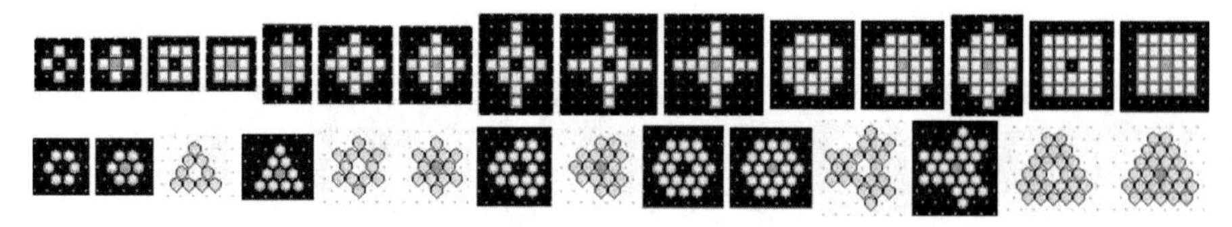

2D neighborhoods k=4–25: top row square, bottom row hex – black indicates the default.

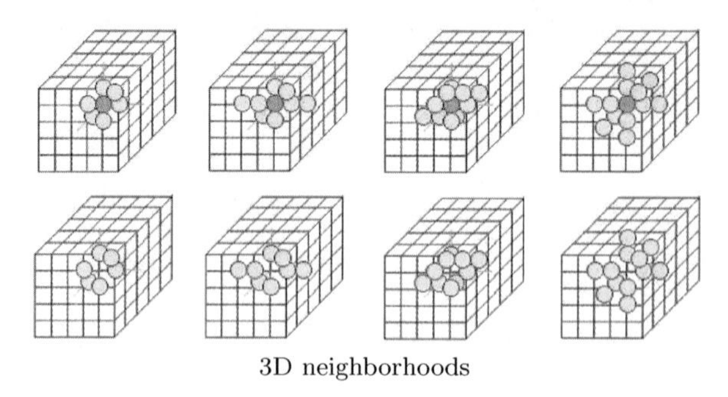

3D neighborhoods

Fig. 8.21 Predefined 1D, 2D, and 3D neighborhood templates. For 1D and 2D, k-max≤25 in TFO-mode, otherwise k-max≤13. For 3D, k-max≤13. For 2D, the neighborhood/lattice can be either square or hexagonal.

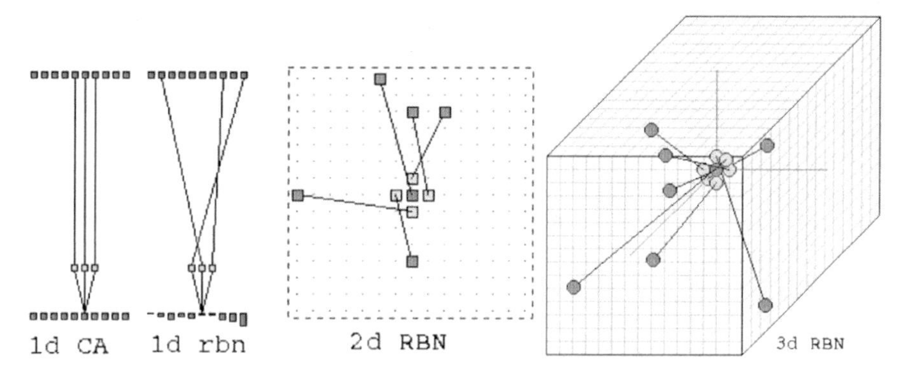

Fig. 8.22 The wiring of a cell: For random wiring, cells anywhere in the network are wired back to each position in a "pseudo-neighborhood," which corresponds to the CA neighborhood. Each cell's wiring can be shown (and amended) by moving around the network with the mouse or keyboard. *Left:* 1D: The wiring is shown between two time-steps. *Center:* 2D: $k = 5$. *Right:* 3D: $k = 7$.

8.3.4 Dimensions

Fig. 8.21 shows some of the predefined neighborhoods templates in 1D, 2D, or 3D, where each neighbor is indexed from 0 to $k-1$. The templates, which are designed to maximize symmetry, are assigned automatically when the dimensions are selected (the default is 1D) and conform to k or the k-mix. In 2D, there is a choice of square or a hexagonal neighborhood template. A hexagonal template results in a lattice with hexagonal tiling (Figs. 8.19 and 8.14). The choice of dimensions also sets the network's *native* geometry, with periodic boundary conditions, where lattice edges wrap around to their opposite edges. The *presentation* geometry can be different from the native geometry. The neighborhood templates also serve as the pseudo-neighborhoods for random wiring, because the rule requires an indexed neighborhood.

8.3.5 Wiring or Connections

The wiring scheme, the network connections, defines the location of the output terminals of each cell's input wires (Fig. 8.22), which connect to each neighbor in the indexed neighborhood or pseudo-neighborhood. For CA, the output terminals correspond to the neighborhood, but for random wiring, they could be anywhere within the network.

Wiring is perhaps the least studied aspect of the dynamics *on* networks. DDLab makes it possible to weave the tapestry.

The 1D, 2D, or 3D networks, whether CA or randomly wired, can be set up automatically, with a wide variety of constraints and biases. For example, random wiring can be confined within a local patch of cells with a set diameter in 1D, 2D or 3D (Fig. 8.24, *Right*). Some wires can be released from the neighborhood or patch. Part of the network only can be designated to accept a particular type of wiring scheme, for example rows in 2D and layers in 3D. The wiring can be biased to connect designated rows or layers.

A wiring scheme can be set and amended just for a predefined sub-network within the network, which may be saved/loaded, and hybrid networks of CA/DDN can be created, or a network of sub-networks in any combination.

Random wiring can be set or amended by hand. The wiring may be hypercube if $k = \log_2 n$ or $\log_2 n + 1$; for example, $k = 3$ or 4, $n = 8$.

The network parameters including wiring can be displayed and amended in a 1D, 2D, or 3D graphic format (Figs. 8.22 and 8.26) or in a "spreadsheet" (Fig. 8.27). The wiring can also be shown in a network-graph which can be rearranged in many ways, including dragging nodes with the mouse as in Figs. 8.1, 8.2, 8.4, and 8.28.

8.3.6 Rule or Rule Mix

A network may have one rule or a rule-mix. The rule mix can be confined to a subset of preselected rules. Rules may be set and modified in a wide variety of ways: in decimal, hex, as a rule-table bit or value string, at random, or loaded from a file. A rule scheme can be set and amended just for a predefined sub-network within the network, and may be saved/loaded.

In its most general form of update logic, a rule is expressed as a full look-up table, but there are sub-sets of the general case, two types of totalistic rules, and their corresponding outer-totalistic rules.

The simplest, t-totalistic rules, depend on the sum of values in the neighborhood. k-totalistic rules depend on the frequency of each value (color) in the neighborhood (see Fig. 8.19). If $k = 2$, these two types are identical.

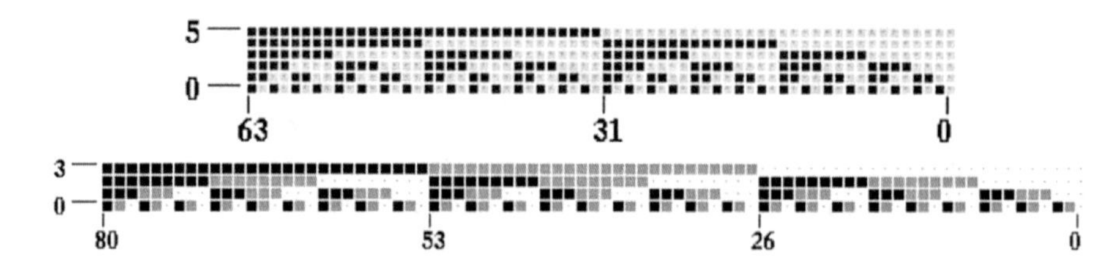

Fig. 8.23 Examples of the neighborhood matrix of full look-up tables (rule-tables), showing all possible neighborhoods (vertically). The position of each neighbor is indexed from $k - 1$ to 0. *Above:* $[v, k] = [2, 6]$ all 64 neighborhoods, from all-1s to all-0s. *Below:* $[v, k] = [4, 3]$ all 81 neighborhoods, from all-3s to all-0s.

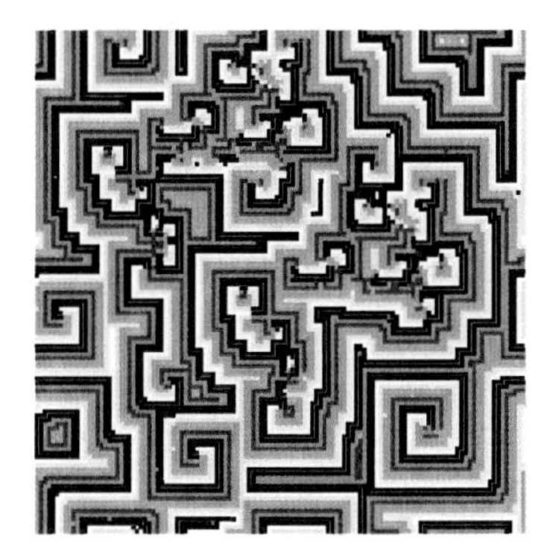

Fig. 8.24 Reaction-Diffusion or excitable media [3] dynamics. *Left*: $[v, k] = [8, 8]$, threshold interval is 1 to 6, $n = 122 \times 122$. *Right*: $[v, k] = [8, 11]$, threshold interval is 2 to 7, $n = 255 \times 255$, random connections within a 24-diameter local zone.

Both types of totalistic rules can be made into outer-totalistic rules, where a different rule applies for each value of the central cell; the Game-of-Life is one such rule (Fig. 8.13).

Totalistic and outer-totalistic rules can be run in either SEED-mode or TFO-mode – transformations and mutations then apply to to either the full or the totalistic look-up table.

There are also subsets of rules that can be automatically selected at random, including isometric rules – where rotated and reflected nhoods (1D, 2D or 3D) have the same output, chain-rules [28] described in Sect. 8.8, and Altenberg rules (Fig. 8.35).

Rules can be biased by various parameters, λ [9], Z [16, 23], canalizing inputs [4], and Post functions [12]. The Game-of-Life, majority (Fig. 8.25), and other predefined rules can be selected.

Rules may be changed into their equivalents (by reflection and negative transformations) and transformed into equivalent rules with larger or smaller neighborhoods. Rules transformed to larger neighborhoods are useful to achieve finer mutations (Fig. 8.33). Rule parameters λ and Z and the frequency of canalizing inputs in a network can be set to any arbitrary level.

DDlab is also able to implement reaction-diffusion or excitable media dynamics [3] (Fig. 8.24), which can be set as an outer-k-totalistic rule in TFO-mode or as a full look-up table in SEED-mode.

8.3.7 Initial State, the Seed

An initial network state, the seed, is required to run a network forward and generate space-time patterns. A seed is also required to generate a sub-tree or a single basin by first running forward to find the attractor, then backward to generate the sub-tree from each attractor state. A basin of attraction field

Fig. 8.25 *Top Left*: A 2D seed, $n = 88 \times 88$, drawn with the mouse and keyboard, $[v, k] = [8, 4]$. Colors 0 to $(v - 1)$ can be selected. The image/seed can be moved, rotated, and complimented. Sub-patterns saved earlier can be loaded into specified positions within the main pattern. In this example the seed is then transformed by applying a CA majority rule for three time-steps.

does not require setting a seed, because appropriate seeds are automatically provided.

To generate a sub-tree from a random seed it is usually necessary to run forward by a few steps to penetrate the sub-tree before running backward (Fig. 8.12). This is because for most DDN and CA (except chain-rules [28] Sect. 8.8), most states in state space have no predecessors; they are the sub-tree leaves or Garden-of-Eden states (Fig. 8.40).

As in setting a rule, there are a wide variety of methods for defining the seed: in decimal or hex, drawing bits or values in 1D, 2D, or 3D – a mini-paint program (Fig. 8.25), at random (with various constraints or biases), or loaded from a file.

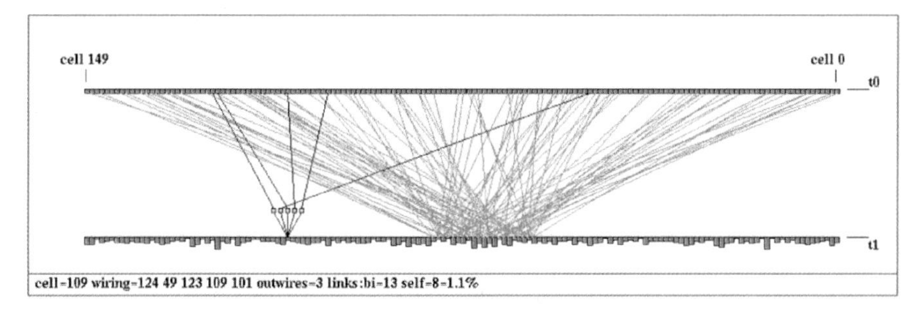

Fig. 8.26 The 1D wiring graphic, showing wiring to a block within a 1D network. $k = 5$, $n = 150$. The block was defined from cells 60–80. Revisions to rules and wiring can be the confined just to the block. The "active cell" is still visible and can be moved as usual. The 1D wiring graphic can also be shown as a circle.

8.4 Network Architecture

Once the network is defined, DDLab provides tabular and graphical methods for examining, amending, duplicating, unraveling, and displaying the network. These functions can be accessed from the initial sequence of prompts, or during an output run if interrupted, or when the run stops.

8.4.1 Wiring Graphic

The wiring graphic (Figs. 8.22 and 8.26) can be displayed in 1D, 2D, or 3D. The network's wiring and rules can be examined, changed, and tailored to requirements, including biased random settings to predefined parts of the network. Individual cells can be examined by moving around the diagram with the keyboard, and the cell's rule or individual wires modified. A block of cells can be defined and its wiring randomized or relocalized. These are very flexible methods, and for RBN/DDN it is usually easier to set up a suitable dummy network initially, and then tailor it in the wiring graphic.

From the wiring graphic it is possible to create a system of independent or weakly coupled sub-networks either directly, or by saving smaller networks to a file, and then loading them at appropriate positions in a base network. Thus, a 2D network can be tiled with sub-networks, and 1D, 2D, or 3D sub-networks can be inserted into a 3D base network.

The parameters of the sub-networks can be different from the base network, provided the base network is set up appropriately, with the right attributes to accommodate the sub-network. For example, to load a DDN into a CA, the CA may need be set up as if it were a DDN. To load a k-mix sub-network into a homogeneous-k base network, k in the base network needs to be at least as big as the biggest k in the sub-network. Options are available

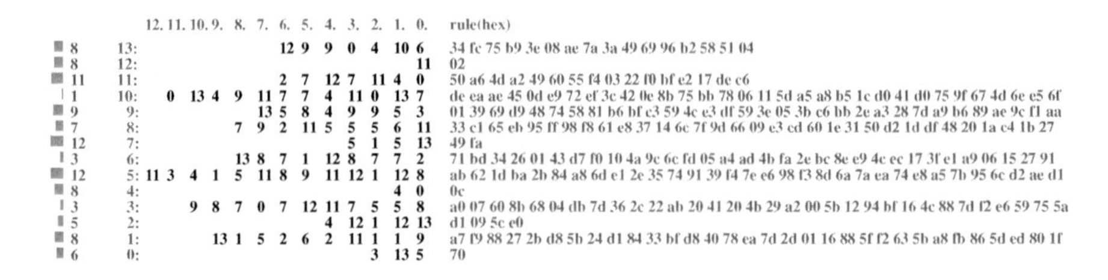

Fig. 8.27 The wiring matrix for a mixed k network with random wiring. $n = 14$, $k = 2$–13. Rows 13. . . 0 show the wiring and rule for each cell. Columns 12. . . 0 show from where each pseudo-neighborhood position is wired. The far left column shows the "out-degree" of each cell, the number of output wires that link to it, as a histogram. The far right column shows the rules in hex (as much as will fit). It is possible to move around the wiring matrix as in a spreadsheet to change wiring settings.

to set up networks in this way. Once loaded, the wiring can be fine-tuned to interconnect the sub-networks.

A network can be automatically duplicated to create a total network made up of two identical sub-networks. There is a function to see the difference pattern (or damage-spread) between two networks from similar initial states.

Connectivity measures from the wiring graphic include the following:

- Average k (inputs), and the number of reciprocal links and self-links.
- Histograms of the frequency distribution of inputs k, of outputs, or both (all connections) in the network.
- The recursive inputs/outputs to/from a network element, whether direct or indirect, showing the "degrees of separation" between cells.

8.4.2 Wiring Matrix

The wiring matrix (Fig. 8.27) displays the wiring and rule information in a tabular format, like a spread-sheet. For each cell tabulated against its pseudo-neighborhood index, the position of the input cell is shown. You can move around the matrix with the keyboard and amend this position.

8.4.3 Network-Graph

Another method of reviewing network architecture is an adjacency matrix and network-graph (Figs. 8.1, 8.2, 8.4, and 8.28) that looks just at the network connections, nodes linked by directed edges. It does not allow changes to the underlying network, but includes flexible methods for representing the network and rearranging and unraveling its graph.

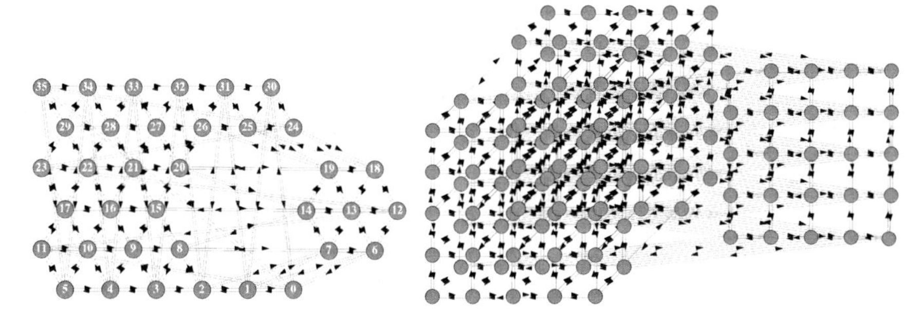

Fig. 8.28 Network-graphs of a 2D and 3D CA. *Left*: A 2D hexagonal network, showing a node and its 1-step inputs dragged out. *Right*: A 3D network shown as an axonometric projection seen from below as if looking up into a cage. A vertical slice has been defined and dragged from the graph.

The methods and options are similar to those of the jump-graph (Sect. 8.5.3) – in fact the functions below apply equally to the jump-graph.

For example, single nodes, connected fragments, or whole components can be dragged with the mouse to new positions with "elastic band" edges. Fragments depend on inputs, outputs, or both, and the link distance of a fragment from a node can be defined.

Dragging can include the node + its immediate links (step 1), the node + immediate links + their immediate links (step 2), etc. The average directed shortest path and non-directed small-world distance can be calculated. Arbitrary 1D, 2D, and 3D blocks can be dragged. Nodes with the fewest links can be automatically moved to the outer edges. This makes it possible to unravel a graph. The pre-programmed graph layouts available are a circle of nodes, a spiral, or 1D, 2D, or 3D, jiggled, and random. The graph can be rotated, expanded, contracted, and various other manipulations can be performed. The graph layout can be saved/loaded. An "ant" can be launched into the network that moves according to the link probabilities (as in a Markov chain) keeping a count of node hits.

8.5 Running Space-Time Patterns and Attractor Basins

Once the network is defined, a further series of prompts set the presentation and analysis in advance of the final output. The prompts depend on the context and previous selections and are presented in a succession of pop-up windows. They are arranged into the following categories:

- For attractor basins: various, layout, display, pause/data and mutation.
- For space-time patterns: various, analysis, and histograms.

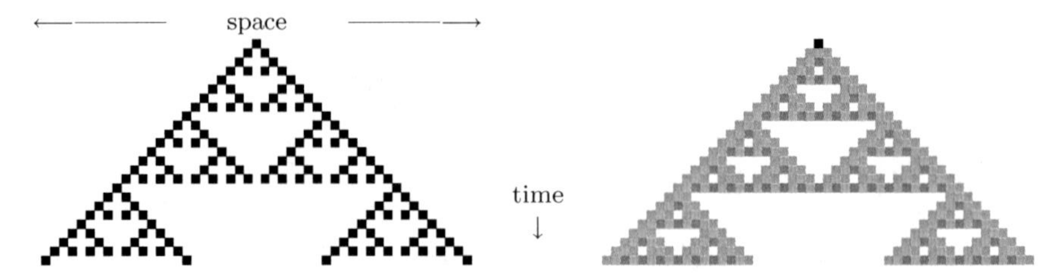

Fig. 8.29 Space-time patterns of a 1D binary CA, $[v, k] = [2, 3]$, $n = 51$, rule 90. Twenty four time-steps from an initial state with a single central 1. Two alternative presentations are shown. *Left*: cells by value. *Right*: cells colored according to their look-up neighborhood.

These options all have default setting, so you can skip them entirely or skip just the remaining categories at any point with **d**, or backtrack with **q** and skip to the required category.

After these prompts, space-time patterns start. Attractor basins start after some further final options. During the running process there are many interrupt and on-the-fly options, especially for space-time patterns. These are too numerous to list in full, but the sub-sections below provide some flavor.

8.5.1 Options for Space-Time Patterns

There are many alternative forms of presentation of space-time patterns and these can be changed on-the-fly, with a key press, while the space-time patterns are running. Cells in space-time patterns are colored according to their value or, alternatively, according to their neighborhood at the previous time-step, the entry in the look-up table that determined the cell's value (Fig. 8.29). Space-time patterns can be filtered to suppress cells that updated according to the most frequently occurring neighborhoods, thus exposing "gliders" and other structures (Figs. 8.16 and 8.30).

The presentation can highlight cells that have not changed in the previous x generations, where x can be set to any value. The emergence of such frozen elements are a sign of order, which can be induced by "canalizing inputs," applied in Kauffman's RBN model of gene regulatory networks [6, 4]. A 1D space-time pattern is presented scrolling vertically, or by successive vertical sweeps. The 2D networks can be toggled between square and hexagonal layout, and presented with a time dimension (2D+time), scrolling either vertically or diagonally (Fig. 8.13). The 3D networks are presented within a 3D "cage" (Fig. 8.20), but can also be shown in slices. The presentation of space-time patterns can be switched on-the-fly among 1D, 2D, 2D+time, and 3D, irrespective of their native dimensions. DDLab automatically unravels or bundles up the dimensions.

Space-time patters may iterate too fast for comfort, especially for 1D. They can be slowed down progressively on-the-fly or restored to full speed with the arrow keys.

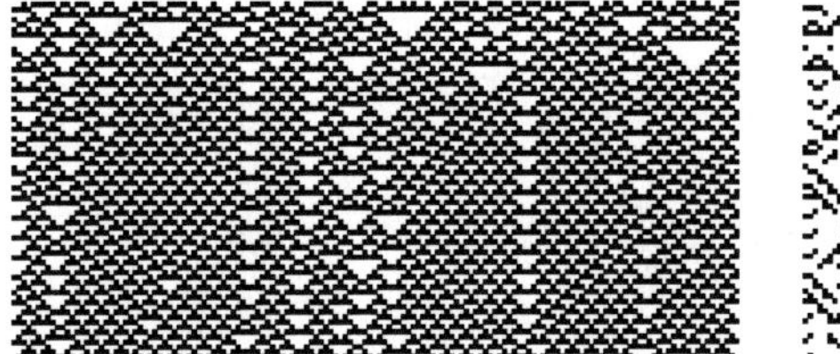

Fig. 8.30 Filtering a binary 1D space-time pattern with interacting gliders embedded in a complicated background. *Left*: unfiltered. *Right*: the same space-time pattern filtered. Filtering is done on-the-fly for any rule. In this example, $k = 3$ rule 54 was first transformed to its equivalent $k = 5$ rule (hex: 0f3c0f3c). $n = 150$, 75 time-steps from a random initial state.

There are many other on-the-fly options which do not interrupt the running space-time patterns, in whatever form of presentation. These include skipping time-steps, pausing after x time-steps, reversing to previous time-steps, changing the scale, changing or perturbing the current state, changing rules, toggling wiring between local and random, and showing cell stability. Bits/values in a rule can be progressively flipped and unflipped. Various qualities of noise can be added. Synchronous updating can be toggled with asynchronous in a random or a predefined partial order.

Concurrently with the standard presentations, space-time patterns can be displayed in a separate window according to the network-graph layout, which can be rearranged in any arbitrary way, including various default layouts. For example, a 1D space-time pattern can be shown in a circular layout and then scrolled diagonally (Fig. 8.9).

8.5.2 Options for Attractor Basins

Many options are provided for the layout of attractor basins, their size, position, spacing, rotation, and type of node display – as a spot, in decimal, hex, or a 1D or 2D bit pattern. Just the transient nodes, Garden-of-Eden nodes (or none) can be displayed. The in-degree spread or fan-angle can be adjusted.

Sub-trees, single basins, and the basin of attraction field can be draw for a range of network sized (Fig. 8.17). The basin of attraction field can be generated within the movable nodes of a jump-graph or layout-graph to achieve any arbitrary layout (Figs. 8.31 and 8.32).

Regular 1D and 2D CA produce attractor basins where sub-trees and basins are equivalent by rotational symmetry [16]. This allows "compression" of basins into just nonequivalent prototypes and also their sub-trees. As attractor basins are generating, the reverse space-time pattern can be simultaneously displayed.

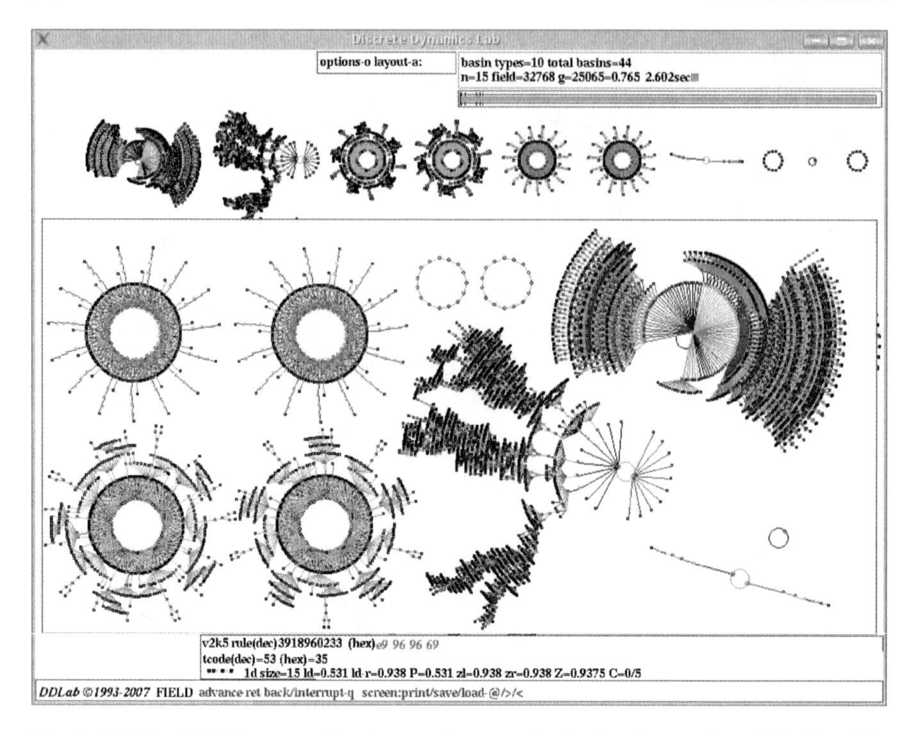

Fig. 8.31 The DDLab screen showing two layouts of a basin of attraction field. *Top*: The normal layout but with spacing adjusted after each basin. *Inset below*: The layout-graph where nodes (thus basins) can be dragged into any position. Both results can be saved as vector PostScript files. This example is for a binary 1D CA, $k = 5$, totalistic rule 53, $n = 15$ [2.4 second].

An attractor basin run can be set to pause to see data on each transient tree, each basin, or each field. Any combination of these data, including the complete list of states in basins and sub-trees, can be displayed in the terminal window or saved to a file. When pausing, the default position, spacing, rotation, and other properties of the next basin can be amended.

When an attractor basin has finished, a key-press will generate a "mutant" according to the mutant setting – for example, flipping a bit or value in the rule table, or moving a wire. The pause may be turned off to create a continuous demo. A "screensave" option shows mutant basins continually growing at random positions.

Attractor basins can be generated according to synchronous updating or asynchronous updating in a pre-defined partial order.

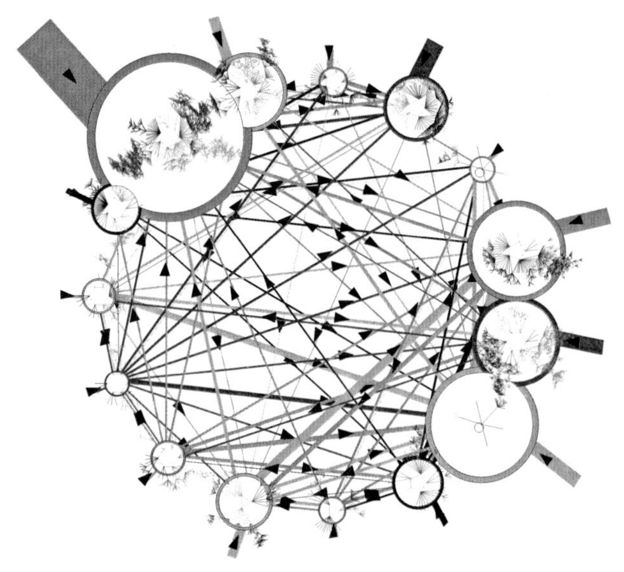

Fig. 8.32 The jump-graph with the basins of attraction field (RBN $k = 3$, $n = 13$, in Fig. 8.5) redrawn within its nodes. The jump-graph shows the probability of jumping between basins due to single bit-flips to attractor states. Nodes representing basins are scaled according the number of states in the basin (basin volume), and can be rearranged and dragged. Links are scaled according to both basin volume and the jump probability. Arrows indicate the direction of jumps. Short stubs are self-jumps. When jump-graph links are eliminated, this becomes a layout-graph with mobile nodes where basins can be redrawn at node positions.

8.5.3 Attractor Jump-Graph and Layout-Graph

The stability of basins of attraction is especially relevant to the stability of cell types in biology, following the genetic regulatory network approach [5, 11].

DDLab provides an analysis of the basin of attraction field, tracking where all possible 1-bit flips (or 1-value flips) to attractor states end up, whether to the same or to another basin. The information is presented in two ways, as a jump-table – a matrix showing the jump probabilities between basins, and graphically as the attractor jump-graph – a graph with weighed vertices and edges giving a graphic representation of the jump-table (Fig. 8.32).

The jump-graph can be analyzed and manipulated in many ways, and rearranged and unraveled, including dragging vertices and defined components to new positions with "elastic band" edges; the same methods as for the network-graph (Sect. 8.4.3).

Basins of attraction can be redrawn within the jump-graph nodes (Fig. 8.32). If the jump links are eliminated, this becomes a layout-graph (Fig. 8.31) where dragging/rearranging nodes provides an alternative flexible method for positioning basins – the resulting graphics can be saved as vector PostScript – for example, Figs. 8.7, 8.10, and 8.15 (center).

Fig. 8.33 Mutant basins of attraction of the $[v, k] = [2, 3]$ rule 60 ($n = 8$, seed all 0s). *Top left*: The original rule, where all states fall into just one very regular basin. The rule was first transformed to its equivalent $k = 5$ rule (f00ff00f in hex), with 32 bits in its rule table. All 32 one-bit mutant basins are shown. If the rule is the genotype, the basin of attraction can be seen as the phenotype [0.02 seconds]

8.5.4 Mutation

As well as on-the-fly changes to presentation, a wide variety of on-the-fly network "mutations" can be made.

When running forward, key-press options allow mutations to wiring, rules, and current state. A number of "complex" CA rules (with glider interactions) are provided as files with DDLab, and these can be activated on-the-fly.

When attractor basins are complete, a key-press will regenerate the attractor basin of a mutant network. Various mutation options can be preset including random bit/value-flips in rules and random rewiring of a given number of wires. Sets of states can be specified and highlighted in attractor basins to see how mutations affect their distribution. The complete set of one-bit mutants of a rule can be displayed together as in Fig. 8.33.

8.5.5 Learning and Forgetting

Learning and forgetting algorithms allow attaching and detaching sets of states as predecessors of a given state by automatically mutating rules or altering wiring couplings [18, 20]. This allows "sculpting" the basin of attraction field to approach a desired scheme of hierarchical categorization, the network's content addressable memory. Because any such change, especially in a small network, usually has significant side effects, the methods are not good at designing categories from scratch, but are useful for fine-tuning a network that is already close to where it is supposed to be.

More generally, a very preliminary method for reverse engineering a network, also known as the inverse problem, is included in DDLab, by pruning the connections in a fully connected network to satisfy an exhaustive map (for network sizes $n \leq 13$). The inverse problem is to find a minimal network that will satisfy a full or partial mapping, fragments of attractor basins, or trajectories.

8.5.6 Sequential Updating

By default, network updating is synchronous, in parallel. DDLab also allows sequential updating, both for space-time patterns and attractor basins. Sequential orders are forward, backward, or a random order, but any specific order can be set from the $n!$ possible sequential orders for a network of size n. The order can be saved/loaded from a file.

An algorithm in DDLab computes the neutral order components (limited to network size $n \leq 12$). These are sets of sequential orders with identical dynamics. DDlab treats these components as sub-trees generated from a root order and can generate a single component sub-tree or the entire set of components sub-trees making up sequence space (the neutral field) which are drawn in an analogous way to attractor basins [21].

DDLab can also set a sequential partial order, where cells are allocated to groups which update sequentially in a given order, but with synchronous updating within each group. This allows the updating to be tuned between fully synchronous and fully sequential.

Setting an update probability can also achieve asynchronicity, where only a part of the network, randomly assigned, updates at each time-step.

8.5.7 Vector PostScript of Graphic Output

Graphic output in DDLab can be captured as vector PostScript, which is preferable for publication quality images over bitmap graphics, as they can be scaled without degrading the resolution. Many of the figures in this chapter are vector images; others are bitmap graphics captured with a screen grabber. At the time of writing, vector PostScript files are available for the following: space-time patterns in 1D and 2D, attractor basins of all kinds, the network-graph, attractor jump-graph, and layout-graph.

Vector PostScript files are ASCII files in the PostScript language and so can be edited for changes and additions. If selected, functions in DDLab generate these files automatically according to the the current presentation of the DDLab graphic, but also with various options for the PostScript image itself such as color or gray scale. There is no limit on the size of a PostScript file, so for large attractor basins with many links and nodes, proceed with caution to avoid excessively large files. If an attractor basin is interrupted, the PostScript file of part of the graphic generated up to that point will be valid.

8.5.8 Filing Network Parameters, States, and Data

Network parameters and states can be saved and loaded for the following: k-mix, wiring schemes, rules, rule schemes, wiring/rule schemes, and network states. Data on attractor basins, at various levels of detail, can be automatically saved. A file of "exhaustive pairs," made up of each state and its successor, can be created.

Various data including mean entropy and entropy variance of space-time patterns can be automatically generated and saved, allowing a sorted sample of CA rules to be created, discriminating between order, complexity, and chaos [23], as in Fig. 8.37. A large collection of complex rules, those featuring "gliders" or other large-scale emergent structures, can be assembled. Pre-assembled files of CA rules sorted by this method are provided with DDLab.

8.6 Measures and Data

DDLab provides static measures for rules, wiring and various quantitative, statistical, and analytical measures, and data on both space-time patterns and attractor basin topology. The measures and data are shown graphically in most cases as well as numerically.

Section 8.4 listed connectivity measures from the wiring graphic. Other measures are listed in the following subsections.

8.6.1 Static Rule Measures

DDLab provides various static measures depending just on the rule-table, or rule-tables in mixed-rule networks. These measures or parameters are generalized for multi-value.

- The λ-parameter [9] which measures the density of non-quiescent values in the rule-table. λ approximates the Z-parameter.
- The λ-ratio [16] equivalent to λ, but defined differently.
- The P-parameter [4] equivalent to λ for binary systems, but defined differently again.

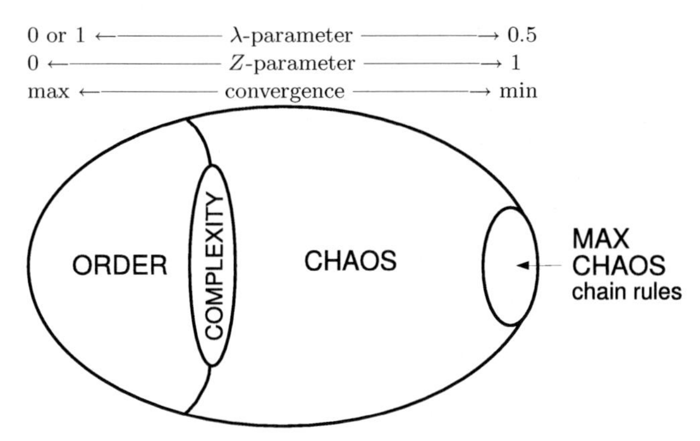

Fig. 8.34 A view of rule-space after Langton and his λ-parameter [9]. Tuning the Z-parameter from 0 to 1 moves the dynamics from maximum to minimum convergence, from order to chaos, traversing a phase transition where complexity might be found. The chain-rules, added on the right, are maximally chaotic and have the very least convergence, decreasing with system size, making them suitable for dynamical encryption.

- The Z-parameter [16], a probability measure, which predicts the convergence, bushiness, of sub-trees in attractor basins, thus order/chaos in the dynamics (Fig. 8.34).
- The (weighted) average λ and Z in a rule-mix.
- Canalizing inputs [6, 4] and the frequency of canalizing "genes" and inputs, which can be tuned to required values and displayed graphically. Canalizing inputs induce order in RBN/DDN (Fig. 8.38).
- Post functions [12], which also induce order in RBN, and where the dynamics are closed under composition.

Single rules or a rule-mix can be tuned to adjust any of these measures to any arbitrary level.

8.6.2 Measures on Space-Time Patterns

Some measures on space-time patterns are listed below:

- The rule-table look-up-frequency histogram in a moving window of time-steps, and its input-entropy plot (Fig. 8.16). This is the basis of the method for automatically filtering space-time patterns [23] (Figs. 8.16 and 8.30).
- The space-time color density in a moving window of time-steps (Fig. 8.35).
- The variance or standard deviation of the entropy, and an entropy/density scatter plot [23], where complex rules have their own distinctive signatures (Fig. 8.36).

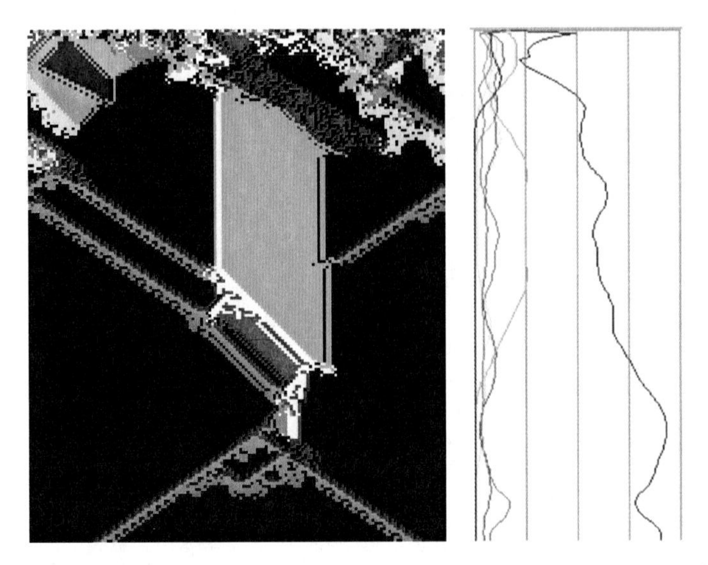

Fig. 8.35 *Left*: A 1D CA space-time pattern, from a random initial state, of an Altenberg k-totalistic rule $[v, k] = [8, 7]$, $n = 150$, where the probability of a rule-table output depends on the fraction of colors in its neighborhood, resulting inevitably in emergent structure. *Right*: The color density plotted for each of the 8 colors, relative to a moving window of 10 time-steps.

- A scatter plot of mean entropy against the standard deviation (or an alternative "min-max" measure) of the entropy for an arbitrarily large sample of CA rules, which allows ordered, complex, and chaotic rules to be classified automatically [21, 23], also shown as a 2D frequency histogram (Fig. 8.37). Ordered, complex, and chaotic dynamics are located in different regions, allowing a statistical measure of their frequency. The rules can be sorted by the various measures, allowing complex rules to be found automatically.
- Various methods for showing the activity/stability of network cells, or frozen "genes" [4].
- The damage-spread [21], or pattern-difference, between two networks, in 1D or 2D. A histogram of damage-spread frequency can be automatically generated for identical networks with initial states differing by 1 bit.
- The Derrida plot [4, 6], and Derrida coefficient, analogous to the Liapunov exponent in continuous dynamical systems, which measures how pairs of network trajectories diverge/converge in terms of their Hamming distance, indicating if a RBN/DDN is in the ordered or chaotic regime (see Fig. 8.38).
- A scatter plot of successive iterations in a 2D phase plane, the "return map" (Fig. 8.39), which has a fractal structure for chaotic rules.

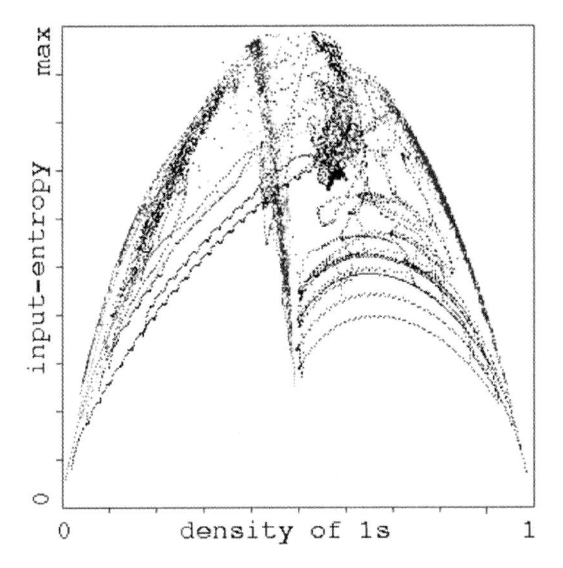

Fig. 8.36 Entropy/density scatter plot [23]. Input-entropy is plotted against the density of 1s relative to a moving window of 10 time-steps. Plots for a number of $[v,k] = [2,5]$ complex rules ($n = 150$) are show superimposed in different colors, each of which has its own distinctive signature, with a marked vertical extent (i.e., high entropy variance). About 1000 time-steps are plotted from several random initial states for each rule.

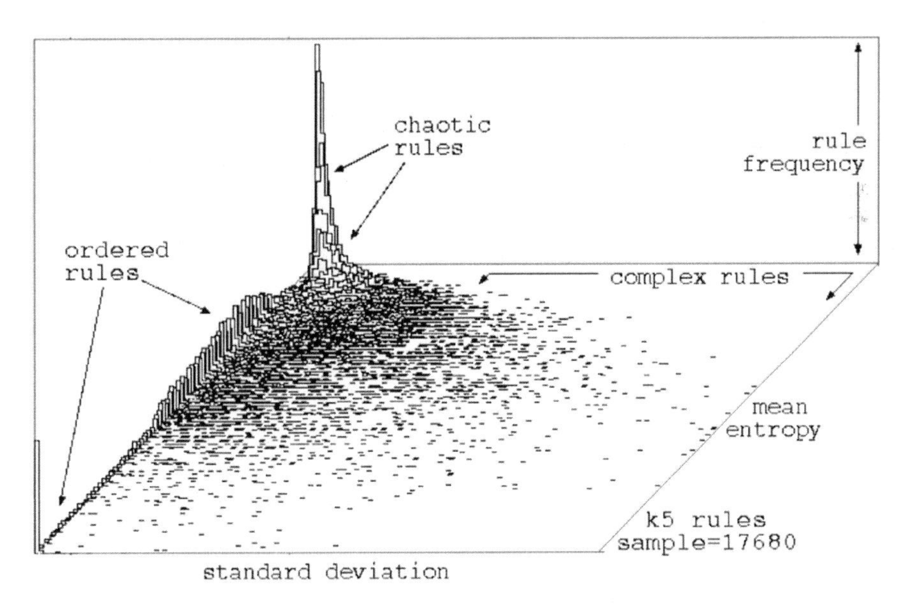

Fig. 8.37 Classifying a random sample of $[v,k] = [2,5]$ rules automatically by plotting mean entropy against the standard deviation of the entropy. The 2D histogram shows the frequency of rules falling within a 128×128 grid. The rules can be sorted, and their space-time patterns can be selectively examined.

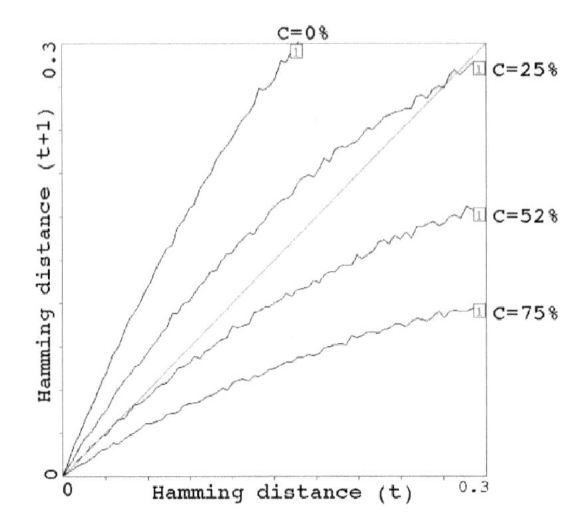

Fig. 8.38 Derrida plots for random Boolean networks ($[v, k] = [2, 5]$, $n = 36 \times 36$). This is a statistical measure of how pairs of network trajectories diverge/converge in terms of their Hamming distance. A curve above the main diagonal indicates divergence and chaos, below the diagonal – convergence and order. A curve tangential to the diagonal indicates balanced dynamics. This example shows four plots where the the percentage of canalizing inputs in the randomly biased network is 0%, 25%, 52%, and 75%, showing progressively greater order, with the 52% curve balanced at the edge of chaos.

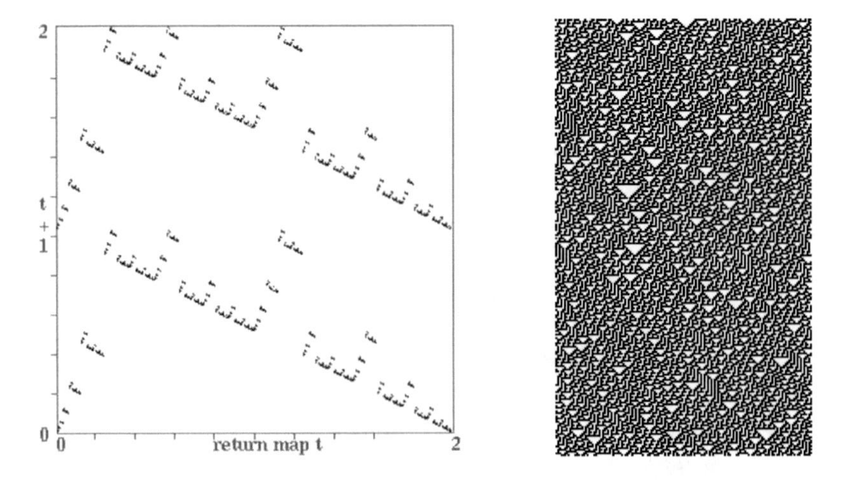

Fig. 8.39 *Left*: The return map for the space-time patterns of the binary 1D $k = 3$ rule 30, which is a maximally chaotic chain-rule, $n = 150$. The space-time pattern was run for about 10,000 time-steps. Note the fractal structure. Each state (bit-string) $B_0, B_1, B_2, B_3, \ldots, B_{n-1}$ is converted into a decimal number 0–2 as follows: $B_0 + B_1/2 + B_2/4 + B_3/8 + \cdots + B_{n-1}/2^{n-1}$. As the network is iterated, this value at time-step t (x-axis) is plotted against the value at time-step t+1 (y-axis). *Right*: The space-time pattern itself, showing 250 time-steps.

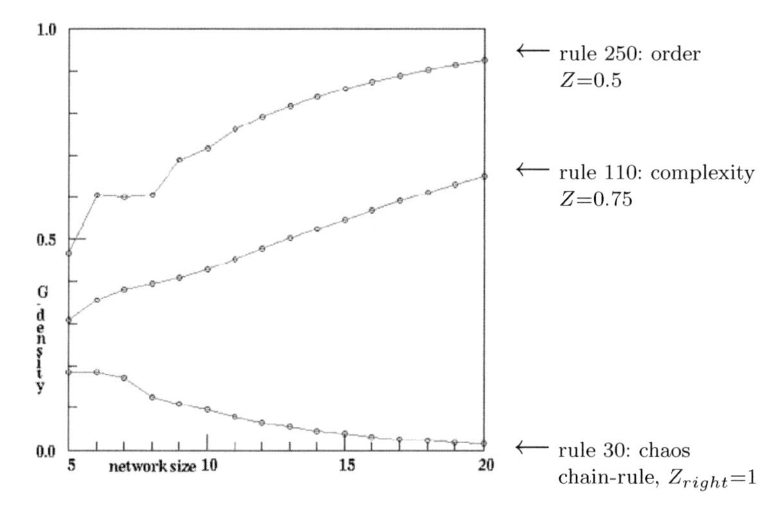

Fig. 8.40 A plot of leaf (Garden-of-Eden) density with increasing system size, for an ordered, complex, and maximally chaotic chain-rule, $n = 5$ to 20. The measures are for the basin of attraction field, and so for the entire state-space. For rule-space in general, leaf-density increases with greater n, but at a slower rate for increasing Z, but uniquely, leaf-density decreases for chain-rules (Sect. 8.8) where either $Z_{left} = 1$ or $Z_{right} = 1$, but not both.

8.6.3 Measures on Attractor Basins

Some measures on sub-trees, basins of attraction, and the basin of attraction field are listed below:

- The number of basins in the basin of attraction field, their size, attractor period, and branching structure of transient trees. Details of states belonging to different basins, sub-trees, their distance from attractors or the sub-tree root, and their in-degree.
- A histogram showing the frequency of arriving at different attractors from random initial states. This provides statistical data on the basin of attraction field for large networks. The number of basins, their relative size, period, and the average run-in length are measured statistically. The data can be used to automatically generate an attractor jump-graph as in Fig. 8.32. An analogous method shows the frequency of arriving at different "skeletons," consisting of patterns that have frozen to a specified degree.
- Garden-of-Eden or leaf-density plotted against the λ and Z parameters, and against network size as in Fig. 8.40.
- A histogram of the in-degree frequency in attractor basins or sub-trees.

- The state-space matrix, a plot of the left half against the right half of each state bit string, using color to identify different basins, or attractor cycle states.
- The attractor jump-graph (Fig. 8.32) as described in Section 8.5.3.

8.7 Reverse Algorithms

There are three different reverse algorithms for generating the pre-images of a network state. These have all been generalized for multi-value networks.

- An algorithm for 1D CA, or networks with 1D CA wiring but mixed rules.
- A general algorithm for RBN/DDN, which also works for the above.
- An exhaustive, brute force, algorithm that works for any "random mapping," including the two cases above.

The first two reverse algorithms generate the pre-images of a state directly; the speed of computation decreases with both neighborhood size k and network size. The speed of the third, exhaustive, algorithm is largely independent of k, but is especially sensitive to network size.

The method used to generate pre-images will be chosen automatically, but can be overridden. For example, a regular 1D CA can be made to use either of the two other algorithms for benchmark purposes and for a reality check that all methods agree. The time taken to generate attractor basins is displayed in DDLab. For the basin of attraction field, a progress bar indicates the proportion of states in state space used up so far.

The 1D CA algorithm [16, 23] is the most efficient and fastest. Furthermore, compression of 1D CA attractor basins by rotation symmetry speeds up the process [16].

Any other network architecture, RBN or DDN, with non-local wiring, will be handled by the slower *general* reverse algorithm [17, 23]. A histogram revealing the inner workings of this algorithm can be displayed (Fig. 8.41).

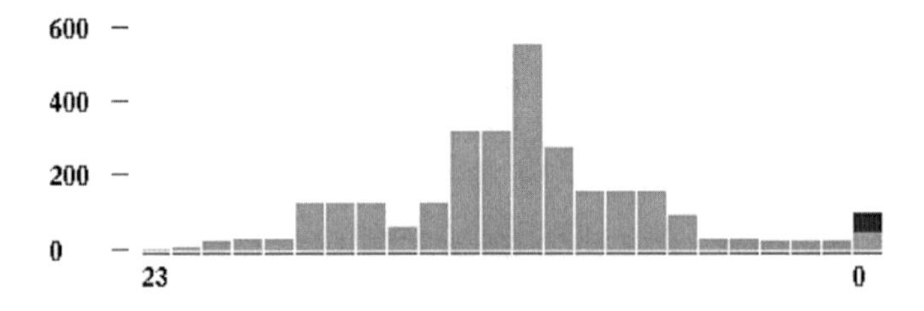

Fig. 8.41 Computing RBN pre-images. The changing size of a typical partial pre-image stack at successive elements. $n = 24$, $k = 3$. This histogram can be automatically generated for a look at the inner workings of the RBN/DDN reverse algorithm.

Regular 2D or 3D CA will also use this general reverse algorithm. Compression algorithms come into play in orthogonal 2D CA to take advantage of the various rotation symmetries on the torus.

The third, brute-force, exhaustive reverse algorithm first sets up a mapping, a list of "exhaustive pairs," of each state in state space and its successor (this can be saved/loaded). The pre-images of states are generated by reference to this list. The exhaustive method is restricted to small systems because the size of the mapping increases exponentially as v^n, and scanning the list for pre-images is slow compared to the direct reverse algorithms for CA and DDN. However, the method is not sensitive to increasing neighborhood size k and is useful for small but highly connected networks. The exhaustive method is also used for sequential and partial order updating.

A random mapping routine can assign a successor to each state in state space, possibly with some bias. Attractor basins can then be reconstructed by reference to this random map with the exhaustive algorithm. The space of random maps for a given system size corresponds to the space of all possible basin of attraction fields and is the super-set of all other deterministic discrete dynamical networks.

8.8 Chain-Rules and Dynamical Encryption

The CA reverse algorithm is especially efficient for a sub-set of maximally chaotic 1D CA rules, the "chain-rules," which can be automatically generated in DDLab for any $[v, k]$. These are rules based on special values of the Z-parameter, where either Z_{left}=1 or Z_{right}=1, but not both [16, 28]. The approximate number of chain-rules for binary CA is the square root of rule-space $2^{2^{k-1}}$.

Chain-rules are special because in contrast to the vast majority of rule-space (Fig. 8.34), the typical number of predecessors of a state (in-degree) is extremely low, *decreasing* with system size. For larger systems, the in-degree is likely to be exactly 1. Consequently, the leaf-density is also very low, also decreasing with system size (Fig. 8.40), becoming vanishingly small in the limit. This means nearly all states have predecessors, embedded deeply along chain-like transients.

Large 1D CA can be therefore run backward very efficiently with chain-rules, generating a chain of predecessors. As the rules rapidly scramble patterns, they allow a method of encryption [28] which is available in DDLab; run backward to encrypt, and forward to decrypt (Figs. 8.42 and 8.43). When decrypting, the information pops out suddenly from chaos and merges back into chaos. A chain-rule (of which there is a virtually inexhaustible supply) is the cypher key and, to a lesser extent, the number of backward steps.

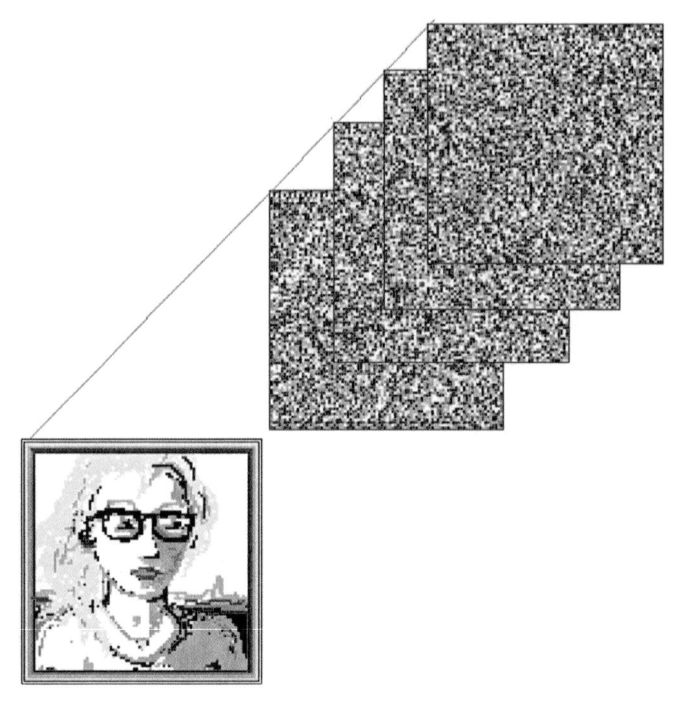

Fig. 8.42 The portrait was made with the drawing function in DDLab (as Fig. 8.25). However, this is a 1D pattern displayed in 2D, $n = 7744$, 88×88. As $v = 8$ there are 3 bits per cell, so the size of the bit string=23,232, which could be an ASCII file, or any other form of information. With a $[v, k] = [8, 4]$ chain-rule constructed at random, and the portrait as the root state, a sub-tree was generated with the CA reverse algorithm and set to stop after four backward time-steps. The sub-tree has no branching, and branching is highly unlikely to occur. The state reached is the encryption [50.5 second].

Fig. 8.43 To decrypt, starting from the encrypted state in Fig. 8.42, the CA with the same chain-rule was run forward to recover the original image. This figure shows time-steps -2 to $+7$, to illustrate how the portrait was scrambled both before and after time-step 4.

8.9 DDLab Website, Manual, and Reviews

DDlab continues to be developed with updates at irregular intervals. A new manual for the multi-value version is in the process of being written, and parts are already posted on DDLab's website. This, together with the existing manual, will provide a useful guide.

The latest information, downloads, manuals, updates, examples, lecture slides, galleries, reviews, publications, and preprints can be found in [25].

Some reviews of DDLab and "The Global Dynamics of Cellular Automata" [16]:

- John E. Myers in *Complexity*, Vol. 3, No. 1, Sept./Oct. 1997.
- Andrew Adamatzky in *Kybernetes*, Vol. 28, No. 8 and 9, 1999.
- Stuart Kauffman in *Complexity* Vol. 5, No. 6, July/Aug. 2000.
- H. Van Dyke Parunak in JASSS, *The Journal of Artificial Societies and Social Simulation*, Vol. 4, Issue 4, Oct. 2001.

Acknowledgements Many people have influenced DDLab by contributing ideas, suggesting new features, providing encouragement, criticism, and helping with programming. I reserve all the blame for its shortcomings. I would like to thank Mike Lesser, Grant Warrel, Crayton Walker, Chris Langton, Stuart Kauffman, Wentian Li, Pedro P.B. de Oliveira, Inman Harvey, Phil Husbands, Guillaume Barreau, Josh Smith, Raja Das, Christian Reidys, the late Brosl Hasslacher, Steve Harris, Simon Frazer, Burt Voorhees, John Myers, Roland Somogyi, Lee Altenberg, Andy Adamatzky, Mark Tilden, Rodney Douglas, Terry Bossomaier, Ed Coxon, Oskar Itzinger, Pietro di Fenizio, Pau Fernandez, Ricard Sole, Paolo Patelli, Jose Manuel Gomez Soto, Dave Burraston, and many other friends and colleagues (to whom I apologize for not listing); also DDLab users who have provided feedback, and researchers and staff at the Santa Fe Institute in the 1990s.

This chapter is dedicated to the memory of
Brosl Hasslacher

References

1. Adamatzky A (1994) Identification of Cellular Automata. Taylor and Francis, London.
2. Burraston D and Martin A (2006) Digital Behaviors and Generative Music, Wild Nature and the Digital Life, Special Issue, Leonardo Electronic Almanac 14(7–8), http://leoalmanac.org/journal/Vol_14/lea_v14_n07-08/dburastonamartin.asp
3. Greenberg JM and Hastings SP (1978) Spatial patterns for discrete models of diffusion in excitable media, SIAM J. Appl. Math. 34(3):515–523.
4. Harris SE, Sawhill BK, Wuensche A, and Kauffman SA, (2002) A model of transcriptional regulatory networks based on biases in the observated regulation rules, Complexity 7(4):23–40.
5. Kauffman SA (1969) Metabolic stability and epigenesis in randomly constructed genetic nets, Theoretical Biology 22(3):439–467.
6. Kauffman SA (1993) The Origins of Order. Oxford University Press.
7. Kauffman SA (2000) Book and Software Reviews, The global dynamics of cellular automata, by Andrew Wuensche and Mike Lesser, Complexity 5(6):47–48. http://www.cogs.susx.ac.uk/users/andywu/kauffman_rev.html

8. Langton CG (1986) Studying artificial life with cellular automata, Physica D, Vol. 22, 120–149.
9. Langton CG (1990) Computation at the edge of chaos: phase transitions and emergent computation, Physica D, Vol. 42, 12–37.
10. Langton CG (1992) Foreword to The Global Dynamics of Cellular Automata [16]; available at http://www.cogs.susx.ac.uk/users/andywu/gdca.html.
11. Somogyi R and Sniegoski CA (1996) Modeling the complexity of genetic networks: understanding multigene and pleiotropic regulation, Complexity, Vol. 1, 45–63.
12. Shmulevich I, Lahdesmaki H, Dougherty ER, Astola J, and Zhang W (2003) The role of certain Post classes in Boolean network models of genetic networks, Proceedings of the National Academy of Sciences of the USA, 100(19):10734–10739.
13. Wolfram S (1983) Statistical mechanics of cellular automata, Rev. Mod. Phys. 55(3):601–664.
14. Wolfram S (ed.) (1986) Theory and Application of Cellular Automata, World Scientific, Singapore.
15. Wolfram S (2002) A New Kind of Science, Wolfram Media, Champaign, IL.
16. Wuensche A and Lesser MJ (1992) The Global Dynamics of Cellular Automata. Santa Fe Institure Studies in the Sciences of Complexity, Reference Vol 1, Addison-Wesley, Reading, MA. http://www.cogs.susx.ac.uk/users/andywu/gdca.html
17. Wuensche A (1994) Complexity in One-D Cellular Automata: Gliders, Basins of Attraction and the Z parameter, Santa Fe Institute Working Paper 94-04-025.
18. Wuensche A (1994) The ghost in the machine: Basin of attraction fields of random Boolean networks. In: Artificial Life III, ed. Langton CG, Addison-Wesley, Reading, MA, 496–501.
19. Wuensche A (1995) Discrete Dynamics Lab: Cellular Automata – Random Boolean Networks (DDLab DOS manual). ftp://ftp.cogs.susx.ac.uk/pub/users/andywu/dd_older_versions/dd_1995_july/ddlab_10.zip
20. Wuensche A (1996) The emergence of memory: Categorisation far from equilibrium, in Towards a Science of Consciousness: The First Tuscon Discussions and Debates, eds. Hameroff SR, Kaszniak AW, and Scott AC. MIT Press, Cambridge, MA, 383–392.
21. Wuensche A (1997) Attractor basins of discrete networks: Implications on self-organisation and memory. Cognitive Science Research Paper 461, DPhil Thesis, University of Sussex.
22. Wuensche A (1998) Genomic regulation modeled as a network with basins of attraction, in Pacific Symposium on Biocomputing '98, eds. Altman RB, Dunker AK, Hunter, and Klien TE. World Scientific, Singapore, 89–102.
23. Wuensche A (1999) Classifying cellular automata automatically: Finding gliders, filtering, and relating space-time patterns, attractor basins, and the Z parameter. Complexity 4(3):47–66.
24. Wuensche A (2000) Basins of attraction in cellular automata; order-complexity-chaos in small universes, Complexity 5(6):19–25; based on Complexity in Small Universes, in an art exhibition and events Objective Wonder: Data as Art, and Merged Realities Exposition'99, University of Arizona, in collaboration with Chris Langton, details at http://www.cogs.susx.ac.uk/users/andywu/Exh2/Exh3.html.
25. Wuensche A (2001) Discrete Dynamics Lab: the DDLab website with the software, manual and other resources: http://www.ddlab.org and http://www.cogs.susx.ac.uk/users/andywu/ddlab.html.
26. Wuensche A (2005) Glider dynamics in 3-value hexagonal cellular automata: The beehive rule, Int. J. of Unconventional Computing, 1(4):375–398.
27. Wuensche A and Adamatzky A (2006) On spiral glider-guns in hexagonal cellular automata: Activator-inhibitor paradigm, Int. J. of Mod. Phys. C, 17(7):1009–1026.
28. Wuensche A (2008) Encryption using cellular automata chain-rules, in Automata-2008: Theory and Applications of Cellular Automata, eds. Adamatzky A, Alonso-Sanz R, and Lawniczak A. Luniver Press, Beckinton, UK, 126–136.

Chapter 9
EINSTein: A Multiagent-Based Model of Combat

Andrew Ilachinski

Artificial Life techniques – specifically, multiagent-based models and evolutionary learning algorithms – provide a powerful new approach to understanding some of the fundamental processes of war. This chapter introduces a simple artificial "toy model" of combat called EINSTein. EINSTein is designed to illustrate how certain aspects of land combat can be viewed as self-organized, emergent phenomena resulting from the dynamical web of interactions among notional combatants. EINSTein's bottom-up, synthesist approach to the modeling of combat stands in stark contrast to the more traditional top-down, or reductionist, approach taken by conventional military models, and it represents a step toward developing a complex systems theoretic toolbox for identifying, exploring, and possibly exploiting self-organized, emergent collective patterns of behavior on the real battlefield. A description of the model is provided, along with examples of emergent spatial patterns and behaviors.

9.1 Background

"War is... not the action of a living force upon lifeless mass... but always the collision of two living forces."

— Carl von Clausewitz, Prussian military strategist (1780–1831)

In 1914, F. W. Lanchester introduced a set of coupled ordinary differential equations – now commonly called the Lanchester equations (LEs) – as models of attrition in modern warfare [1]. In the simplest case of directed fire, for example, the LEs embody the intuitive idea that one side's attrition rate is proportional to the opposing side's size:

$$\begin{cases} \frac{dR}{dt} = -\alpha_B B(t), & R(0) = R_0, \\ \frac{dB}{dt} = -\alpha_R R(t), & B(0) = B_0, \end{cases} \tag{9.1}$$

where R_0 and B_0 are the initial red and blue force levels, respectively, and α_R and α_B represent the effective firing rates at which one unit of strength on one side causes attrition on the other side's forces. The closed-form solution of these equations is given in terms of hyperbolic functions as

$$\begin{cases} R(t) = R_0 \cosh\left(t\sqrt{\alpha_B \alpha_R}\right) - B_0 \sqrt{\alpha_B/\alpha_R} \sinh\left(t\sqrt{\alpha_B \alpha_R}\right), \\ B(t) = B_0 \cosh\left(t\sqrt{\alpha_B \alpha_R}\right) - R_0 \sqrt{\alpha_R/\alpha_B} \sinh\left(t\sqrt{\alpha_B \alpha_R}\right) \end{cases} \tag{9.2}$$

and satisfies the simple "square-law" state equation

$$\alpha_R \left[R_0^2 - R(t)^2\right] = \alpha_B \left[B_0^2 - B(t)^2\right]. \tag{9.3}$$

Similar ideas were proposed around that time by Chase [2] and Osipov [3]. These equations are formally equivalent to the Lotka–Volterra equations used for modeling the dynamics of interacting predator–prey populations [4]. Despite their relative simplicity, the LEs have since served as the fundamental mathematical models upon which most modern theories of combat attrition are based and are to this day embedded in many "state-of-the art" military models of combat. Taylor [5] provided a thorough mathematical discussion.

On the one hand, there is much to like about the LEs, since they are very intuitive and therefore easy to apply, and they provide relatively simple closed-form solutions. On the other hand, as is typically the case in the more general setting of nonlinear dynamical system theory, knowing the "exact" solution to a simplified problem does not necessarily imply that one has gained a deep insight into the problem. Moreover, almost all attempts to correlate LE-based models with historical combat data have proven inconclusive, a result that is in no small part due to the paucity of data. Most data consist only of initial force levels and casualties, and typically for one side only. Moreover, the actual number of casualties is usually uncertain because the definition of "casualty" varies (killed, killed + wounded, killed + missing, etc.).

Two noteworthy battles for which detailed daily attrition data and daily force levels do exist are the battle of Iwo Jima in World War II and the Inchon-Seoul campaign in the Korean War. While the battle of Iwo Jima is frequently cited as evidence for the efficacy of the classic LEs, the conditions under which it was fought were very close to the ideal list of assumptions under which the LEs themselves are derived. A detailed analysis of the Inchon-Seoul campaign has also proved inconclusive [6]. Weiss [7], Fain [8], Richardson [9], and others analyzed attrition in battles fought from 200 B.C. to World War II.

While the LEs may be relevant for the kind of static trench warfare and artillery duels that characterized most of World War I, they lack the spatial degrees of freedom to realistically model modern combat. They are certainly too simple to adequately represent the more modern vision of combat, which depends on small, highly trained, well-armed autonomous teams working in concert, continually adapting to changing conditions and environments. The fundamental problem is that the LEs idealize combat much in the same way as Newton's laws idealize physics.

The two most significant drawbacks to using LEs to model land combat are that (1) they are unable to account for any spatial variation of forces (no link is established, for example, between movement and attrition) and (2) they do not incorporate the human factor in combat (i.e., the uniquely individual, often imperfect, psychology and decision-making capability of the human soldier.) While there have been many extensions to and generalizations of the LEs over the years, all designed to minimize the deficiencies inherent in their original formulation (including reformulations as stochastic differential equations and partial differential equations), most existing models remain essentially Lanchesterian in nature, the driving factor being force-on-force attrition.

9.2 Land Combat as a Complex Adaptive System

To address all of these shortcomings, the Center for Naval Analyses and the Office of Naval Research are exploring developments in a complex adaptive systems theory – particularly the set of agent-based models and simulation tools developed in the artificial life community – as a means of understanding land warfare in a fundamentally different way.

Military conflicts, particularly land combat, possess the key characteristics of complex adaptive systems (CASs) [10, 11, 12, 13]: Combat forces are composed of a large number of nonlinearly interacting parts and are organized in a command and control hierarchy; local action, which often appears disordered, induces long-range order (i.e., combat is self-organized); military conflicts, by their nature, proceed far from equilibrium; military forces, in order to survive, must continually adapt to a changing combat environment; and there is no master "voice" that dictates the actions of each and every combatant (i.e., battlefield action effectively proceeds according to a decentralized control).

A number of recent papers discuss the fundamental role that nonlinearity plays in combat. See, for example, Beckerman [14], Beyerchen [15], Hedgepeth [16], Ilachinski [17, 18], Miller and Sulcoski [19], Saperstein [20], and Tagarev and Nicholls [21]. The general approach of the EINSTein project is to extend these largely conceptual and general links that have been drawn between properties of land warfare and properties of complex systems into a set of practical connections and analytical research tools.

9.3 Agent-Based Modeling and Simulation

Models based on differential equations homogenize the properties of entire populations and ignore the spatial component altogether. Partial differential equations – by introducing a physical space to account for movement – fare somewhat better, but still treat the agent population as a continuum. In contrast, agent-based models (ABMs) consist of a discrete heterogeneous set of spatially distributed individual agents, each of which has its own characteristic properties and rules of behavior. These properties can also change as agents evolve in time.

Agent-based models of CASs are becoming an increasingly popular exploratory tool in the artificial life community and are predicated on the basic idea that the (often complicated) global behavior of a real system derives, collectively, from simpler, low-level interactions among its constituent agents [22]. Insights about the real-world system that an ABM is designed to model can then be gained by looking at the emergent structures induced by the interactions taking place within the simulation, as well as the feedback that these patterns might have on the rules governing the individual agents' behavior.[1] Agent-based simulations engender a significant shift in the kinds of questions that are asked of the real system being simulated. For example, where traditional models ask, effectively, "How can I characterize the system's top-level behavior with a few (equally top-level) variables?" ABMs instead ask, *"What low-level rules and what kinds of heterogeneous, autonomous agents do I need to have in order to synthesize the system's observed high-level behavior?"*

Perhaps the most important benefit of using an agent-based simulation to gain insight into why a system behaves the way it does – whether that system is a collection of traders on the stock market floor, neurons in a brain, or soldiers on the battlefield – is that once the simulation is used to generate the desired behavior, the researcher has immediate and simultaneous access to both the top-level (i.e., generated) behavior of the system and a low-level description of the system's underlying dynamics. Because they take an actively generative, or synthesist, approach to understanding a system, from the bottom up, ABMs are thus a powerful methodological tool for not just describing behaviors but also explaining why specific behaviors occur. While an analytical solution may provide an accurate description of a phenomenon, it is only with an agent-based simulation that one can fine-tune one's understanding of the precise set of conditions under which certain behaviors emerge.

---

[1] Two excellent recent texts on agent-based modeling, as applied to a variety of disciplines, are by Ferber [23] and Weiss [24]. Collections of papers focusing on systems that involve aspects of "human reasoning" are by Gilbert and Troitzsch [25], Gilbert and Conte [26], and Conte et al. [27]. More recently, ABMs have been applied successfully to traffic pattern analysis [28] and social evolution [29, 30].

In the context of modeling combat, agent-based simulations represent a fundamental shift from focusing on simple force-on-force attrition calculations to considering how complex, high-level properties and behaviors of combat emerge out of (sometimes coevolving) low-level rules of behaviors and interactions. The final outcome of a battle – as defined, say, by measuring the surviving force strengths – takes second stage to exploring how two forces might coevolve as a series of firefights and skirmishes unfold. ABMs are designed to allow the user to explore the evolving patterns of macroscopic behavior that result from the collective interactions of individual agents, as well as the feedback that these patterns might have on the rules governing the individual agents' behavior.

9.4 EINSTein

EINSTein (*Enhanced ISAAC Neural Simulation Tool*) is an adaptive ABM of combat and is an outgrowth of a more far-reaching project to develop a complexity-based fundamental theory of warfare [31]. EINSTein [32] builds upon and extends an earlier proof-of-concept, DOS-based combat simulator called ISAAC (*Irreducible Semi-Autonomous Adaptive Combat*), which was developed for the US Marines Corps [33]. All approved-for-public-release documents, project reports and summaries, tutorials, sample runs, and an auto-install program for Windows-based PCs may be downloaded from [70]. Details of the EINSTein toolkit are provided in [34].

EINSTein represents one of the first systematic attempts, within the military operations research community, to simulate combat – on a small to medium scale – by using autonomous agents to model individual behaviors and personalities rather than specific weapons. Because agents are all endowed with a rudimentary form of "intelligence," they can respond to a very large class of changing conditions as they evolve during battle. Because of the relative simplicity of the underlying dynamical rules, EINSTein can rapidly provide outcomes for a wide spectrum of tunable parameter values defining specific scenarios and can thus be used to effectively map out the space of possible behaviors.

Fundamentally, EINSTein addresses the basic question: *"To what extent is land combat a self-organized emergent phenomenon?"* Or, more precisely, "What are the conditions under which high-level patterns (such as penetration, flanking maneuvers, attack, etc.) emerge from a given set of low-lying dynamical primitive actions (move forward, move backward, approach/retreat-from enemy, etc.)." As such, EINSTein's intended use is not as a full system-level model of combat but as an interactive toolbox – or "conceptual playground" – in which to explore high-level emergent behaviors arising from various low-level (i.e., individual combatant and squad-level) interaction rules. The idea behind developing this toolbox is emphatically not to model in detail a specific piece of hardware (an M16 rifle or M101 105mm howitzer, for example). Instead, the idea is to explore the middle ground between – at one extreme – highly realistic models that provide little insight into basic processes and – at the other extreme – ultraminimalist models that strip away all but the simplest dynamical variables and leave out the most interesting real behavior that is, to explore the fundamental dynamical trade-offs among a large number of notional variables.

The underlying dynamics is patterned after mobile cellular automata rules [35] and are somewhat reminiscent of Braitenberg's vehicles [36]. Mobile cellular automata have been used before to model predator–prey interactions in natural ecologies [37]. They have also been applied to combat modeling [38], but in a much more limited fashion than the one used by EINSTein.

9.4.1 Features

EINSTein's major features include the following:

- Dialog-driven I/O, using a Windows graphical user interface (GUI) front-end
- Object-oriented C++ source code base
- Integrated natural terrain maps and terrain-based adaptive decisions
- Context-dependent and user-defined agent behaviors
- Multiple squads, with intersquad communication links
- Local and global command-agent dynamics
- Genetic algorithm toolkit to tailor agent rules to desired force-level behaviors
- Data collection and multidimensional visualization tools
- Mission fitness-landscape profilers
- Over 250 user-programmable functions on the source code level

Fig. 9.1 provides a screenshot of a typical run-session in EINSTein. The screenshot contains three active windows: main battlefield view (which includes passable and impassable terrain elements), trace view (which shows color coded territorial occupancy), and combat view (which provides a gray-scaled filter of combat intensity). All views are simultaneously updated during a run. Toward the right-hand side of the screenshot appear two data dialogs that summarize red and blue agent parameter values. Appearing on the lower left side and along the bottom of the figure are time-series graphs of red and blue center-of-mass coordinates (as measured from the red flag) and the average number of agents within red and blue agent's sensor ranges, and a dialog that allows the user to define communication relays among individual squads.

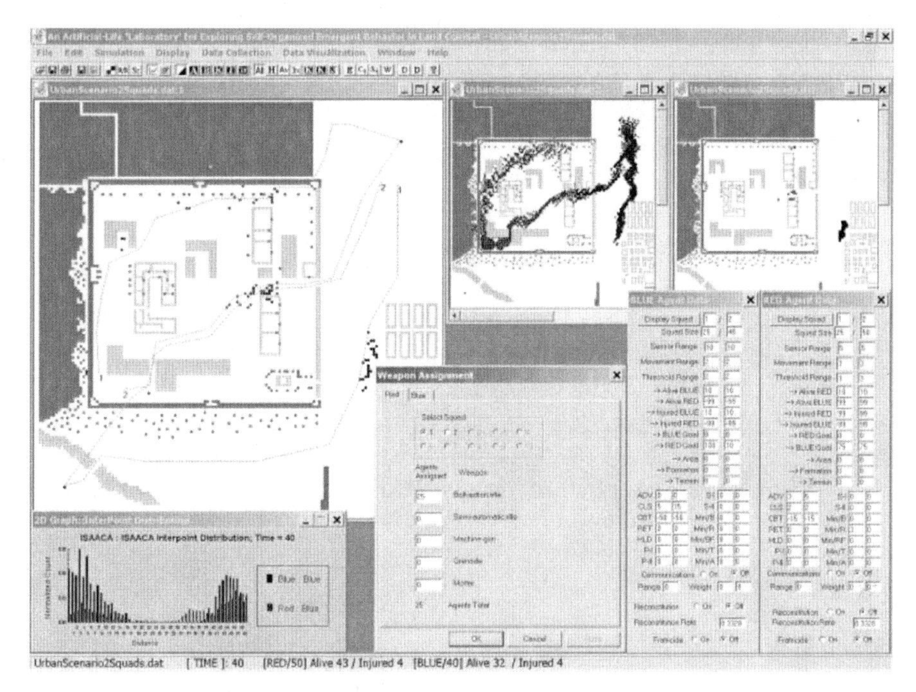

Fig. 9.1 Screenshot of EINSTein's GUI front-end.

9.4.2 Source Code

EINSTein is written and compiled using Microsoft's Visual C++ and makes use of Pinnacle Publishing Inc.'s Graphics Server[2] for displaying time-series plots and three-dimensional (3D) fitness-landscapes. EINSTein consists of roughly 100K lines of object-oriented source code.

The source code is divided into three basic parts: (1) the *combat engine* (parts of which are summarized below); (2) the *graphical user interface*; and (3) the *data-collection/data-visualization functions*. These parts are essentially machine (i.e., CPU and/or operating system) independent and may be compiled separately. EINSTein's source code base is thus highly portable and is relatively easy to modify to suit particular problems and interests. For example, an EINSTein-based combat environment may be developed as a stand-alone program on a CPU platform other than the original MS Windows target machine used for EINSTein's original development. Any developer/analyst interested in porting EINSTein over to some other machine and/or operating system is tasked only with providing his own machine-specific GUI as a "wrap-around" to the stand-alone combat and data-visualization engines (that may be provided as dynamic-link libraries – DLLs). Moreover, it is very

[2] Graphics Server is a commercial plug-in, licensed from Pinnacle Publishing, Inc. [68].

easy to add, delete, and/or change the existing source code, including making complicated changes that significantly alter how agents decide their moves.

At the heart of EINSTein lies the combat engine (discussed later). The combat engine processes all run-time, combat-related logical decisions and is the core script upon which multiple time-series data collection, fitness-landscape sweeps over the agents' parameter space, and genetic algorithm searches all depend.

9.4.3 Design Philosophy

"Things should be as simple as possible, but not simpler." — Albert Einstein

EINSTein's design is predicated upon two guiding principles: (1) to keep all dynamical components and rules as simple as possible (with a view toward optimizing the trade-off between run time and realism) and (2) to treat all forms of information (and the way in which all information is processed locally by agents) in a contextually consistent manner. The meaning of this second principle will become clear in the exposition below.

9.4.3.1 Simplicity

The first guiding principle is to keep things *simple*. Specifically, EINSTein is designed to make it as intuitive as possible for the user to program specific agent behaviors. This is done by deliberately keeping the set of combat and movement rules small and by defining those rules as simply as possible. Thus, the power projection rule is essentially "target and fire upon any enemy agent within a threshold fire range" rather than some other, more complicated (albeit, possibly more physically realistic) prescription. The idea is to qualitatively probe the behavioral consequences of the interaction among a large number of notional variables, not to provide an explicit detailed model of the minutiae of real-world combat.

9.4.3.2 Consistency

The second guiding principle is keep things *consistent*. All dynamical decisions – whether they are made by individual agents, by local or global commanders, or by the user (when scripting a scenario's objectives) – consist of boundedly rational (i.e., locally optimal) penalty assessments. Actions are based on an agent's *personality* (see later), which consists of numerical weights that attach greater or lesser degrees of relative importance to each factor relevant to selecting a particular move in a given local context (from the point of view of a given agent). It is in this sense that all forms of information, on various levels, are treated on a consistent basis.

The decisions taking place on different levels of the simulation all follow the same general template of probing and responding to the environment. Each decision consists of a personality-mediated "answer" to the following three basic questions:

- What are my immediate and long-term goals?
- What do I currently know about my local environment?
- What must I do to attain my goals?

As we will see in detail ahead, at the most primitive level, each agent cares only about "moving toward" or "moving away from" all other agents and/or his own and the enemy's flag. An agent's personality prescribes the relative weight assigned to each of these immediate "goals." On the other hand, a global commander must weigh such features as overall force strength, casualty rate, rate of advance, and so on in order to attain certain long-term goals. Local and supreme commanders have their own unique concerns. While the actual decisions are different in each case and on each information level – for example, an individual agent's decision to "stay put" in order to survive is quite different and uses a different form of information, from a global commander's drive to "get to the enemy flag as quickly as possible" – the general manner in which these decisions are made is the same.

9.4.4 Program Flow

A typical sequence of programming steps during an interactive run consists of multiple loops through the following basic steps:

1. *Initialize battlefield and agent distribution parameters.*
2. *Initialize time-step counter.*
3. *Adjudicate combat.*
4. *Refresh battlefield graphics display.*
5. *Find context-dependent personality weight vector for each red and blue agent.*
6. *Compute local penalty function to determine best move.*
7. *Move agents to their newly selected position (or leave them where they are).*
8. *Refresh graphics display and loop through steps 3–7.*

The most important parts of this skeletal structure are the adjudication of combat, the adaptation of personality weights, and the decision-making process that each agent goes through in choosing its next move. Before describing the details of what each of these steps involves, we must first discuss how each agent partitions its local information. During interactive runs (i.e., whenever the fitness-landscape profiler and genetic algorithm breeder batch modes are both inactive), the user can pause the simulation at any time to make on-the-fly changes to any, or all, agent parameters (including adding or subtracting "playing" agents) and then resume the run with the changed values.

9.5 Combat Engine

9.5.1 Agents

The basic element of EINSTein is an agent, which loosely represents a primitive combat unit (infantryman, tank, transport vehicle, etc.) that is equipped with the following characteristics:

- *Doctrine:* a default local-rule set specifying how to act in a generic environment
- *Mission:* goals directing behavior
- *Situational awareness:* sensors generating an internal map of environment
- *Adaptability:* an internal mechanism to alter behavior and/or rules

Each agent exists in one of three states: alive, injured, or killed. Injured agents can (but are not required to) have different personalities and offensive/defensive characteristics from when they were alive. For example, the user can specify that injured agents are able to move half as far, and shoot half as accurately, as their "alive" counterparts. Up to 10 distinct groups (or "squads") of personalities, of varying sizes, can be defined. The user can also specify how agents from one squad react to agents from other squads.

Each agent has associated with it a set of ranges (sensor range, fire range, communications range, etc.), within which it senses and assimilates various forms of local information, and a personality, which determines the general manner in which it responds to its environment. A global rule set determines combat attrition, reconstitution, and (in future versions) reinforcement. EINSTein also contains both local and global commanders, each with their own command radii and obeying an evolving command-and-control (C2) hierarchy of rules.

9.5.2 Battlefield

The putative combat battlefield is represented by a two-dimensional lattice of discrete sites. Each site of the lattice may be occupied by one of two kinds of agents: red or blue. The initial state consists of user-specified formations of red and blue agents positioned anywhere within the battlefield. Formations may include squad-specific bounding rectangles or may be completely random. Red and blue flags are also typically (but not always) positioned in diagonally opposite corners. A typical goal, for both red and blue agents, is to reach the enemy's flag.

EINSTein includes an option to add terrain elements. Terrain can be either impassable or passable. If passable, the user can also tune an agent's behavior to a particular terrain type. For example, if an agent is positioned within "heavy brush," its movement range and visibility (from other nearby agents) may be curtailed.

9.5.3 Agent Personalities

Each agent is equipped with a user-specified personality – or internal value system – nominally defined by a six-component personality weight vector, $\mathbf{w} = (w_1, w_2, \ldots, w_6)$, where $-1 \leq w_i \leq 1$ and $\Sigma_i |w_i| = 1$. The components of $\mathbf{w}$ specify how an individual agent responds to specific kinds of local information within its sensor range.

The personality weight vector may be health dependent; that is, w_{alive} need not, in general, be equal to w_{injured}. The components of $\mathbf{w}$ can also be negative – in which case they signify a propensity for moving away from, rather than toward, a given entity.

9.5.4 Penalty Function

An agent's personality weight vector is used to rank each possible move according to a penalty function. The simplest penalty function effectively measures the total distance that the agent will be from other agents (including both friendly and enemy agents) and from its own and enemy flags, weighing each component distance by the appropriate component of the personality weight vector, $\mathbf{w}$. An agent moves to the position that incurs the least penalty; that is, an agent's move is the one that best satisfies its personality-driven desire to "move closer to" or "farther away from" other agents in given states and either of the two flags. The general form of the penalty function is given by:

$$Z(B_{x,y}) = \frac{1}{\sqrt{2}r_S} \begin{bmatrix} \frac{\omega_{AF}}{N_{AF}} \sum_{i \in AF}^{N_{AF}} D_{i,B_{x,y}} + \frac{\omega_{AE}}{N_{AE}} \sum_{j \in AE}^{N_{AE}} D_{j,B_{x,y}} \\ + \frac{\omega_{IF}}{N_{IF}} \sum_{i \in IF}^{N_{IF}} D_{i,B_{x,y}} + \frac{\omega_{IE}}{N_{IE}} \sum_{j \in IE}^{N_{IE}} D_{j,B_{x,y}} \end{bmatrix} \tag{9.4}$$

$$+\omega_{FF} \frac{D^{new}_{FF,B_{x,y}}}{D^{old}_{FF,B_{x,y}}} + \omega_{EF} \frac{D^{new}_{EF,B_{x,y}}}{D^{old}_{EF,B_{x,y}}},$$

where $B_{x,y}$ is the (x, y) coordinate of battlefield B; AF, IF, AE, and IE represent respectively the sets of alive friends, injured friends, alive enemies, and injured enemies within the given agent's sensor range, r_S; w_i are the components of the personality weight vector; $\sqrt{2}r_S$ is a scale factor; N_X is the total number of elements of type X within the given agent's sensor range (e.g., N_F is the number of alive friends within range r_S); $D_{A,B}$ is the distance between elements A and B; FF and EF denote the friendly and enemy flags, respectively; and represent distances computed using the given agent's new (candidate move) position and old (current) position, respectively.

A penalty is computed for each possible move; that is, for each of the $N = (2r_m + 1)^2$ possible sites to which an agent can "step" in one time step: $Z_1(B_{x,y})$, $Z_2(B_{x+1,y})$, $Z_3(B_{x-1,y}), \ldots, Z_N(B_{x+n,y+n})$. The actual move is the one that incurs the least penalty. If there is a set of moves (consisting of more than one possible move) all of whose penalties are within $\epsilon_{\text{Penalty}} \geq 0$ of the minimal penalty, an agent randomly selects the actual move among the candidate moves making up that set. Users can also define paths near which agents must try to stay while maneuvering toward their ultimate goal.

The penalty function shown above includes only a few relative-proximity-based weights. In practice, the penalty function is more complicated and incorporates more terms, although its basic form is the same. Additional terms can include the propensity for maintaining the minimum distance from friendly or enemy agents, staying near a designated patrol area, the cost of traversing terrain, desire for finding local cover (from fire) and/or concealment (from enemy sensors), and combat intensity (see Table 9.1).

9.5.5 Meta-rules

An agent's personality may be augmented by a set of meta-rules that tell it how to alter its default personality according to dynamic environmental contexts. A typical meta-rule consists of altering a few of the components of an agent's personality vector according to a set of associated local threshold constraints. The three simplest meta-rule classes effectively define the

Weight	Meaning = Relative Weight for...
w_{AF}	...moving toward/away from alive friendly agents
w_{IF}	...moving toward/away from injured friendly agents
w_{AE}	...moving toward/away from alive enemy agents
w_{IE}	...moving toward/away from injured enemy agents
w_{FF}	...moving toward/away from friendly flag
w_{EF}	...moving toward/away from enemy flag
w_{BB}	...moving toward/away from the boundary of battlefield
w_{area}	...staying near some (squad-specific) area
w_{squad}	...maintaining formation with own squad-mates
$w_{\mathrm{fire-team}}$	...maintaining formation with own fireteam-mates
S_{ij}	...how agents from squad S_i react to agents from squad S_j
SS'_{ij}	...how agents from squad S_i react to agents from enemy squad S_j
w_{LC}	...moving toward/away from local commander
w_{obeyLC}	...obeying orders issued by local commander
w_{terrain}	...moving toward/away from terrain elements
$w_{\mathrm{enemy-fire}}$	...moving toward/away from enemy agents that have fired on agent

Table 9.1 A partial list of EINSTein's *primitive weight* set

local conditions under which an agent is allowed to *advance toward enemy flag* (class 1), *cluster with friendly forces* (class 2), and *engage the enemy in combat* (class 3).

For example, a class-1 meta-rule prevents an agent from advancing toward the enemy flag unless it is locally surrounded by a threshold number of friendly agents; that is, it is a notional indicator of local combat support. A class-2 meta-rule can be used to prevent an agent from moving toward friendly agents once it is surrounded by a threshold number. Finally, a class-3 meta-rule can be used to fix the local conditions under which an agent is allowed to move toward or away from possibly engaging an enemy agent in combat. Specifically, an agent is allowed to engage an enemy if and only if the difference between friendly and enemy force strengths locally exceeds a given threshold.

Other meta-rule classes include *retreat*, *pursuit*, *support*, and *hold position*. A global rule set determines combat attrition (see later), communication, reconstitution, and (in future versions) reinforcement. EINSTein also contains both local and global commanders, each of which is equipped with its own unique command-personality and area of responsibility, and obeys an evolving command and control hierarchy of rules. Table 9.2 summarizes some of EINSTein's meta-rules.[3]

[3] Note that threshold constraints (τ_{Advance}, τ_{Cluster}, and Δ_{Combat}) are explicitly defined only for the first three meta-rules. These meta-rules are used in the sample runs discussed later. In fact, each of the meta-rules appearing in Table 9.2 has one or more threshold constraints associated with it, and the set also requires additional logic to dynamically resolve ambiguities as they arise during the course of a run. Details are in [31].

Meta-rule	Description
w_{AF}	...moving towards/away from alive friendly agents
Advance	Advance to enemy flag if the number of friends $\geq \tau_{\text{Advance}}$
Cluster	Stop seeking friends if number of friends $\geq \tau_{\text{Cluster}}$
Combat	Engage enemy if the $N_{\text{friends}} - N_{\text{enemies}} \geq \Delta_{\text{Combat}}$
Hold	Hold current position
Pursuit-I	Temporarily turn off pursuit of enemy agents
Pursuit-II	Temporarily turn exclusive pursuit on
Retreat	Retreat toward own flag
Run Away	Run away, fast, from enemy agents
Support-I	Provide support for nearby injured
Support-II	Seek support from nearby friends
Min-D Friend	Maintain minimum distance from all friendly agents
Min-D Enemy	Maintain minimum distance from all enemy agents
Min-D Flag	Maintain minimum distance from all friendly flags
Min-D Terrain	Maintain minimum distance from terrain
Min-D Area	Maintain minimum distance from a fixed area on battlefield

Table 9.2 A partial list of EINSTein's *meta-rule* set

9.5.6 Combat

During the combat phase of an iteration step for the whole system, each agent X (on either side) is given an opportunity to fire at all enemy agents Y_i that are within a fire range r_F of X's position. If an agent is shot by an enemy agent, its current state is degraded either from alive to injured or from injured to dead. Once killed, an agent is permanently removed from the battlefield. The probability that a given Y_i is shot is fixed by user-specified single-shot probabilities. Weapons are assigned to individual agents and are either point-to-point (i.e., rifles) or area destruction (i.e., grenades).

By default, all enemy agents within a given agent's fire range are targeted for a possible hit. However, the user has the option of limiting the number of enemy targets that can be engaged simultaneously. If this option is selected and the number of enemy agents within an agent's fire-range exceeds a user-defined threshold number (say N), then N agents are randomly chosen among the agents in this set. Grenades include additional targeting logic (to maximize expected inflicted damage on the enemy).

This basic combat logic may be enhanced by three additional functions: (1) *defense*, which adds a notional ability to agents to be able to withstand a greater number of "hits" before having their state degraded; (2) *reconstitution*, which adds a provision for previously injured agents to be reconstituted to their alive state; and (3) *fratricide* ("friendly fire"), which adds an element of realism by making it possible to inadvertently hit friendly forces.

9.5.7 Run Modes

EINSTein can be run in three basic modes (see EINSTein's *User's Guide* [31]):

- *Interactive mode*, in which the combat engine is run interactively using a fixed set of rules. This mode, which allows the user to make on-the-fly changes to the values of any (or all) parameters defining a given run, is particularly well suited for playing simple *"What if?"* scenarios. The interactive mode also makes it easy to search for interesting emergent behavior.
- *Data-collection mode*, in which the user can (1) generate time series of various changing quantities describing the step-by-step evolution of a battle and (2) keep track of certain measures of how well mission objectives are met at a battle's conclusion. Additionally, the user can generate behavioral profiles on two-dimensional slices of EINSTein's N-dimensional parameter space.
- *Genetic algorithm "breeder" mode*, in which a genetic algorithm is used to breed an agent force that is optimally suited for performing a specific mission against a fixed enemy force. This mode is designed to suggest ways in which ABMs may eventually be used to evolve real-world tactics and strategies.

9.6 Sample Patterns and Behavior

EINSTein possesses a large repertoire of emergent behaviors: *forward advance*, *frontal attack*, *local clustering*, *penetration*, *retreat*, *attack posturing*, *containment*, *flanking maneuvers*, and *"Guerrilla-like" assaults*, among many others. Moreover, behaviors frequently arise that appear to involve some form of intelligent division of red and blue forces to deal with local firestorms and skirmishes, particularly those forces whose personalities have been bred (via a genetic algorithm) to perform a specific mission. It is important to point out that such behaviors are not hard-wired but are, rather, an emergent property of a decentralized, but dynamically interdependent, swarm of agents.

Fig. 9.2 shows screen captures of spatial patterns resulting from 16 different rules and illustrates the diversity of behaviors that emerges out of a relatively simple set of rules. (Note that the sample patterns shown here are for clashing red and blue forces consisting of a *single* squad. Multisquad scenarios, in which agents belonging to different squads obey different rules, and interact with one another according to an additional layer of micro-rules, often result in considerably more complicated emergent behaviors.) An important long-term goal is for EINSTein to be flexible enough to serve as a general tool (that transcends the specific notional combat environment to which it is obviously tailored) for exploring the still very poorly understood mapping between micro-rules and emergent macro-behaviors in complex adaptive systems.

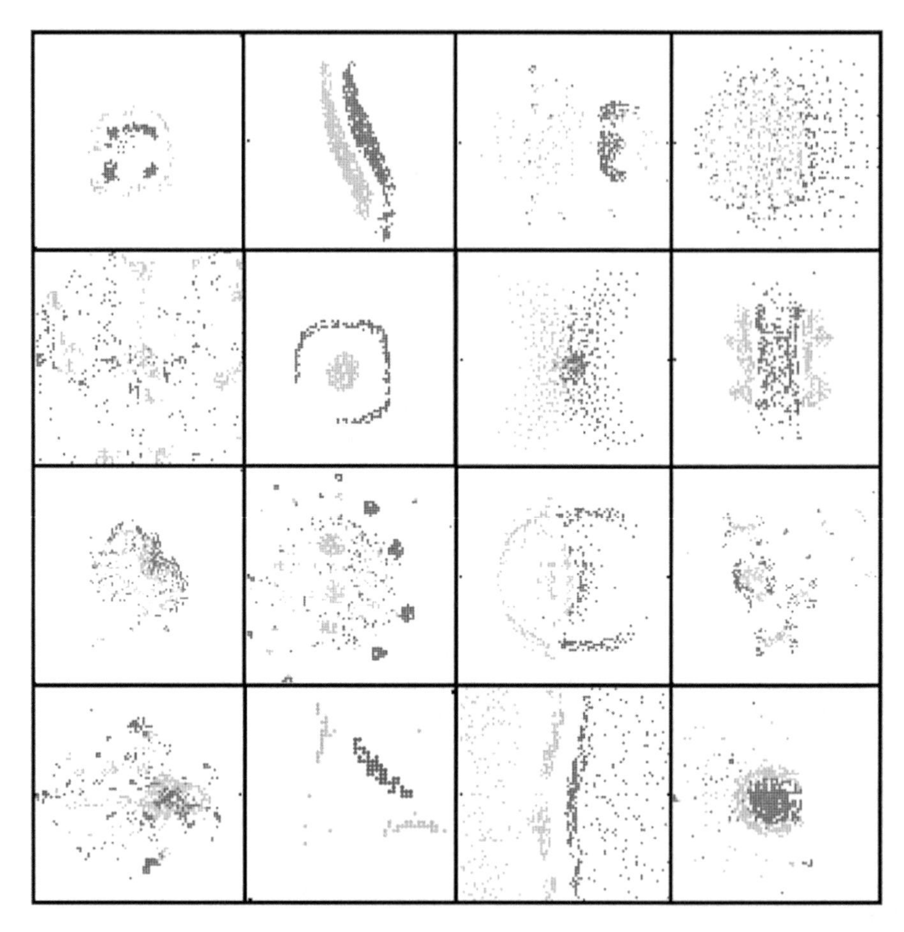

Fig. 9.2 A sampling of emergent spatial patterns of agents obeying EINSTein's micro-rules. Each of the 16 squares represents a different rule and contains a single snapshot of a typical run.

9.6.1 Qualitative Classes of Behavior

Simulations run for many different scenarios and initial conditions suggest that EINSTein's collective behavior generally falls into one of six broad qualitative classes (labeled, suggestively, according to different kinds of fluid flow):

- *Laminar flow*, which typically consists of one (or, at most, a few) well-defined "linear" battlefronts. This class is so named because it is visually suggestive of laminar fluid flow of two fluids and is reminiscent of static trench warfare in World War I. Laminar rules can actually be divided into two types of behaviors, characterized according to a system's overall stability (i.e., according to whether the system is stable, or not stable, to initial conditions).
- *Viscous flow*, in which the unfolding battle typically consists of a single tight cluster (or, at most, a few clusters) of interpenetrating red and blue agents.
- *Dispersive flow*, in which – as soon as red and blue agents maneuver within view of the opposing side's forces – the battle unfolds as a single, explosive, dispersion of forces. Dispersive systems exhibit little, if any, of the "front-like" linear structures that form for laminar-flow rules.
- *Turbulent flow*, in which combat consists of either spatially distributed, but otherwise confined and/or clustered individual combat zones, or a series of close-to space-filling local firestorms. In either case, there is almost always a significant degree of local maneuvering.
- *Autopoeitic flow*, in which agents self-organize into persistent dissipative structures. These formations typically maintain their integrity for long times (on the scale of individual agents entering and leaving the structure) and undergo "higher-level" maneuvering, including longitudinal motion and rotation.[4]
- *Swarming*, in which agents self-organize into nested swarms of attacking and/or defending forces.

We should be quick to point out that this taxonomy is neither complete nor well defined, in a mathematical sense. Because of the qualitative distinctions among classes, there is considerable overlap among them. Moreover, a given scenario, as it unfolds in time, usually consists of several phases of behavior during which one class predominates at one time and other classes at other times. Indeed, for such cases, which occur frequently, it is of considerable interest to understand the nature of the transition between distinct behavioral phases. For example, the initial stages of a scenario may unfold in typically laminar fashion and suddenly transition over into a turbulent phase.

A finer distinction among these six classes can be made on the basis of a more refined statistical analysis of emergent behavior. There is strong evidence to suggest, for example, that while attrition rates for certain classes of rules display smooth Gaussian statistics, other classes (overlapping with viscous-flow and turbulent-flow rules) display interesting fractal power-law scaling behaviors [40]. Insofar as the "box-counting" fractal dimension [41] is useful for describing the degree of agent clustering on the battlefield, it

---

[4] *Autopoiesis* refers to dynamical systems that are simultaneously self-creating and self-maintaining. It was introduced as an explanatory mechanism within biology by Maturana and Varela [39].

can also be used as a simple discriminant between laminar and turbulent classes of behavior. Measuring temporal correlations in the time series of various statistical quantities describing combat is also useful in this regard. The case studies presented here are selected mainly to highlight the qualitative behavioral classes described previously.

9.6.2 Lanchesterian Combat

On the simplest level, EINSTein is an interactive, exploratory tool that allows users to take conceptual excursions away from Lanchesterian oversimplifications of real combat. It is therefore of interest to first define a Lanchesterian scenario within EINSTein that can subsequently be used as a test bed to which the outcomes of other, non-Lanchesterian, scenarios can be compared. The set of simulation parameters that are appropriate for simulating a maneuverless, Lanchester-like combat scenario in EINSTein includes a red/blue movement range of $r_m = 0$ (so that the position of all agents is fixed) and a red/blue sensor range that is large enough so that all agents have all enemy agents within their view (for the example below, $r_S = 40$).

Fig. 9.3 shows several snapshots of a typical run. Initial conditions consist of 100 red and 100 blue agents (in a tightly packed block formation, with block-centers 15 units distant on a 60-by-60 battlefield) and a red/blue single-shot probability of hit $P_{\text{hit}} = 0.005$. Note that the outcome of the battle is a function of the initial sizes of red and blue forces and P_{hit} alone and does not depend on maneuver or any other agent, squad, or force characteristics.

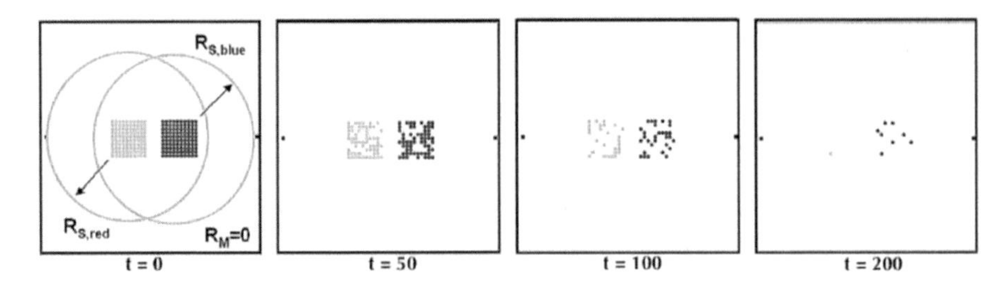

Fig. 9.3 Screenshots of a typical run using an EINSTein rule-set that approximates LE-like combat.

While the Lanchester scenario shown here is highly unrealistic, of course, it is important to remember that most conventional military models (even those that include some form of maneuvering) adjudicate combat by effectively sweeping over a series of similarly idealized maneuverless skirmishes until one side, or both sides, of the conflict decide to withdraw after sustaining a threshold number of casualties. Most models are still almost entirely attrition driven. The only substantive role that maneuver and adaptability play is in getting the individual combatants into position to fight.

A typical signature of such Lanchesterian-like combat scenarios is a linear dependence of the *mean attrition rate* – defined as the average number of combatants lost, $\langle \alpha \rangle$, during some specified time interval, $\Delta\tau = t - t_0$ – on the *single-shot kill* (or, in our case here, *single-shot hit*) probability, P_{ss}:

$$\langle \alpha \rangle = \left\langle \frac{\Delta n}{\Delta \tau} \right\rangle = \left\langle \frac{n(t_0 + t) - n(t_0)}{\Delta \tau} \right\rangle = \sum_{i=1}^{N} P_{ss}(i) = N P_{ss}, \tag{9.5}$$

where N is the total number of agents, $n(t)$ is the number of agents at time t, $P_{ss}(i)$ is the single-shot hit probability of the ith agent, and we have assumed, for the final expression on the right, that $P_{ss}(i) = P_{ss}$ for all i.

What happens if agents are allowed to maneuver? If the maneuver is in any sense "intelligent" (i.e., if agents react reasonably intelligently to changing levels of combat intensity as a battle unfolds), intuitively we should not expect the same linear dependence between $\langle \alpha \rangle$ and P_{ss} to hold. In the extreme case of infinitely timid combatants that run away at the slightest provocation, no fighting at all will occur. In the case where one side applies sophisticated targeting algorithms to maximize enemy casualties but minimize friendly casualties, we might expect a marked increase in that force's relative fighting ability.

To illustrate these ideas, consider an "explosive skirmish" scenario, which is characterized by a rapid, explosive burst of agents as they collide and maneuver close-in during a series of local firefights and skirmishes as the battle slowly dissipates (see screenshots in Fig. 9.4). Table 9.3 lists the parameters values used for these runs.

	Sample 1		Sample 2	
Parameter	Red	Blue	Red	Blue
Agents	250	250	100	100
r_S	5	5	5	5
r_F	3	3	3	3
r_M	2	2	1	1
ω_1	10	10	10	10
ω_2	40	40	40	25
ω_3	10	10	10	10
ω_4	40	40	40	25
ω_5	0	0	0	0
ω_6	25	25	5	50
$\tau_{Advance}$	3	3	3	3
$\tau_{Cluster}$	8	3	5	8
Δ_{Combat}	−99	−3	−10	−3

Table 9.3 Parameter values used for scenarios shown in Fig. 9.4

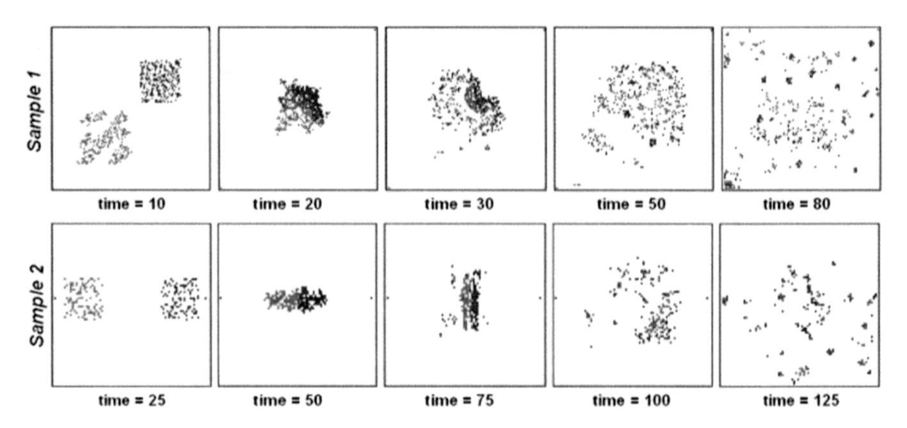

Fig. 9.4 Screenshots of typical runs of the "explosive skirmish" scenario using parameters given in Table 9.3.

9.6.2.1 Fractal Dimensions of Spatial Dispositions

If one were to plot attrition rate, $\langle\alpha\rangle$, versus single-shot probability of hit, P_{ss} – for either of the two "explosive skirmish" scenarios shown in Fig. 9.4 – one would find that $\langle\alpha\rangle \propto P_{ss}^n$, where $n \approx 1/2$. Moreover, careful analysis of the spatial patterns, as they emerge in multiple runs, suggests that what lies at the core of a certain class of non-Lanchesterian scenarios is a fractal scaling of combat forces.

Recall that fractal dimensions – such as the *capacity* dimension (or "box-counting" dimension) – are measures that provide useful structural (and/or, in the case of *information* dimension, statistical information) about a given point set. Fractals are geometric objects characterized by some form of self-similarity; that is, parts of a fractal, when magnified to an appropriate scale, appear similar to the whole. Fractals are thus objects that harbor an effectively infinite amount of detail on all levels. Coastlines of islands and continents and terrain features are approximate fractals. A magnified image of a part of a leaf is similar to an image of the entire leaf. Strange attractors also typically have a fractal structure. Loosely speaking, a fractal dimension specifies the minimum number of variables that are needed to specify an object. For a one-dimensional line, for example, say the x-axis, one piece of information, the x-variable, is needed to specify any position on the line. The fractal dimension of the x-axis is said to be equal to 1. Similarly, two coordinates are needed to specify a position on a two-dimensional plane, so that the fractal dimension of a plane is equal to 2. Fractals are objects whose fractal dimension is noninteger-valued.

How might fractals relate specifically to combat? Intuitively, since real combat consists of anything but a series of random skirmishes in which opposing sides constantly shoot at each other, we expect attrition data to contain spa-

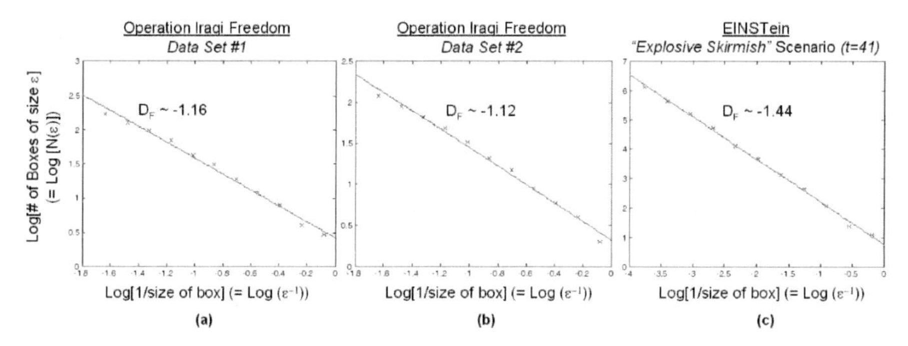

Fig. 9.5 Single-time estimates of D_F for two spatial dispositions of real-world forces and a snapshot of notional forces as arrayed in EINSTein. The (x, y) locations of the real-world data are the longitudes and latitudes of Coalition ground forces during *Operation Iraqi Freedom* in 2003. See the text for additional details.

tiotemporal correlations. Because EINSTein includes rules for maneuver, we expect to see spatiotemporal correlations emerge as a consequence of maneuver (as we would expect to also see in other multiagent-based combat simulations that contain intelligent maneuver).

Spatial distributions of agents on a battlefield are nothing more than sets of abstract points on a lattice. The degree of clustering and maneuver can therefore be measured by calculating a fractal dimension for the distribution in exactly the same way one typically calculates the fractal dimension for point sets representing various *one-* and *two*-dimensional attractors of dynamical systems. The only real difference between using the (x, y) positions of agents and points of an attractor, in practice, is that when dealing with agents, we are naturally limited in the number of "data points" with which we have to work. Even large scenarios are limited to about 500 agents or so to a side. Moreover, agents are allowed to sit only on integer-valued sites, so that the set of possible (x, y)'s is also more limited. Nonetheless, the actual calculation of fractal dimensions proceeds in exactly the same manner for the two cases; just as for continuous dynamical systems, the measures provide an important insight into the geometry of the spatial distributions.

As a concrete example, consider the *capacity* dimension, which is defined by $D_F = \lim_{\epsilon \to 0} \ln [N(\epsilon)] / \ln(1/\epsilon)$, where $N(\epsilon)$ is the number of d-dimensional boxes of side ϵ that contain at least one agent. Alternatively, we can write that $N(\varepsilon) = \varepsilon^{-D_F}$ and call D_F the power-law *scaling exponent*. For a solid block of maneuverless agents, such the solid red and blue blocks of agents dueling it out in the Lanchesterian scenario shown in Fig. 9.3, $D_F \sim 2$; D_F will be *less than* 2 if the agents occupy only a portion of the entire battlefield, as the whole battlefield is be used to estimate D_F.

Fig. 9.5 compares single-time estimates of D_F for (1) two spatial dispositions of real-world forces (Figs. 9.5a and 9.5b) and (2) a snapshot of notional forces as arrayed during the dispersive-flow phase of the explosive skirmish scenario in EINSTein (Fig. 9.5c). The (x, y) locations of the real-world data consists of the longitudes and latitudes of Coalition ground forces during *Operation Iraqi Freedom* in 2003.[5] In each case, the (x, y) locations are first scaled so that the real and notional battlefields assume the same effective "size" of 1-by-1 sites and the scaled battlefield is then divided into a total of $N_{total} = \varepsilon^{-2}$ boxes of length ε. Several different values of ε are chosen, and for each ε, the number of boxes, $N(\varepsilon)$, that contain at least one agent are counted. Since we are limited by how finely we are able to partition the battlefield (as well as by the relatively limited number of agents that define our set of (x, y) locations: ~ 500), the fractal dimension is estimated by the slope of a linear fit on a plot of $\log [N(\varepsilon)]$ versus $\log [1/\varepsilon]$. (Note that we are not in any way suggesting that a finite distribution of either real or notional combatants represents a genuine fractal in a mathematically rigorous sense. We are only suggesting that for limited domains (in battlefield size, duration of conflict, and number of agents) their distribution is such that it can reasonably well be characterized by a fractal-like power-law scaling.)

The point of Fig. 9.5 is not to compare the absolute values of D_F for the different cases – which we could have anticipated as being different, particularly since EINSTein's explosive skirmish example does not intentionally model any real-world scenario – but rather to illustrate the important fact that EINSTein is able to reproduce a spatial fractal scaling *at all* (at least for the limited spatial ranges being considered). In EINSTein, as in the real world, "intelligent maneuvering" implies an agent's position at time t is strongly correlated with local combat conditions. While it has for a long time been known, on the basis of empirical evidence, that real-world forces tend to arrange themselves in self-organized fractal fashion (see [9] and [44]), no satisfactory generative explanation for why this is so has yet appeared.

How does the fractal dimension change during combat? To illustrate the kinds of spatial configurations that can arise in different combat scenarios, consider Fig. 9.6. It shows three plots each (using different initial configurations of agents) of D_F as a function of time for (1) the *Lanchesterian* scenario, (2) a scenario in which the agents are allowed to move but do so completely randomly (and are initially distributed randomly on the battlefield), and (3) the Sample 1 explosive skirmish scenario defined by the parameters appearing in Table 9.3.

These three cases are defined by first using the explosive skirmish scenario to select a single-shot probability of hit, P_{ss}, for which the mean attrition after 100 time steps is equal to 20%. That same value is then used for the other two cases as well. Also, in order to better place the random scenario in between the Lanchester scenario (in which all agents "see" and "fire at" all other agents at all times) and the explosive skirmish scenario (in which agents' sensor and fire ranges are relatively small), agents in the *random* sce-

[5] These scaled geo-locations are from software databases maintained, and kindly provided to the author, by Dr. Michael Shepko and Dr. David Mazel of the Center for Naval Analyses, Alexandria, Virginia.

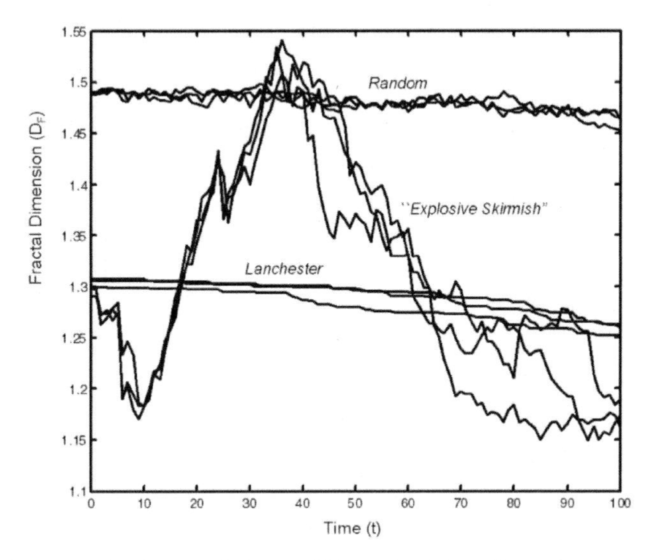

Fig. 9.6 Three sample plots, each (using different initial configurations of agents) of the fractal dimension D_F as a function of time for the *Lanchesterian* (i.e., maneuverless) scenario, a scenario in which the agents are allowed to move but do so completely *randomly*, and the Sample 1 explosive skirmish scenario defined by the parameters appearing in Table 9.3.

nario are assigned sensor and fire ranges equal to three times their value for the explosive skirmish scenario. Because these three cases represent qualitatively different combat scenarios that range from maneuverless, all-seeing/all-shooting agents to intelligently maneuvering agents able to sense, and adapt to, only local streams of information, it is instructive to use them as a basis for comparing simulation output. Except for a small drift of values among different runs for the same scenario and/or differences in precise values at a given time for a given run, the scenarios are each characterized by the unique manner in which the fractal dimension of its associated spatial distribution evolves in time; that is, the time evolution of D_F represents a kind of *behavioral signature* of a given scenario. There is also evidence to suggest a deeper coupling between D_F and combat dynamics.

9.6.2.2 Power-Law Scaling of Attrition

Lauren [40] has used EINSTein (and other ABMs of combat; see [42]) to identify some significant differences between agent-based attrition statistics and results derived from stochastic LE-based models. In particular, he has found evidence to suggest that the *intensity of battles* obeys a fractal power-law dependence on frequency and displays other traits characteristic of high-dimensional chaotic systems, such as fat-tailed probability distributions and intermittency. Specifically, the attrition rate appears to depend on the cube root of the kill probability, which stands in marked contrast to results obtained for stochastic variants of LE-based models, in which, typically, the attrition rate scales linearly with an increase in kill probability.[6] If the ABM more accurately represents real combat processes, an $\sim 1/3$ power-law scaling implies that a relatively "weak" force, with a small kill probability, may actually constitute a much more potent force than a simple LE-based approach suggests. The potency of the force comes from its ability to maneuver (which is never explicitly modeled by LE-based approaches) and to selectively concentrate firepower on the enemy while maneuvering. This deceptively simple result has an important consequence for peacekeeping activities in the Third World, in which a strong, modern force may (and often, does) significantly underestimate the ability of ostensibly poorly trained and/or poorly armed militia to inflict damage.

The appearance of fractal power-law scaling in EINSTein (and other agent-based combat models) is particularly interesting in light of the fact that it has been observed before in real combat [44]. While it has been previously argued, on intuitive grounds, that this must be due to the dynamical coupling between local information processing and maneuver – features that are completely ignored by Lanchesterian models – no generative "explanation" for why fractal power-law scaling appears in combat has heretofore existed. It is therefore tempting to speculate that there are phases of real combat that are poised at *self-organized critical states* (see e.g. [45, 46]).

9.6.3 A Step Away from Lanchester

With an eye toward exploring non-Lanchesterian scenarios, consider an example that includes both simple maneuver and terrain. Fig. 9.7 shows the initial state, consisting of 12 red and 12 blue agents positioned near their respective "flags" (in the lower left and upper right corners, respectively). The red agents are arrayed along a berm (i.e., a permeable terrain element, which appears green in the figure), whose dynamical effect is to reduce their visibility to the approaching blue enemy agents to 15% of the nominal value. As blue agents approach the red flag, red agents remain fixed at their posi-

---

[6] The key observation is that the attrition rate generally depends not just on P_{ss} (as in Eq. 9.5), but on both P_{ss} and D_F, the latter measure representing the spatial distribution of agents [34, 42]. To derive Eq. 9.5, for Lanchesterian combat, one assumes that one side's attrition rate is proportional to the opposing side's size (and *nothing else*); in the general case, one must assume that the attrition rate also depends on the probability that an agent actually "sees" an enemy (or cluster of enemy agents) in a given period of time. The likelihood of this happening, in turn, may be expressed in terms of D_F. Lauren et al. [43] have recently introduced the generalized Lanchester equation $\langle \Delta B/\Delta t\rangle \propto k^{D_F/2} \cdot \Delta t^{D_F/2-1} \cdot R(t)$, where B and R are the number of red and blue agents, k is the rate at which red (or blue) kill blue (or red) agents, and t is the time.

tions (simulating a notional "hunkered-down" condition). The red and blue weapon characteristics (probability of hit and range) are equal.

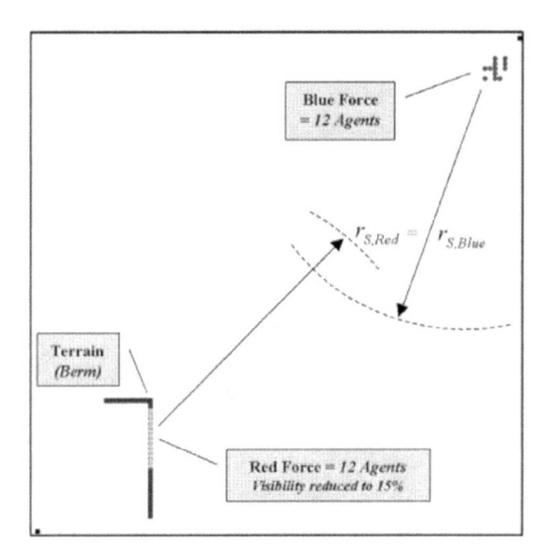

Fig. 9.7 Initial state for simple non-Lanchesterian scenario; see text for details.

Runs typically proceed as follows. Because of the stealth afforded the dug-in red agents by the berm, red agents are targeted and engaged with a much lower probability than the approaching blue force. The attrition of the attacking force (blue) is significantly higher than the attrition of the defending force (red). When the attackers are able to survive (with some of their force intact) – on some particular run of the scenario – it is because they are able to maneuver out of range (which occurs when the force strength drops below the combat effective threshold of 50% and attempts to withdraw) and red is unable to pursue. (As an aside, EINSTein's ability to prescribe retreat conditions adds a certain realism to the model. Faced with mounting attrition, real squads fall back and regroup.)

The red force usually remains at full strength after the engagement (the probability of zero red casualties is about 80%). This result is intuitively satisfying, since, historically (all other factors being equal), defending forces have the advantage over an attacking force traversing open ground. An obvious question to ask is, *"How large must the blue force be in order to overcome the advantage of the red's terrain?"* Fig. 9.8 plots the fraction of the initial forces that remain at the end of the engagement (150 steps) versus the attacker-to-defender force-size ratio (the lines are simple fits to the data to guide the eye). In the runs used to generate this graph, the size of the blue force ranges from 12 to 40 agents, while the red force remains at 12. Note that the red and blue survival curves merge at roughly a 2.8:1 ratio; which is interesting in light of the well-known "rule of thumb" that attackers require a 3:1 force ratio against a defended position [47].

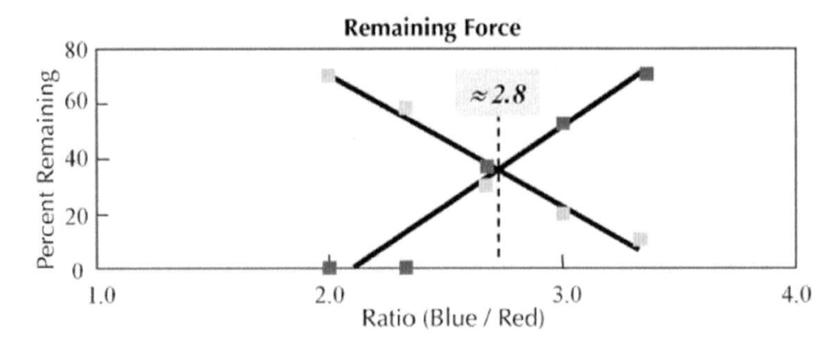

Fig. 9.8 Impact of attacker-to-defender force ratio on survival for the simple non-Lanchesterian scenario shown in Fig. 9.7. The red and blue survival curves merge at about a 2.8:1 ratio, which compares favorably to the well-known "rule of thumb" that attackers require a 3:1 force ratio against a defended position [47].

9.6.4 Swarming Forces

One of the first detailed studies of swarming, as a major theme in military history, was recently conducted by Sean Edwards, as part of the Swarming and the Future of Conflict project at RAND [48]. Edwards' report focuses on 10 carefully selected historical examples of swarming, includes a series of important lessons-learned distilled from these examples about the advantages and disadvantages of swarming, and provides some examples of successful countermeasures that have been used against swarming in the past.

Edwards noted that swarming consists of four overlapping stages: (1) *location*, (2) *convergence*, (3) *attack*, and (4) *dispersion*. Moreover, swarming forces must be capable of a sustainable pulsing; that is, networks of swarming agents must be able to come together rapidly and stealthily on a target, then redisperse, and, finally, recombine for a new pulse:

> The swarm concept is built on the principles of complexity theory, and it assumes that blue units have to operate autonomously and adaptively according to the overall mission statement ... It is important that swarm units converge and attack simultaneously. Each individual swarm unit is vulnerable on its own, but if it is united in a concerted effort with other friendly units, overall lethality can be multiplied, because the phenomenon of the swarm effect is greater than the sum of its parts. Individual units or incompletely assembled groups are vulnerable to defeat in detail against the larger enemy force with its superior fire-power and mass.

The report noted that swarming scenarios have already played a role in certain high-level war-gaming exercises, such as at the *Dominating Maneuver Game*, held at the US Army War College in 1997. Edwards concluded his survey by speculating about the feasibility of a future "swarming doctrine" that would consist of small, distributed, highly maneuverable units converging rapidly on specific targets.

Because of its decentralized rule-base and rich space of behavioral primitives, EINSTein is an ideal test bed with which to explore the nature of battlefield swarming and the efficacy of swarm-like tactics. Typically, but not always, one side appears to swarm the other when there is a significant mismatch in firepower, total force strength, and/or maneuvering ability. (Swarming also occasionally emerges as a useful "tactic" to use against certain opponents when EINSTein's built-in genetic algorithm is tasked with finding optimal attack strategies.) While it is common to find swarm-like behavior for personalities that include large cluster meta-rule thresholds, τ_{Cluster} (which increases the likelihood that agents will remain in close proximity to friendly agents), the most interesting "self-organized" examples of swarming are those for which τ_{Cluster} is, at most, a few agents.

Table 9.4 lists some of the parameter values defining four representative swarm scenarios (I–IV). In scenario I, blue attacks red; in scenario II, blue defends. Blue agents are more aggressive than red in all four scenarios (as defined by the values of their respective combat meta-rule thresholds, Δ_{Combat}). Note that in scenarios II and III defending blue agents are able to communicate with other blue agents that are within a range $r_C = 25$ of their position. Fig. 9.9 show snapshots of typical runs using parameters for scenarios I–IV.

	I	I	II	II	III	III	IV	IV
Force	Red	Blue	Red	Blue	Red	Blue	Red	Blue
Size	150	225	90	125	25	100	200	200
r_S	5	5	5	10	3	7	3	7
r_F	3	3	3	7	2	5	2	5
r_M	1	1	1	2	1	1	1	1
w_{AF}	25	10	10	0	5	0	5	0
w_{AE}	25	50	40	99	40	5	40	5
w_{IF}	75	0	10	0	5	0	5	0
w_{IE}	25	99	40	99	90	50	90	50
w_{FF}	0	0	0	0	0	0	0	0
w_{EF}	75	25	50	0	0	0	0	0
τ_{Advance}	5	1	3	N/A	N/A	N/A	N/A	N/A
τ_{Cluster}	15	3	3	12	5	5	5	5
Δ_{Combat}	5	-7	0	-15	-5	-10	-5	-10
Comms	no	no	no	yes, $r_C = 25$	no	yes, $r_C = 25$	no	no

Table 9.4 Agent parameter values for scenarios I–IV shown in Fig. 9.9

9.6.5 Nonmonotonicity

For a fixed set of force characteristics, number, type, and lethality of weapon systems, and tactics, one might intuitively expect that as one side's capability is unilaterally enhanced – say, by increasing sensor range or its ability to maneuver – the other side's ability to perform its mission ought to be com-

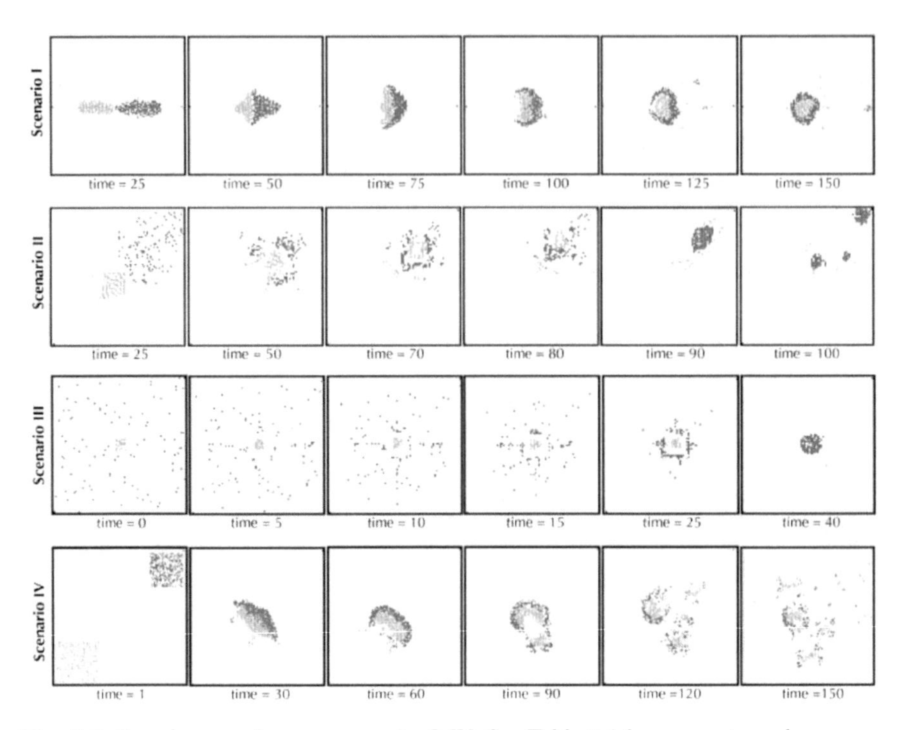

Fig. 9.9 Sample runs of swarm scenarios I–IV. See Table 9.4 for parameter values.

mensurately diminished. In other words, our expectations are that mission success scales monotonically with force capability.

In fact, nonmonotonicities abound in both real-world behavior and simulations. With respect to models and simulations, of course, one must always be on guard against the possibility that nonmonotonic scaling is an artifact of the code and therefore does not represent real processes. As pointed out by a RAND study that addressed this issue [49], "a combat model with a single decision based on the state of the battle ... can produce nonmonotonic behavior in the outcomes of the model and chaotic behavior in its underlying dynamics."

Fig. 9.10 shows an instructive example of genuinely nonmonotonic behavior; genuine in the sense that the nonmonotonicity emerges directly out of the primitive rule set. The three rows in Fig. 9.10 contain snapshots of three separate runs in which red's sensor range is systematically increased in increments of 2: $r_{S,\text{red}} = 5$ for the top sequence; $r_{S,\text{red}} = 7$ for the middle sequence; $r_{S,\text{red}} = 9$ for the bottom sequence. Blue's sensor range, $r_{S,\text{blue}}$, remains fixed at $r_{S,\text{blue}} = 5$ throughout all three runs. The values of other pertinent red and blue agent parameters are given in Table 9.5.

In each of the runs, there are 100 red and 50 blue agents. Red is also the more the aggressive force. Blue engages red in combat if the number of

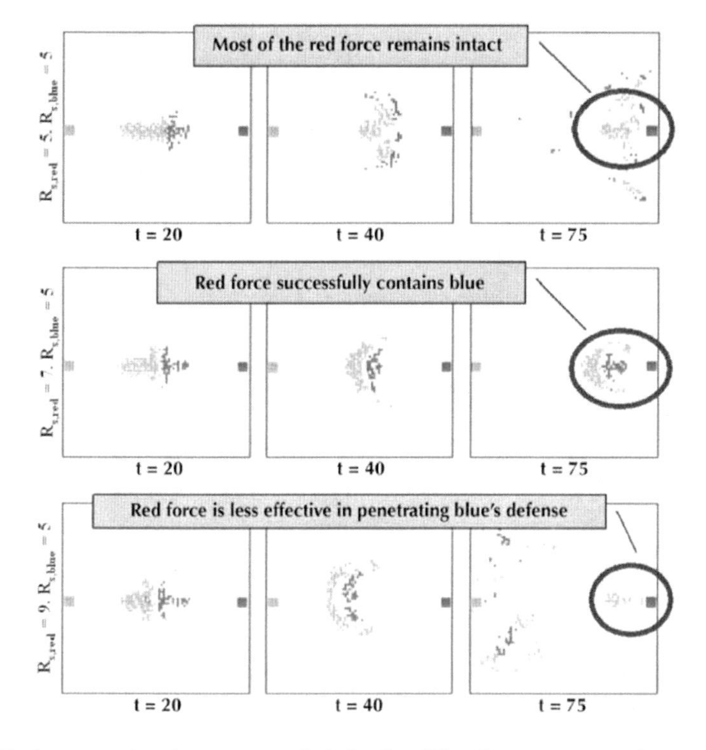

Fig. 9.10 An example of nonmonotonic behavior. The three rows contain snapshots of three separate runs in which red's sensor range is increased in increments of 2 (from $r_{S,\text{red}} = 5$ on the top row to $r_{S,\text{red}} = 9$ on the bottom). Blue's sensor range is fixed at $r_{S,\text{blue}} = 5$ throughout. Comparing the bottom row to the top two rows, we see that increasing red's sensor appears to have a detrimental effect on red's overall ability to penetrate blue's defense.

friendly and enemy agents is locally about even, while red will fight blue even if outnumbered by four enemy combatants. Both sides have the same fire range ($r_F = 4$), and the same single-shot probability ($P_{\text{hit}} = 0.005$) and can simultaneously engage the same maximum of three enemy targets. (Note that the flags for this run are near the middle of the left and right edges of the notional battlefield rather than at the corners.)

The top row of Fig. 9.10 shows screenshots of a run in which red's sensor range is equal to blue's. Here the red force easily penetrates the blue defense as it moves toward the blue flag. During red's advance, a number of agents

	N	r_S	r_F	r_M	$\mathbf{w} = (w_{\text{AF}}, w_{\text{AE}}, w_{\text{IF}}, w_{\text{IE}}, w_{\text{FF}}, w_{\text{EF}})$	τ_{Adv}	τ_{Cluster}	Δ_{Combat}
Red	100	5, 7, 9	4	1	$\mathbf{w}_{\text{Red}} = (10, 90, 10, 50, 0, 99)$	2	4	−4
Blue	50	5	4	1	$\mathbf{w}_{\text{Blue}} = (10, 90, 10, 50, 0, 99)$	2	4	0

Table 9.5 Agent parameter values for *nonmonotonic* run appearing in Fig. 9.10

are "stripped" away from the main red-blue cluster in the center as they respond to the presence of nearby blue agents. The snapshots in the middle row of Fig. 9.10 show that when red's sensor range is two units greater than blue's, red is not only able to mass almost its entire force on the blue flag (by $t = 90$ – not shown – blue's flag is completely enveloped by red forces), but to also defend its own flag from all blue forces as well. In this instance, the red force knows enough about and can respond quickly enough to enemy action such that it is able to march into enemy territory effectively unhindered by enemy forces and "scoop up" blue agents as they are encountered.

What happens as red's sensor range is increased still further? One might intuitively guess that red can only do at least as well, certainly no worse; that is, red's mission performance scales monotonically with the amount of information that each red agent is allowed to have about the engagement. However, as the snapshots for bottom row of Fig. 9.10 reveal, when red's sensor range is increased to $r_{S,\text{red}} = 9$ – *so that all red agents are locally aware of more information* – red, as a force, turns in an objectively weaker mission performance than on the preceding runs. "Weaker" here meaning that red is less effective in (1) establishing a presence near the blue flag and (2) defending blue's advance toward the red flag.

The nonmonotonic behavior is immediately obvious from Fig. 9.11, which shows a 3D fitness landscape for mission objective = *maximize number of red agents near blue flag* (where "near" is defined as anywhere within 10 battlefield units). The landscape sweeps over $r_{S,\text{red}}$ $(= 1, 2, \ldots, 16)$ and the red combat meta-rule threshold Δ_{Combat} $(= -15, -14, \ldots, +15)$. Higher-valued fitness values translate to mean better performance.

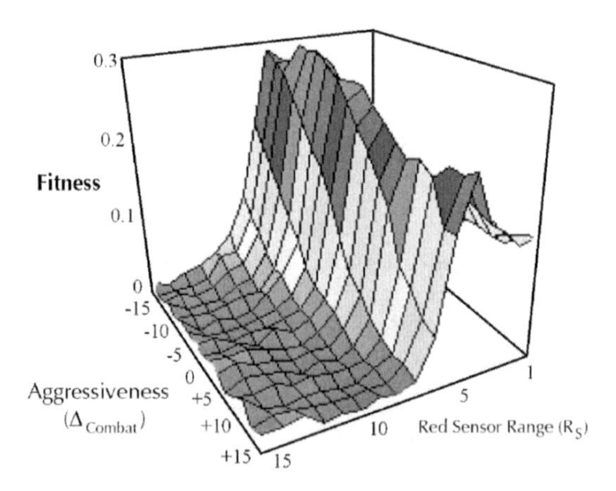

Fig. 9.11 Fitness landscape for mission = *maximize number of red agents near blue flag*, as a function of combat aggressiveness (Δ_{Combat}) and red sensor range ($r_{S,\text{red}}$). Higher-valued fitness values translate to mean better performance. Note that (this particular fitness measure) does not scale monotonically with sensor range.

This example illustrates that when the resources and personalities of both sides remain fixed in a conflict, how well side X does over side Y does not necessarily scale monotonically with X's sensor capability. As one side is forced to assimilate more and more information (with increasing sensor range), there will inevitably come a point when the available resources will be spread too thin and the overall fighting ability will therefore be curtailed. Agent-based models such as EINSTein are well suited for providing insights into more operationally significant questions such as, *"How must X 's resources and/or tactics (i.e., personality) be altered in order to ensure at least the same level of mission performance?"*

9.7 Genetic Algorithm Breeding

One of EINSTein's most powerful built-in features is a genetic algorithm "breeder" run-mode. Genetic algorithms (GAs) are a class of heuristic search methods and computational models of adaptation and evolution based on natural selection. In nature, the search for beneficial adaptations to a continually changing environment (i.e., evolution) is fostered by the cumulative evolutionary knowledge that each species possesses of its forebears. This knowledge, which is encoded in the chromosomes of each member of a species, is passed on from one generation to the next by a mating process in which the chromosomes of "parents" produce "offspring" chromosomes. GAs mimic and exploit the genetic dynamics underlying natural evolution to search for optimal solutions of general combinatorial optimization problems. They have been applied to the traveling salesman problem, VLSI circuit layout, gas pipeline control, the parametric design of aircraft, neural net architecture, models of international security, and strategy formulation [50].

Fig. 9.12 illustrates how GAs are used in EINSTein. Chromosomes define individual agents. Genes encode the components of the personality weight vector, sensor range, fire range, meta-rule thresholds, and so forth. The initial GA population consists of a set of randomly generated chromosomes. The fitness function represents a user-specified mission "fitness" (see later). The target of the GA search is, by default, the red force. The parameter values defining the blue force – once they are defined at the start of a search – are held fixed.

9.7.1 Search Space

EINSTein uses up to 80 genes to conduct a GA search; the actual number depends on the particular region of the parameter space the user wishes to explore. Some genes are integer-valued (such as the agent-to-agent commu-

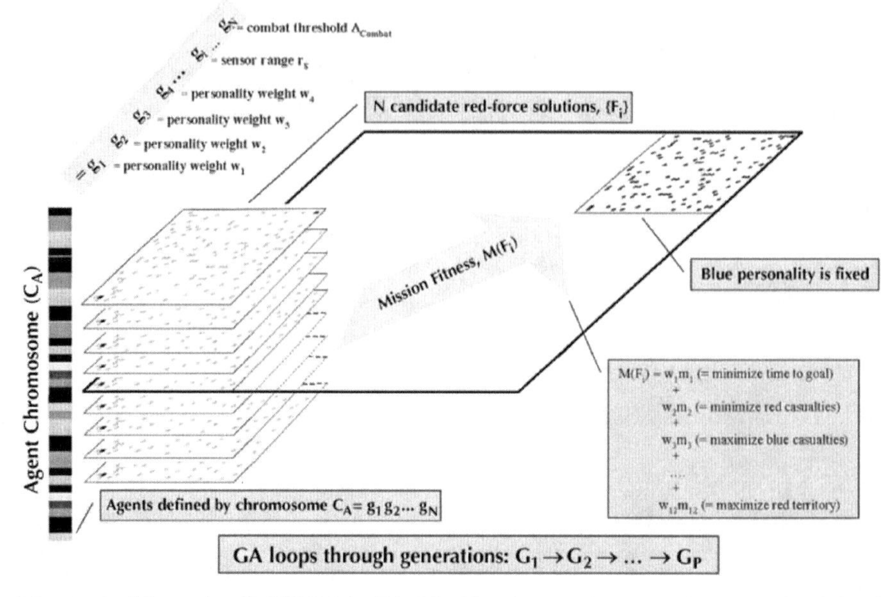

Fig. 9.12 Schematic of EINSTein's GA. The blue force and mission fitness are both fixed by the user. The GA encodes components of the agents' personality weight vector, sensor range, fire range, meta-rule thresholds, and so forth and breeds the "best" red force using populations of N red force "candidate" solutions; see the text for details.

nication links), while others are real-valued. All appropriate translations to integer values and/or binary toggles (*on/off*) are performed automatically by the program. Typically, each gene encodes the value of a basic parameter defining the red force. For example, g_1 encodes red's sensor range when an agent is in the alive state, g_3 encodes red's alive-state fire range, and so on. Some special genes encode the sign ($+$ or $-$) of an associated parametric gene. Thus, the actual value of each of the components of red's personality weight vector, for example, is actually encoded by *two genes*: one gene specifying the component's absolute value, and the other gene specifying its sign.

EINSTein's GA can conduct its search over five spaces:

- *Single-squad personality:* GA searches over the personality-space defining a single squad.
- *Multiple-squad personality:* GA searches over the personality-space defining multiple squads. The number of squads and the size of each squad remain fixed throughout this GA run mode.
- *Squad composition:* GA searches over squad composition space. The personality parameters defining squads 1 through 10 are fixed according to the values defined in the default input data file used to start the interactive run. The GA searches over the space defined by the number of squads

(1–10) and size of each squad (constrained by the total number of agents as defined by the data file).

- *Intersquad communications connectivity:* GA searches over the zero-one entries defining the communications matrix. The number of squads and the number of agents per squad are kept fixed at the values defined in the default input data file used to start the interactive run.
- *Intersquad weight connectivity:* GA searches over (real-valued) entries defining the squad interconnectivity matrix. The number of squads and the number of agents per squad are kept fixed at the values defined in the default input data file.

9.7.2 Mission Fitness

The mission fitness (MF) is a measure of how well agents perform a user-defined mission. Typical missions are *"Get to blue flag as quickly as possible,"* *"Minimize red casualties,"* and *"Maximize the ratio of blue to red casualties,"* or some combination of these. MFs are always defined from red's perspective. The user assigns weights ($0 \leq w_i \leq 1$)[7] to represent the relative degree of importance of each mission-fitness primitive, m_i (see Table 9.6). While the mission primitives are relatively few in number and simple, they can be combined to define more complicated multiobjective functions.

Weight	Primitive	Description
w_1	m_1	Minimize time to goal
w_2	m_2	Minimize friendly casualties
w_3	m_3	Maximize enemy casualties
w_4	m_4	Maximize friendly-to-enemy survival ratio
w_5	m_5	Minimize friendly center-of-mass distance to enemy flag
w_6	m_6	Maximize enemy center-of-mass distance to friendly flag
w_7	m_7	Maximize N_{friends} within distance D of enemy flag
w_8	m_8	Minimize N_{enemy} within distance D of friendly flag
w_9	m_9	Minimize number of friendly fratricide hits
w_{10}	m_{10}	Maximize number of enemy fratricide hits
w_{11}	m_{11}	Maximize friendly territorial possession
w_{12}	m_{12}	Minimize enemy territorial possession

Table 9.6 EINSTein's GA mission-fitness primitives

The mission-fitness function, M, used by the GA, is a weighted sum of mission primitives: $M = \sum_i m_i$. (It is left up to the user to ensure that mission objectives are both logically consistent and amenable to a "solution.")

[7] *Mission fitness* weights must not be confused with the *personality* weights; agent personalities discussed earlier.

Future versions of EINSTein will include a richer set of mission-fitness primitives, including: locate and kill enemy squad leaders, stay close to friends, stay away from enemies, have combat efficiency (as measured by cumulative number of hits on enemy), clear specified area of enemy agents, occupy area for specified period of time, take the enemy flag under specific conditions (e.g., the user is asked to specify the number of agents that must occupy a given area around the enemy flag for a given length of time), among others.

9.7.3 EINSTein's GA Recipe

The GA uses EINSTein's agent-movement/combat engine to conduct its searches. In pseudocode, the main components of EINSTein GA recipe are as follows:

```
for generation=1,G_max
    for personality=1,P_max
        decode chromosome
        for initial_condition IC=1 toIC_max
            run combat engine
            calculate fitness (for given IC)
        next initial_condition
        calculate mission fitness
    next personality
    find the best personality
    select survivors from population
    perform (single-point) crossover operation
    perform mutation operation
    update progress/status
next generation
write best personality to file
```

In words, the GA uses a randomized pool of chromosomes to define an initial generation of red personalities. For each red personality, and for each of the $IC_{\max}$ initial spatial configurations of red and blue forces, the program then runs EINSTein's combat engine to determine the mission fitness. After looping through all personalities and initial conditions, the GA first sorts and ranks the personalities according to their mission-fitness values, then selects some to be eliminated from the pool and others to breed. The GA then performs the basic operations of crossover and mutation. Finally, after defining a new generation of red personalities, the entire process is repeated until either the user interrupts the evolution or the maximum generation number has been reached (see Fig. 9.12).

9.7.4 Sample GA Breeding Experiment #1

Consider the following mission (as stated from the *red* force's point of view): *"Keep blue agents as far away from the red flag as possible, for as long as possible (up to a maximum 100 iteration steps)"*; that is, set all GA mission weights to zero, except for $w_6 = w_8 = 1/2$; see Table 9.6. This means that the mission fitness M will be close to its maximal value *one* only if red is able to keep all blue agents pinned near their own flag (at a point farthest from the red flag) for the entire duration of the run, and M will be near its minimal value *zero* if red allows blue agents to advance completely unhindered toward the red flag. Combat unfolds on a 40-by-40 battlefield, with 35 agents per side. The GA is run using a pool of 50 red personalities for 50 generations, and each personality is averaged over 25 initial spatial configurations. Blue agents are each assigned (a fixed) personality weight vector $w_{\text{Blue}} = (w_{\text{AF}}, w_{\text{IF}}, w_{\text{AE}}, w_{\text{IE}}, w_{\text{FF}}, w_{\text{EF}}) = (0, 10, 0, 10, 0, 90)$.

Fig. 9.13 shows a typical *learning curve*, where "Best" refers to the fitness of the highest-ranking candidate solution and "Average" refers to the average fitness among all candidate solutions per generation. The GA run described here (using a 1-GHz Pentium IV PC) each requires roughly an hour to complete.

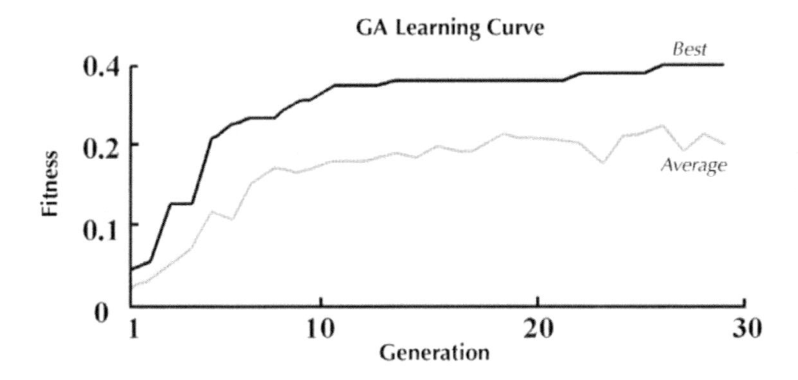

Fig. 9.13 Typical GA learning curve for GA breeding experiment discussed in the text.

Screenshots from a typical run using the highest-ranked red personality (as sampled from "solution" pool representing generation 30) that the GA is able to find for this mission are shown along the top row of Fig. 9.14. They show that red is very successful at keeping blue forces away from its own flag; the closest that red permits blue agents to approach the red flag – during the entire allotted run time of 100 iteration steps – is some point roughly near midfield. In words, the "tactic" here seems to be – from red's

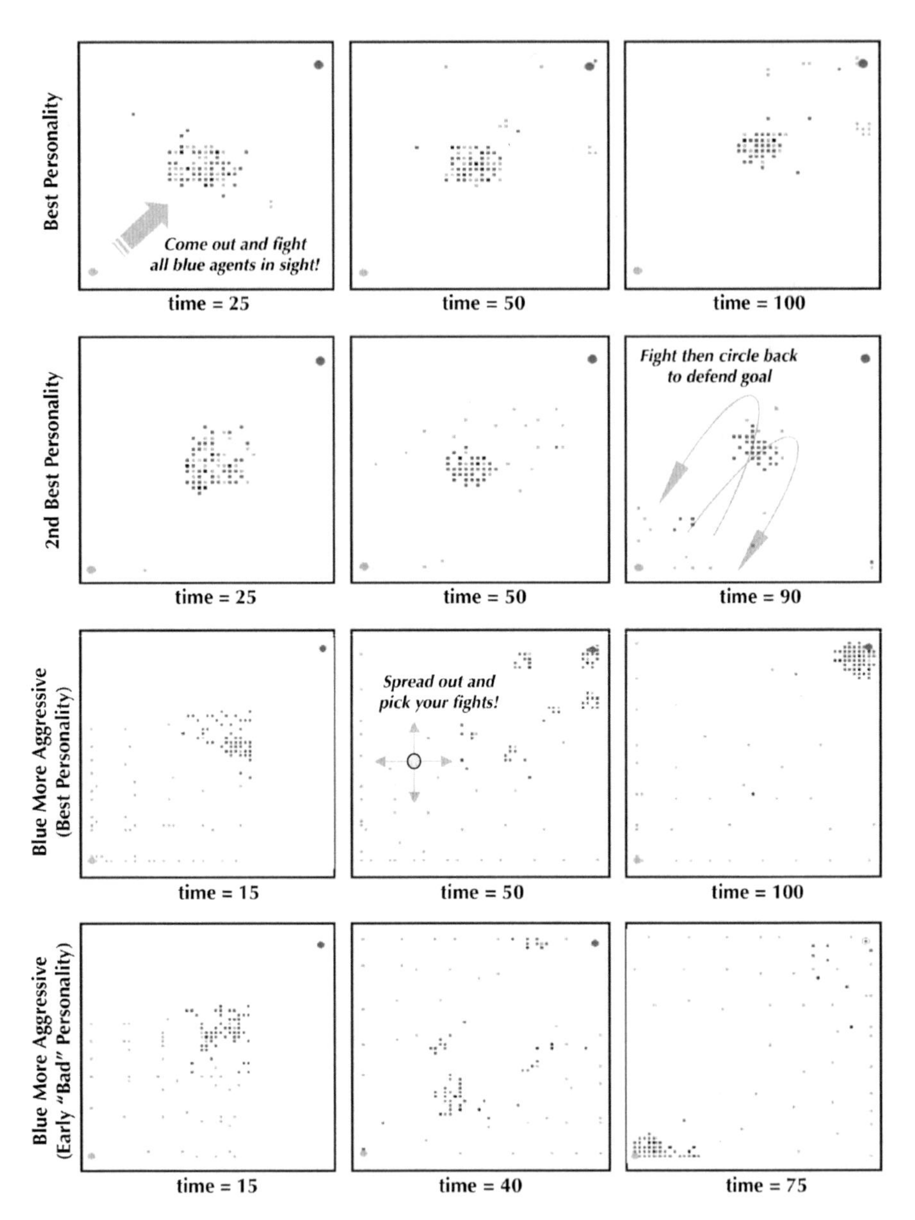

Fig. 9.14 Screenshots from several sample runs of GA breeding experiment #1. The top row shows a run using the highest-ranked red agents after 50 generations. The second row shows the second highest-ranked red force. The third row shows how the red force adapts to a more aggressive blue force. Finally, the fourth row shows an example of how a suboptimal red force performs representing a pool of agents occupying an early portion (generation 10) of the GA's learning curve.

perspective – *"fight all enemy agents within sensor range, and move toward the enemy flag slowly enough to drive the enemy along."* Note that this emergent tactic is also fairly robust, in the sense that if the battle is initialized with a different spatial disposition of red and blue forces (while keeping all personality parameters fixed), red performs this particular mission about as well, on average, as evidenced by these screenshots.

Screenshots from a typical run using the second highest-ranking red personality are shown along the second row of Fig. 9.14. These show a slightly less successful, but nonetheless innovative, alternative tactic. Initially, red agents move away from their own goal to meet the advancing blue forces, just as in the first case (at $t = 25$). Once combat ensues, however, any red agents that find themselves locally isolated now "double back" toward their own flag (positioned in the lower left corner of the battlefield) to regroup with other remaining friendly agents. The red force thus, effectively, forms an impromptu secondary defense against possible blue leakers. Because a few blue agents do manage to fight their way near the red flag at later times (at least in the particular run these screenshots have been taken from; see snapshot for $t = 90$), the red agent parameter values underlying this emergent tactic are not as highly ranked as the parameter values underlying the run shown in the top row.

The series of screenshots appearing in the third row of Fig. 9.14 show the emergent tactic used by the highest-ranked red personality found by the GA after the blue force is made more *aggressive*. For this case, prior to initializing the GA search, blue's personality weight-vector components for moving toward red (i.e., $w_{AE} = w_3$, and $w_{IE} = w_4$) are first increased by 50%. We see that EINSTein's GA discovers an entirely different (and more effective) tactic to use. Here, the red force quickly spreads out to cover as much territory as possible and individual agents attack the enemy as soon as they come within view. As red agents' local territorial coverage is thinned – either through attrition or gradual advance toward the blue flag – other red agents (namely agents that had previously been positioned near the periphery of the battlefield) move closer to the center of the battlefield, thus filling emerging voids. This tactic succeeds in preventing any blue agents from reaching the red flag and also manages to push most of the surviving blue force back toward its own flag (near the top right corner of the battlefield)! As is true of the other cases in this experiment, this tactic is also fairly robust and is not a strong function of the initial spatial disposition of red and blue forces.

The last row of plots in Fig. 9.14 contains snapshots from a run using interim red agent parameter values, *before* the GA has had a chance to evaluate a large number of candidate solutions. This example illustrates how an obviously *sub*-optimal pool of agents behaves differently from their optimized counterparts. The mission parameters and blue force agent personalities are the same as in the case represented by the screenshots in the third row. We see that, initially at least, there does not seem to be much difference in the optimal and sub-optimal behaviors; red agents quickly disperse outward to cover a large area. However, because the GA has not yet had the time to fine-tune all of red's genes, the red force is, in this instance, unable to prevent blue agents from penetrating deeply into its territory. The defensive tactic, however it may be characterized, is obviously ineffective.

9.7.5 Sample GA Breeding Experiment #2

Consider a scenario in which the blue force is tasked with defending its flag against a smaller attacking red force. We use the GA to find a red force that is able to penetrate the blue defense. Table 9.7 lists some pertinent parameter values defining the two forces. The third column of the table (i.e., red trial values) lists baseline red force parameter values (as defined by us, not the GA) used to test the scenario. The fourth and fifth columns (i.e., GA-bred values) list the GA-bred red force "solution." Notice that, in both cases, the number of agents is the same and is fixed (with blue outnumbering red, 100 to 50 in all runs). All baseline red-*trial* alive and injured parameter values are equal.

	Blue Agents	Red *Trial* Agents	Red *GA Bred* *Alive* Agents	Red *GA Bred* *Injured* Agents
N_{Agents}	100	50	50	50
r_S	5	5	8	5
r_F	3	3	8	5
r_M	2	2	2	2
w_{AF}	0	10	3	−22
w_{AE}	100	40	40	95
w_{IF}	0	10	46	−86
w_{IE}	100	40	38	−14
w_{FF}	0	0	−70	14
w_{EF}	0	25	65	31
τ_{Advance}	N/A	3	3	1
τ_{Cluster}	5	10	13	17
Δ_{Combat}	−20	0	−19	+20

Table 9.7 Agent parameter values for GA sample run appearing in Figs. 9.15 and 9.16

Fig. 9.15 shows screenshots from a typical run using the red-trial values. Red agents attack, unsuccessfully, in a tight cluster. The larger blue force (whose agents initially move about randomly around their starting position until a red agent comes within their sensor range) dispels the red force rather easily (within the 30 time steps shown here).

The GA-bred parameters listed along the bottom row in Table 9.7 define the highest-ranked red force that EINSTein's GA is able to find (after 30 generations) with respect to performing the mission = *"maximize the number of red agents able to penetrate within a distance d = 7 units of the blue flag*

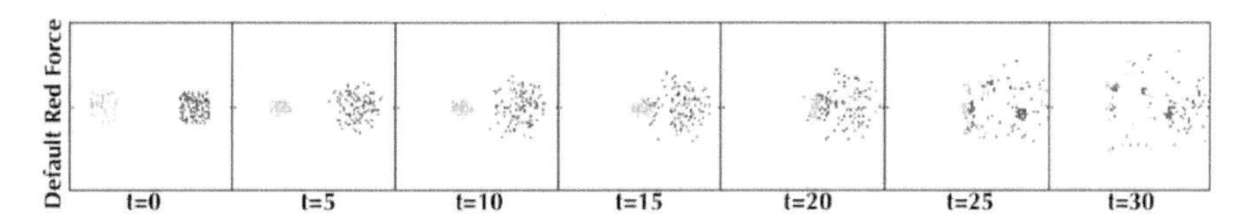

Fig. 9.15 Trial red attacking force (consisting of typical parameter values that are not explicitly tuned for performing any specific mission). Red performance is used simply as a reference for interpreting the output of the sample GA breeding experiment discussed in the text.

within 40 time steps." A population size of 75 was used (i.e., each generation of the GA search consists of 75 red force candidate "solutions") and mission fitness, for a given candidate solution, is averaged over 10 initial configurations of red and blue forces. The fitness equals *one* if a candidate solution performs the specified mission in the best possible manner (i.e., if the red force sustains zero casualties and all agents remain within $d = 7$ of the blue flag starting from the minimal possible time at which they move to within that distance of the flag, for all 10 initial states) and equals *zero* if a candidate solution fails to place a single red agent within $d = 7$ of the blue flag for all 10 initial states (within the mission time limit). Fig. 9.16 shows screenshots from a typical run using the GA-bred red force values. (The arrows are included as visual aids and simply trace the motion of the red agent clusters.) Comparing this sequence of steps to those in the trial run shown in Fig. 9.15, it is obvious that the respective "attack strategies" in the two cases are very different. Indeed, the GA has found just the right mix of agent-agent proximity weights and meta-rules to define a red force that effectively exploits a relative weakness in the randomly maneuvering blue defenders. The emergent "tactic" is to *separate into two roughly equal-sized units, regroup beyond enemy sensor range, and then simultaneously strike, as a pincer, into the heart of the defending enemy cluster.*

Apart from the anecdotal evidence supplied by screenshots of this particular run, the efficacy of this simple GA-bred tactic is illustrated by comparing graphs of the number of agents near the blue flag (averaged over 50 runs)

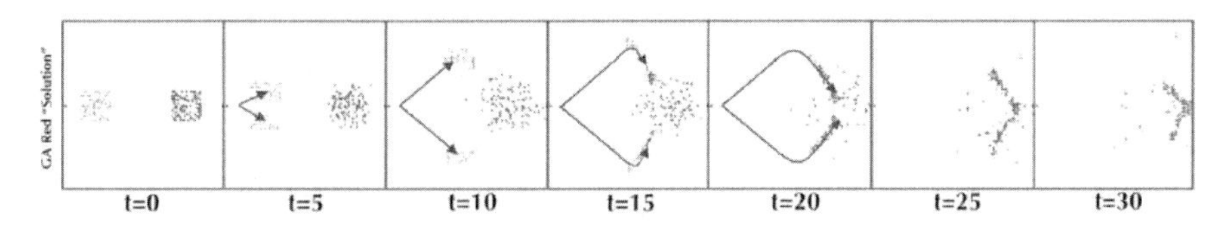

Fig. 9.16 Screenshots from a typical run using the GA-bred red force for the sample GA breeding experiment discussed in the text. Red agents are defined by GA-bred parameter values that are the highest ranked (after 30 generations) with respect to performing the mission = *"maximize the number of red agents able to penetrate within a distance d=7 units of the blue flag within 40 time steps."*

as a function of time for the red-*trial* and *GA-bred* cases. Fig. 9.17 shows that whereas fewer than three red-trial agents, on average, penetrate close to the blue flag (Fig. 9.17a), almost 80% of the entire GA-bred red force is able to do so (Fig. 9.17b) and begins penetrating at an earlier time. Other (well-performing) tactics are possible, of course. A representative sampling is generally provided by looking at the behaviors of some of the higher-ranking red forces remaining at the end of a GA search. It is interesting to run a series of GA runs to systematically probe how red forces "adapt" to different blue personalities. What one observes, typically, is that as the behavior of the blue agents changes, various – often dramatically different – GA-bred red personalities emerge to exploit any new weaknesses in the blue's defensive posture.

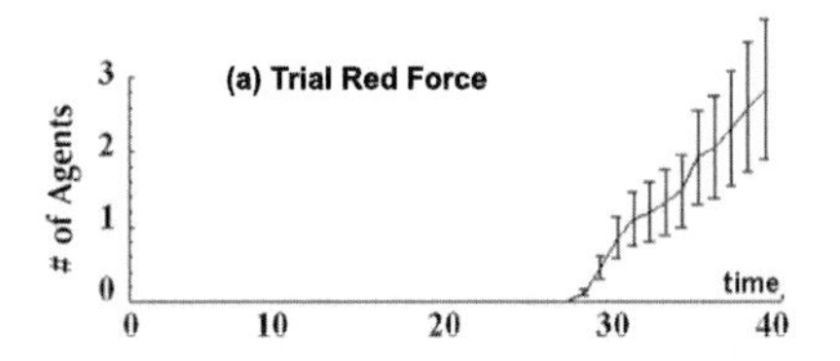

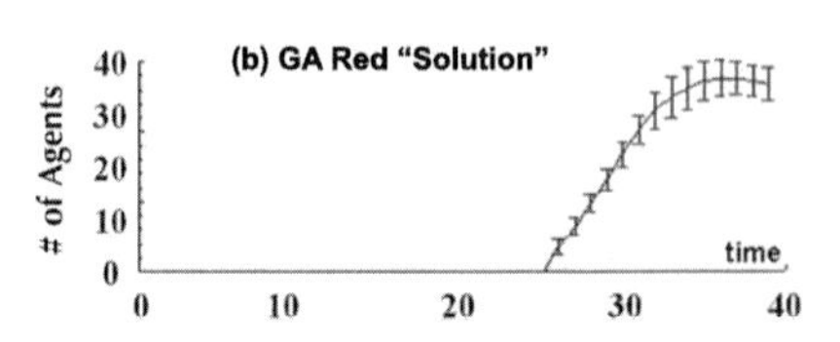

Fig. 9.17 A comparison between the average number of red agents that approach within a distance $d = 7$ of the blue flag for (a) *trial* and (b) *GA-bred* red forces. We see that the GA-bred red force typically performs this mission an order of magnitude more successfully than the trial force.

9.8 Discussion

"The musical notes are only five in number,
but their melodies are so numerous that one cannot hear them all.
The primary colors are only five in number,
but their combinations are so infinite that one cannot visualize them all.
In battle there are only the normal and extraordinary forces,
but their combinations are limitless; none can comprehend them all."

— Sun Tzu, *The Art of War*

The high-level, or poetic, description of EINSTein owes much to the suggestive metaphors appearing in the quote from *The Art of War*. In the same way as, for Sun Tzu, rainbows and melodies are all natural outcomes of combining primary colors and musical notes, EINSTein may be viewed as an "engine" that converts a primitive grammar (i.e., a grammar composed of the basic notes and colors of combat) into the limitless patterns and possibilities of war. The researcher chooses and/or tunes primitive, low-level agents and rules; EINSTein provides the dynamic arena within which these rules interact and spawn high-level patterns and behaviors. On a more practical level, EINSTein was developed with these three important goals in mind:

1. To demonstrate the efficacy of agent-based simulation alternatives to more traditional Lanchester-equation-based models of combat [5].
2. To be used as a general prototype artificial life model/toolkit that can be used as a testbed for exploring self-organized emergent behavior in complex adaptive systems.
3. To provide the military operations research community with an easy-to-use, intuitive agent-based combat-simulation laboratory that – by respecting both the principles of real-world combat and the dynamics of complex adaptive systems – may lead researchers one step closer to a fundamental theory of combat.

To better appreciate how each of these motivations has contributed to EINSTein's (still evolving) architecture, consider the conceptual map of its design, as illustrated schematically in Fig. 9.18. Self-organized patterns emerge out of a set of primitive local rules of combat, both on the individual agent level – via interactions among internal motivations (on the Phenotype-I level, which appears as the middle level in Fig. 9.18) – and squad and force levels (labeled Phenotype-II in Fig. 9.18, and which appears as the topmost level in the figure) – via mutual interactions among many agents in a changing environment.

Of course, a deeper understanding of phenomena governing behaviors on the topmost level can only be achieved by developing a suite of appropriate pattern recognition tools (the need for which is indicated symbolically at the top of Fig. 9.18). Although a number of interesting, and highly suggestive, high-level patterns have already been discovered, much still remains to be done. Consider, for example, the frequent appearance of various power-law scalings and fractal dimensions describing space-time patterns and attrition rates ([34, 40]). The existence of power-law scalings, in particular, strongly suggests that a self-organized, critical-like dynamical mechanism might govern turbulent-like phases of combat. However, the data collection and analysis necessary to rigorously establish the nature of these findings (as well as to establish a mathematically precise set of conditions under which power-law scalings either *do* or *do not* occur) has only just started.

One of the directions in which EINSTein's design is moving (some details of which are described in the next section) is toward a fully developed ontological architecture that assigns specific meaning to the symbolic relationship between *environment* and *action*. The hope is to be able to explore the complementary problem of *reverse behavior engineering*; that is, the problem of finding an appropriate set of primitives (properties and rules) that lead either to empirically observed or desired macroscopic patterns of combat (or, in Fig. 9.18, of finding ways of going from either phenotype level I or II to the genotype level).

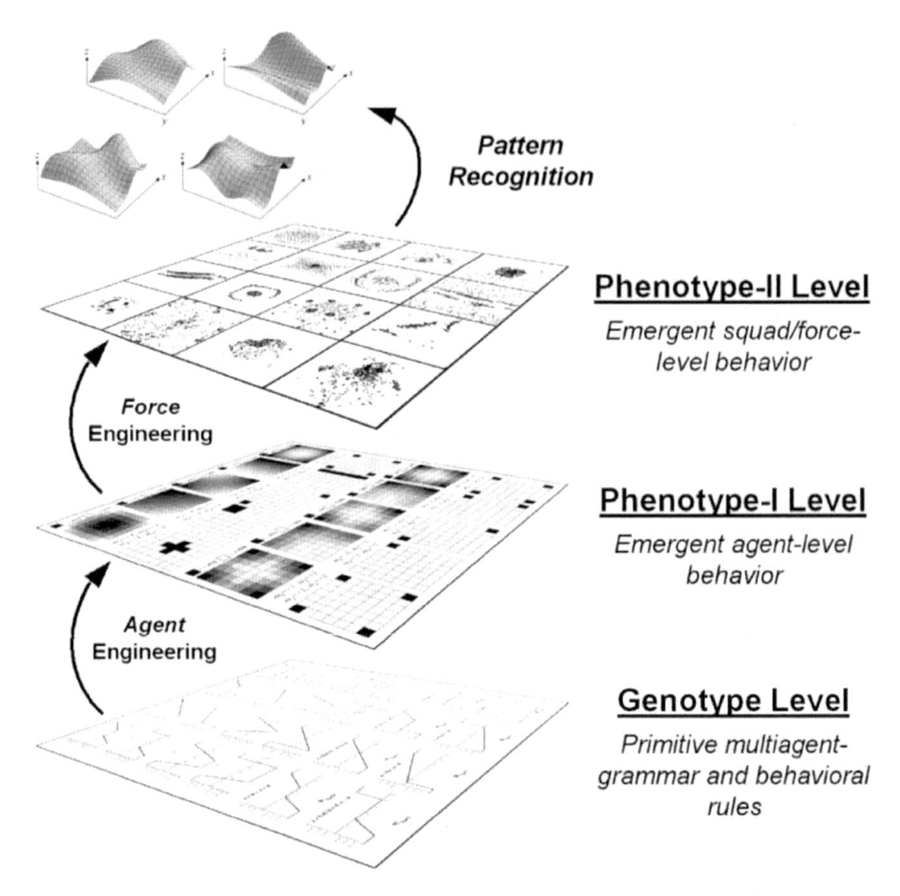

Fig. 9.18 A hierarchy of conceptual levels that illustrate EINSTein's core design.

9.8.1 Why Are Agent-Based Models of Combat Useful?

The most important immediate payoff to using EINSTein is the radically new way at looking at fundamental issues. However, agent-based models are best used to enhance understanding, not as prediction engines. Specifically, EINSTein is being designed to help researchers do the following:

- Understand how all of the different elements of combat fit together in an overall combat phase space: *"Are there regions that are 'sensitive' to small perturbations, and, if so, might there be a way to exploit this in combat (as in selectively driving an opponent into more sensitive regions of phase space)?"*
- Assess the value of information: *"How can I exploit what I know the enemy does not know about me?"*

- Explore trade-offs between centralized and decentralized command-and-control (C2) structures: *"Are some C2 topologies more conducive to information flow and attainment of mission objectives than others?" "What do emergent forms of a self-organized C2 topology look like?"*
- Provide a natural arena in which to explore consequences of various qualitative characteristics of combat (unit cohesion, morale, leadership, etc.).
- Explore emergent properties and/or other "novel" behaviors arising from low-level rules (even combat doctrine if it is well encoded): *"Are there universal patterns of combat behavior?"*
- Provide clues about how near-real-time tactical decision aids may eventually be developed using evolutionary programming techniques.
- Address questions such as *"How do two sides of a conflict coevolve with one another?"* and *"Can one side exploit what it knows of this coevolutionary process to compel the other side to remain 'out of equilibrium?' (or be otherwise trapped in a supoptimal dynamical combat state)."*

EINSTein has been used to explore the following: patrol dynamics, security, and ambush tactics [52]; reconnaissance [53]; counter reconnaissance [54]; communications [55]; distributed operations in open [56] and urban environments [57]; the historical evolution of squad and fire-team composition and weapon-mix [47]; small unit combat [59]; C2 [60]; situational awareness [58]; civil disobedience [61]; peacekeeping operations [62]; and maritime ship stationing [63]. In all, there are some 800 registered users of EINSTein, including researchers from the US Department of Defense, academia, research and development centers and private companies.

9.8.1.1 Command and Control

EINSTein contains embedded code that hardwires in a specific set of C2 functions (i.e., both contain a hierarchy of local and global commanders), so that it can be used to explore the dynamics of a given C2 structure. However, a more compelling question is, *"What is the best C2 topology for dealing with a specific threat, or set of threats?"* One can imagine using a genetic algorithm, or some other heuristic tool to aid in exploring potentially very large fitness landscapes, to search for alternative C2 structures. What forms should local and global command take, and what is the optimal communications matrix among individual combatants, squads, and their local and global commanders?

9.8.1.2 Pattern Recognition

An even deeper issue has to do with identifying the primitive forms of information that are relevant on the battlefield. Traditionally, the role of the combat operations research analyst has been to assimilate and provide useful insights from certain conventional streams of battlefield data: attrition rate, posture profiles, available and depleted resources, logistics, rate of reinforcement, Forward Edge of the Battle Area (FEBA) location, morale, and so forth. While all of these measures are obviously important, and will remain so, having an ABM of combat permits one to ask the following deeper question: *"Are there any other forms of primitive information – perhaps derived from measures commonly used to describe the behavior of nonlinear and complex dynamical systems – that might provide a more thorough understanding of the fundamental dynamical processes of combat?"* We have already mentioned, for example, that evidence suggests that the intensity of battles – both in the real world and in ABMs of combat – obeys a fractal power-law dependence on frequency and displays other traits characteristic of high-dimensional chaotic systems. Are there other, similar but heretofore unexamined, measures that may provide insight into the dynamics of real-world combat?

9.8.1.3 "What If?" Experimentation

The strength of ABMs lies not just in their providing a potentially powerful new general approach to computer simulation but also in their infallible ability to prod researchers into asking a host of interesting new questions. This is particularly apparent when EINSTein is run interactively, with its provision for making quick "on-the-fly" changes to various dynamical parameters. Observations immediately lead to a series of *"What if?"* speculations, which in turn lead to further explorations and further questions. Rather than focusing on a single scenario and estimating the values of simple attrition-based measures of single outcomes ("Who won?"), users of agent-based simulations of combat typically walk away from an interactive session with an enhanced intuition of what the overall combat fitness landscape looks like. Users are also given an opportunity to construct a context for understanding their own conjectures about dynamical combat behavior. The agent-based simulation is therefore a medium in which questions and insights continually feed off one another.

9.8.1.4 Validation

Before any combat model – agent based or not – is judged "useful" to a military operations researcher, it must pass two important tests in the affirmative: (1) Does it provide insight into the specific set of problems the researcher is interested in studying, in a well-defined and self-contained conceptual context? (which is an obvious requirement of even the most basic mathematical model) and (2) Is its output consistent with behavior that is either accepted to be true or has otherwise been observed to occur in the real-world? Most

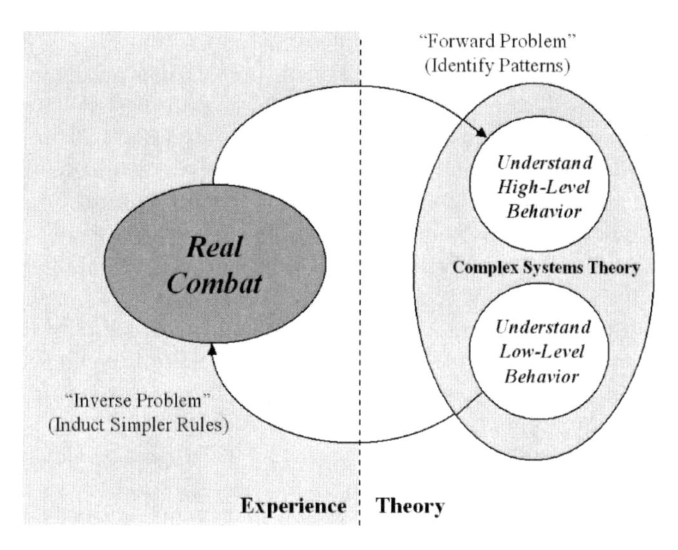

Fig. 9.19 Schematic of the interplay between experience and theory in the *forward* and *inverse* problems of simulating combat.

importantly, a successful model of combat must respect the critical interplay between real-world experience and simulation outcome. That is to say, the model must at some point be *validated.*

Fig. 9.19 illustrates, schematically, the interplay between the *forward* problem of using the model to predict behaviors and identify patterns and the *inverse* problem of using experience and real-world data to add to, change, and/or refine the rules and behaviors that define the model. The forward problem consists of observing real-world behavior, with the objective being to identify any emergent high-level behavioral patterns that the system might possess. The inverse problem deals with trying to induct a set of low-level rules that describe observed high-level behaviors. Starting with observed data, the goal is to find something interesting to say about the properties of the source of the data. Solving the forward problem requires finding the right set of theoretical tools that can be used to identify patterns, while the inverse problem needs tools that can be used to induct low-level rules (or models) that generate the observed high-level behaviors.

Since EINSTein was conceived primarily as an exploratory model, validation is less of an issue than it might be for more ostensibly realistic models. As long as EINSTein's outcomes are intuitive and its inputs are easy to generalize (or translate) for use by more realistic models, EINSTein will remain useful for many different kinds of exploratory analyses. Nonetheless, because EINSTein is already powerful enough to simulate many real-world scenarios, it is proper to ask about how the model may be validated.

One such recent attempt to validate (a beta version of) EINSTein was undertaken at the United States West Point Military Academy's Operations

Research Center for Excellence by Klingaman and Carlton [64], by comparing its output to that of another well-established combat simulation model, called JANUS (a high-resolution conventional – i.e., not agent based – simulation of red and blue forces, with resolution down to the individual platform and soldier). The study endeavored to establish the combat effectiveness of EINSTein agents executing a standard National Training Center combat scenario that consists of a single armored company of 14 blue force tanks engaged versus a similar size force of 14 red battle tanks. One set of blue agents is allowed to gain knowledge (or "learn") by using EINSTein's built-in GA agent breeder function. The behavior of another set of blue agents is kept fixed, and the agents are not allowed to learn. In both cases, EINSTein's combat results for all agent actions are recorded. These *observed actions* are then programmed into JANUS (EINSTein's automatic record of center-of-mass positions are used to define the routes in JANUS), and, for each case, the combat effectiveness resulting from JANUS is compared to the outcome in EINSTein. The validation test consisted of verifying the reasonable expectation that the knowledgeable agents exhibit noticeably different (and, hopefully, improved) behavior and have a significantly better loss-exchange ratio (LER) in both EINSTein and JANUS.

Using a general linear model analysis of two factors (agent type and model) with two levels each (default/GA-bred and EINSTein/JANUS), the study found that although the LER was different for the two models, the LER data in both models follow similar trends. The standard deviations of the mean LER also decrease from the default agents to GA-bred agents in both models. Overall, the study found that EINSTein's agents may be used to portray similarities of combat vehicles reasonably well and that learning as portrayed in EINSTein can be transferred into another combat model. However, citing various limitations due to model-specific constraints on translating agent and environmental characteristics from one model to the other, as well as unavoidable conceptual differences between the two models, Klingaman and Carlton concluded their report by offering three suggestions: (1) that ABMs need increased fidelity in terms of terrain and weapons performance (limitations that are no longer an issue for newer versions of EINSTein; see discussion in the next section), (2) traditional models, such as JANUS, ought to incorporate ABM-like personality traits and decision-making algorithms to allow for more realistic combatant actions, and (3) traditional models ought to incorporate some mechanism to allow for adaptive learning.

One other important suggestion that belongs on Klingaman and Carlton's list of general recommendations is due to Axtell et al. [65], who argued for the need to *align* (or "dock") simulation models. Docking refers to a process of testing to see whether two models can produce the same results, or using one model to check the output of another, but in a more general and systematic manner than was used in the one-time EINSTein-to-JANUS comparison discussed above. The idea is similar to a common practice among technical researchers to use not one, but two or more mathematical packages (such as *Mathematica* [66] and *Maple* [67]) to help check their calculations. The authors illustrate this concept by using, as their testbed simulations, a model of cultural transmission designed by Axelrod [76] and the *Sugarscape* multiagent-based simulation of evolution that takes place on a notional sugar field, developed by Epstein and Axtell [29]. Since the models differ in many ways (and have been designed with quite different goals in mind), the comparison was not an especially easy one to make. Nonetheless, the authors report that the benefits of the sometimes arduous process of alignment far outweighed the hardships. In the end, the user communities of both models benefited from the alignment process by gaining a deeper understanding of how each program works, of their similarities and differences, and of how the inputs and outputs of each program must be modified before a fair comparison of what really happens in either model can be made.

As the list of combat agent models grows beyond those that have already appeared (such as CROCADILE, MANA, SEM, Socrates, and SWarrior; see [34]), the need to "align" the output of these models – in the sense defined by Axtell et al. [65] – can only become more important.

9.8.1.5 Universal Grammar of Combat?

What lies at the heart of an artificial life approach to simulating combat is the hope of discovering a fundamental relationship between the set of higher-level emergent processes (penetration, flanking maneuvers, containment, etc.) and the set of low-level primitive actions (movement, communication, firing at an enemy, etc.). Wolfram [77] has conjectured that the macro-level emergent behavior of all cellular automata rules falls into one of only four universality classes, despite the huge number of possible local rules. While EINSTein's rules are obviously more complicated than those of their elementary cellular automata brethren, it is nonetheless tempting to speculate about whether there exists – and, if so, what the properties are – a *universal grammar of combat.* A step toward achieving this far-reaching goal is being taken with EINSTein's recent integration within *Python*, a scripting language that allows programming-savvy users to extend EINSTein (see discussion below).

9.9 Overview of Features in Newer Versions

EINSTein has evolved considerably beyond the snapshot of the program that appears in the previous sections of this chapter, although the core of the model remains essentially the same and is consistent with everything thus far discussed.[8] This section summarizes the state of the model as it appears

[8] Version 2.1 may be downloaded from [71], and older versions may be downloaded from [72].

at the time of this writing (April 2008) and highlights some of the main changes and additions that have been made to the program since the first edition of this book came out in 2005.

Loosely speaking, EINSTein's emphasis as a CAS-based "combat simulator" has gradually shifted away from describing the mutual interactions among many *simple* agents (although EINSTein's original capability in this regard is still there, of course) to describing interactions among a relatively few, but *complex* agents that are also endowed with a richer internal structure and dynamics. Thus, where earlier versions have focused on the complexity of emergent behaviors on the system level, more recent work adds the ability to explore emergent behaviors on the individual agent level as well. Details are discussed in [34].

Apart from (mostly self-explanatory) changes to the GUI and file I/O (see below), the newest version of EINSTein includes three general classes of enhancements over earlier releases: (1) agent attributes, (2) battlefield environment and behavior functions, and (3) integration within *Python.*

9.9.1 Agent Attributes

An agent's "personality" may now be tuned using many more primitive attributes than previously available. For example (in alphabetical order),[9] (1) ***Acuity***, which regulates the strength of an agent's focus on its priorities and is used to "sharpen" decisions that are made probabilistically (its range of possible values varies from +1, for which an agent only considers options of maximum value, to +1, for which an agent gives equal consideration to all options); (2) ***Camouflage***, which defines how well an agent can hide from other agents in the context of a local environment (if the value is 0, an agent is 100% visible to other agents; if the value is 1, an agent is effectively invisible to both friends and enemies; and intermediate values determine an agent's inherent probability of being detected); (3) ***Frazzle***, which specifies how much an agent's firing accuracy degrades when the agent is presented with multiple simultaneous targets (and increases linearly with the number of simultaneous targets an agent fires at, reaching its maximum value at the weapon's maximum target limit); (4) ***Mass***, which specifies the rate at which agents burn their energy (all agents expend energy while moving, the amount being determined by their "velocity" – i.e., how far they move in a single time step – and by the nature of the terrain over which they are moving); (5) ***Memory***, which regulates how long does an agent retains information about certain internal state variables (and is computed using a time-averaging technique: the contribution of past values decays exponentially, with the value of memory controlling the rate of that exponential decay); (6) ***Recovery_rate***, which

[9] This is only a partial list of agent attributes that have been added to EINSTein since version 1.0. For a complete list, please refer to EINSTein's programming guide [75].

defines how much health an agent regains in a single time step (a value of 0 means an agent never recovers from injury and a value of 1 means an agent fully recovers from all injuries in a single time step; all intermediate values are interpreted as the probability an agent will recover from injuries); (7) ***Resupply_rate***, which specifies how frequently an agent reloads its weapon with new ammunition (a value of 0 means that an agent never reloads; a value of 1 means that an agent reloads at the start of every time step; intermediate values are interpreted as the probability of reload); (8) ***Stamina***, which is the rate at which agents regain their energy (energy expenditure is proportional to agent mass, and the amount recovered is proportional to stamina, if ***mass*** = ***stamina*** then an agent can traverse 100% passable terrain indefinitely); and (9) ***Tremor***, which is the degree to which an agent's "inherent fallibility" reduces its weapon firing accuracy (if the value is 0, then weapon accuracy is unaffected; if the value is 1, then one standard deviation is added to the agent's weapon's average hit distance from an intended target). ***Health*** has also been generalized from a binary variable (healthy/not-healthy) to a continuous one (ranging in value from 0 to 1). And ***movement_range*** is now effectively free to vary from 0 to the size of the battlefield.

9.9.2 Environment and Behavior Functions

There are six major additions to EINSTein's battlefield environment and repertoire of behavior functions and modifiers: (1) ***Pathfinding***, which uses a priority-queue variant of Dijkstra's optimal path algorithm [51] to give agents an innate "intelligence" to find paths between any two points on the battlefield (an ability which, among other things, prevents agents from becoming trapped at corners of an obstacle and generally yields more realistic "flow" around terrain elements); (2) ***Waypoint scripting***, which allows the user to define arbitrarily complex paths (or roads) on the battlefield and tune the way in which agents traverse them (that also allows for far more complex scenarios to be constructed than before)[10]; (3) ***Obstacles***, which are fixed objects in the terrain that interfere with weapon fire and agent movement (and can therefore be used as building blocks to populate a battlefield with notional buildings). A weapon's *loft* and an obstacle's *height* properties determine whether a weapon round reaches an intended target[11]; (4) ***Weapon-construction class***, which allows users to design their own

[10] The "battlefield" that appears on the upper left of the screenshot shown in Fig. 9.1 contains two red and two blue user-defined paths defined using waypoints.

[11] *Impenetrability* defines how much energy an obstacle may absorb from a weapon blast. If an obstacle exists along the straight-line path between a weapon and its target and *height* $>$ *loft*, then all rounds fired by the weapon will be blocked from reaching its target. If *height* $\leq$ *loft*, a weapon's blast energy is reduced by a factor $\propto$ *Impenetrability*. If there are multiple obstacles between a weapon and its kill-zone center, then the reduction in

weapons-of-choice[12] using a palette of 10 primitives (*range, firing rate, capacity, blast radius, power, armor, deviation, reliability, loft*, and *ammunition capacity*); (5) ***Weapon-targeting logic***, which provides agents with an intelligent targeting capability (and with which agents can discriminate targets by weighing the relative potential benefit of firing at the given coordinate on the battlefield). Agents consider factors such as expected blast size, the damage likely to be inflicted on friends and enemies near the target coordinate, and the value or threat that specific enemy agents represent. A targeting penalty function is used to evaluate each of the possible targeting strategies that may be used in a given context and is an analog of the movement penalty function defined in Eq. 9.4; and (6) ***Trigger states***, which generalize EINSTein's older meta-rules (which are still available) by allowing users to associate certain predefined environmental conditions with agent behaviors (i.e., agent actions may now be adaptively triggered by dynamic contexts).

Meta-rules have always allowed agents to tailor their behavior to simple contexts – the Δ_{Combat} meta-rule, for example, defines the conditions under which agents either engage (or do not engage) the enemy (see Table 9.2) – but also constrain the user to making basic *either/or* decisions (and limit an agent's context-specific behavior modification to changing a single component of its default personality weight vector). EINSTein's newer trigger-state logic is vastly more flexible and allows essentially arbitrary modifications of an agent's behavior to be made contingent upon arbitrary environmental conditions. Aside from obviously adding a great deal of realism to scenarios, the new logic also allows analysts to more deeply explore interactions between agent personalities and their dynamic environment.

Fig. 9.20 shows a screenshot of a sample work session in which the user has elected to define a trigger state called "Hunker Down." We see that it depends on three activation conditions (health, energy, and fire density) and that the response (shown along the right-hand side of the figure) consists of changing the value of 7 of 12 personality features (maximize rest and propensity to flock, replace values for alive ally, own flag, and enemy flag with 0, replace the value for alive foe with -100, and replace value for moving toward the local commander with $+100$). The probability that a given condition becomes "active" may be defined as a piecewise discontinuous probability density function using another dialog (not shown) that pops up by pressing the "Define PDF..." button associated with a given condition. The probability, P_A, that a trigger state (in this case, "Hunker Down") is active is the product of the conditional probability distribution functions, P_a^i, applied to

---

blast energy is cumulative (additive) along the straight-line path connecting the site and the kill-zone center.

[12] The user is presented with a default – but modifiable – selection of *bolt-action rifle* (accurate, reliable, low range, low power); *semiautomatic rifle* (higher rate, lower range, less accurate); *machine gun* (high rate, higher power, lower reliability); *grenade* (lower range, larger spread, hard to target); and *mortar* (long range, large spread, high power, accurate).

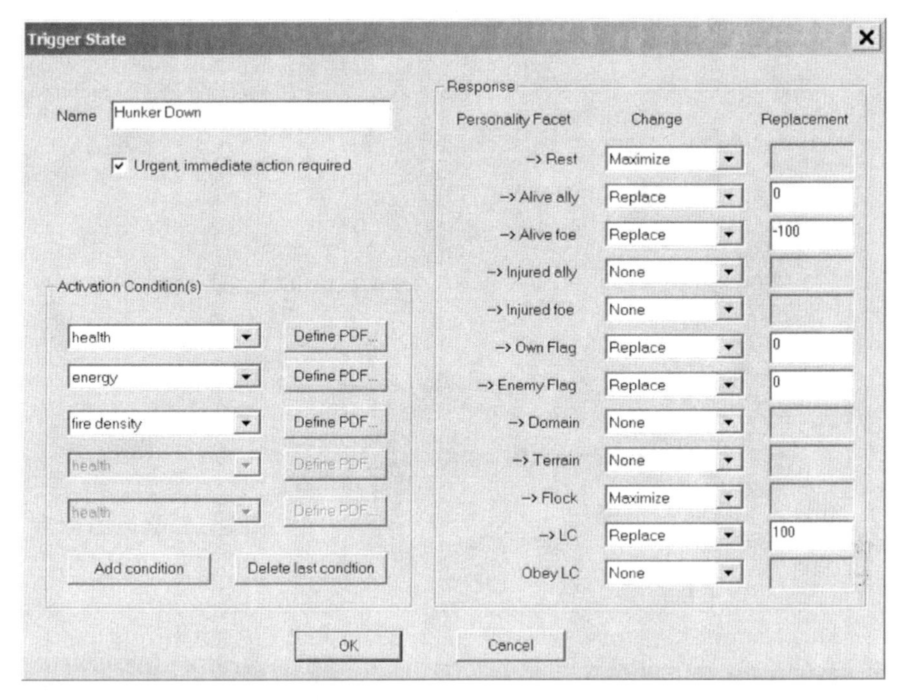

Fig. 9.20 Screenshot of EINSTein's new trigger-state edit dialog.

their corresponding activation conditions, C_i, is $P_A \equiv \prod_i P_a^i(C_i)$. Up to five activation conditions may be active at one time. However, there is no limit on the number of trigger states that may be defined in a scenario for any given agent.

9.9.3 GUI and File I/O

While most of the older elements of EINSTein's GUI (see Fig. 9.1) remain intact, albeit in a modified form to accommodate the growing palette of agent attributes and behaviors, entirely new elements have been added to make it easier for the user to interact with the program as it is running. A click of the *right* mouse button now calls up a series of menus (see Fig. 9.21) that provide quick access to essentially all of EINSTein's key data-entry dialogs. There are also additional options to both probe and alter the environment. For example, the user may choose to "inspect a site" to see the properties of a particular terrain element (if the site is "empty") or to be given a complete list of agent attributes if the site is occupied by an agent. Or, the user can pick up an agent (or a goal), move it from one part of the battlefield to another, and

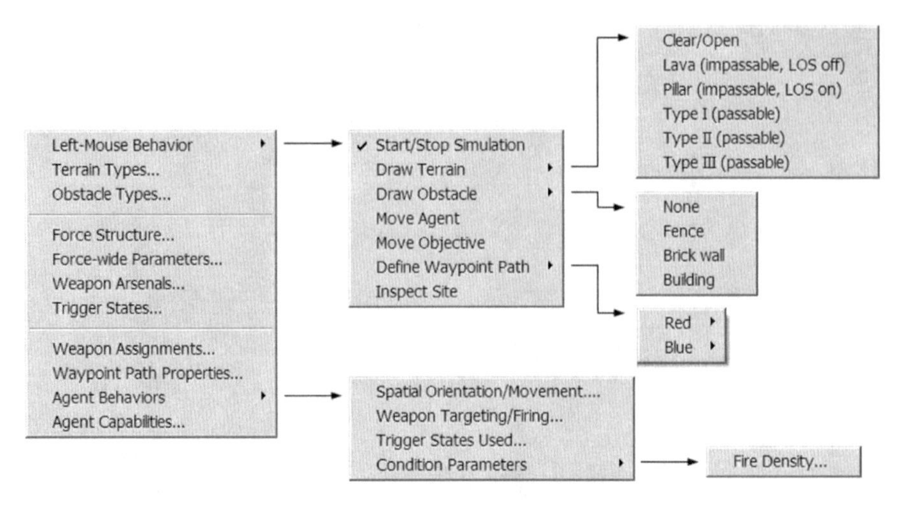

Fig. 9.21 Screenshots of pop-up menus that appear after clicking with the right mouse button in EINSTein versions 2.0 and higher.

continue running the same scenario. Terrain elements, obstacles, waypoints, and paths can also all now be drawn freehand directly on the battlefield.

The types and contents of the scenario and data files depends on the version of the program. Versions 1.0 and older use basic ASCII text files. Starting with versions 1.1 (and newer), EINSTein uses Apache Software's open-source Xerces XML (eXtensible Markup Language) parser in C++ (version 2.3 [73]) to provide XML I/O functionality. Compatibility with EINSTein's older *.dat file format has been retained in the form of I/O functions that appear under the Load and Save options of the main File menu. EINSTein has also been significantly enhanced with data logging and data exporting functions.

9.9.4 Python and wxWidgets

One of the most exciting changes that has been made to EINSTein in recent years is its integration within Python. Python is an interpreted, interactive, object-oriented programming scripting language. It is freely available in source or binary form from the Python developer homepage [69]. Since Python is easily extended with new functions and data types implemented in C or C++ (or other languages callable from C), when combined with the 300+ source-level functions of EINSTein's core engine, Python becomes a powerful programming environment in which to build custom "combat scenario" applications.

PyE (version 2.1) is an extension of EINSTein that allows the simulation to be run directly within the Python programming environment. On the

GUI level, PyE builds on (the pre-Python) EINSTein source code but is without the earlier versions' embedded graphical visualization tools (although the data-extraction routines remain the same). The idea is to first use PyE interactively to generate (and/or fine-tune) desired scenarios and then run PyE in batch mode to generate statistics for follow-on analysis by some other program. A sample Python-script for multiple time-series runs is included in the automatic Windows install file [71].

Future versions of EINSTein (and PyE) will migrate entirely away from EINSTein's original Windows MFC codebase and use wxWidgets [74] as the development platform. wxWidgets is a cross-platform, open-source GUI library that provides a single API for simultaneously developing applications on multiple platforms (such as Win32, Linux/Unix, Mac OS X, X11, and Motif) using a single codebase.

9.10 Next Step

While EINSTein continues to be developed and new capabilities are still being added to the program as of this writing (April 2008), a new direction has recently been taken in a related project. The project's goal is to develop an EINSTein-based model of terrorist-counterterrorist coevolutions called SOTCAC (= *S*elf-*O*rganized *T*errorist-*C*ounterterrorist *A*daptive *C*oevolutions).

SOTCAC leverages and generalizes EINSTein's "combat agents" to a class of "terrorist agents" that live in an abstract social-information space – that is, a mathematical graph whose vertices possess an inner semantic structure, and assimilate, process, and adapt to various forms of information (such as a multidimensional feature space that describes the properties of the Al Qaeda terrorist network, as derived from intelligence sources). SOTCAC uses adaptive agents to describe the self-organized, emergent behavior of terrorist networks – conceived as complex adaptive systems – on three interrelated dynamical levels: (1) *dynamics on networks*, in which notional terrorist agents process and interpret information, search and acquire resources, and adapt to other agents' actions; (2) *dynamics of networks*, in which the terrorist network itself is a fully dynamic, adaptive entity and whose agents build, maintain, and modify the network's local (and therefore, collectively, its global) topology; and (3) dynamics *between networks*, in which the terrorist network and counterterrorist network mutually coevolve; the terrorist network's "goal" is to achieve the critical infrastructure (of manpower, weapons, financial resources, and logistics) required to strike, while the counterterrorist network's mission is to prevent the terrorist network from achieving its goal.

For all of EINSTein's "complexity" on the source code level (EINSTein consists of roughly 150K lines of C++ code, contains over 350 agent functions, and requires a programming manual that is about 400 pages long to define its class structure and primitive behaviors), EINSTein's basic dynamics is – conceptually speaking – very simple: Red and blue agent swarms interpenetrate on a notionally physical battlefield with the "goal" of annihilating the other side and/or "controlling" a certain region of the battlefield (such as the opponent's "flag"). There are effectively two – and only two – kinds of actions that agents must adjudicate as a battle unfolds: (1) *where to move* and (2) *whom to fight*; and everything "happens" in a single notional space (i.e., the "battlefield").

In contrast, the behaviors in SOTCAC depend on (and take place in) four coupled spaces: (1) a *physical space*, analogous to EINSTein's "battlefield"; (2) a *social space*, in which terrorist agents forge/break social bonds and exchange information and materiel resources; (3) an *"image" of the physical space* that represents the counterterrorist network's "best guess" view of the terrorist's activity in the physical space; and (4) an *"image" of the social space* that represents the counterterrorist network's "best guess" view of the terrorist network's activity in its social arena. Not only are the physical and social spaces obviously coupled, but because all counterterrorist actions are fundamentally grounded on what the counterterrorist network "believes" the terrorist network is "doing" at any given time (possibly erroneously), the coupled terrorist physical/social space ("ground truth") is also tightly coupled with the counterterrorist belief network.

SOTCAC's agents must fuse multiple streams of (often conflicting information), decide to whom (and when, and for how long) to establish communication (and/or material flow) links, and – in the case of counterterrorist forces – decide on courses of action based on an incomplete "snapshot" view of what is currently known about (and/or can be inferred from) existing data. Consequently, SOTCAC's effective "space of possible interactions" (i.e., its emergent fitness landscape) is significantly larger than EINSTein's. SOTCAC is currently in development.

Acknowledgements US Marine Corps LtGen (Ret) Paul van Riper's vision of applying the lessons of complexity theory to warfare directly inspired the ISAAC and EINSTein projects. I would like to thank Lyntis Beard (who coined both of the names ISAAC and EINSTein), Rich Bronowitz, David Broyles, Dave Kelsey, David Mazel, Katherine McGrady, Igor Mikolic-Torriera, Mike Shlesinger, Jonathan Schroden, Greg Swider, David Taylor, and Wendy Trickey, all of whom helped form and shape EINSTein over the years. Programming support for the EINSTein project continues to be very skillfully provided by Fred Richards. Funding was provided, in part, by the Office of Naval Research (Contract No. N00014-96-D-0001).

References

1. Lanchester F W (1995) Aircraft in Warfare. Lanchester Press, Sunnyvale, California
2. Chase J V (1902) A Mathematical Investigation of the Effect of Superiority in Combats Upon the Sea. In: Fiske B A (1994) The Navy as a Fighting Machine. U.S. Naval Institute Press, Annapolis, Maryland

3. Osipov M (1995) The Influence of the Numerical Strength of Engaged Forces in Their Casualties. Naval Research Logistics 42:435–490
4. Hofbauer J, Sigmund K (1988) Evolutionary Games and Population Dynamics. Cambridge University Press, Cambridge, United Kingdom
5. Taylor J G (1983) Lanchester Models of Warfare. Operations Research Society of America, Arlington, Virginia
6. Hartley D S, Helmbold R L (1995) Validating Lanchester's Square Law and Other Attrition Models. Naval Research Logistics 42:609–633
7. Weiss H K (1957) Lanchester-Type Nodels of Warfare. Proceedings of 1st Conference on Operations Research, Operations Research Society of America. Arlington, Virginia
8. Fain J (1975) The Lanchester Equations and Historical Warfare. In: Proceedings of the 34th Military Operations Research Symposium. Military Operations Research Society, Alexandria, Virginia
9. Richardson L F (1960) Statistics of Deadly Quarrels. Boxwood Press, Pittsburgh, Pennsylvania
10. Cowan G A, Pines D, Meltzer D (1994) Complexity: Metaphors, Models and Reality. Addison-Wesley, Reading, Massachusetts
11. Kauffman S (1993) Origins of Order. Oxford University Press, New York
12. Langton C G (1995) Artificial Life: An Overview. MIT Press, Cambridge, Massachusetts
13. Mainzer K (1994) Thinking in Complexity. Springer-Verlag, New York
14. Beckerman Linda P (1999) The Non-Linear Dynamics of War. Science Applications International Corporation, `http://www.calresco.org/beckermn/nonlindy.htm`
15. Beyerchen A (1992) Clausewitz, Nonlinearity, and the Unpredictability of War. International Security 17:59–90
16. Hedgepeth W O (1993) The Science of Complexity for Military Operations Research. Phalanx 26:25–26
17. Ilachinski A (1996) Land Warfare and Complexity, Part I. Center for Naval Analyses, Alexandria, Virginia
18. Ilachinski A (1996) Land Warfare and Complexity, Part II. Center for Naval Analyses, Alexandria, Virginia
19. Miller L D, Sulcoski M F (1995) Foreign Physics Research with Military Significance. Defense Intelligence Reference Document, NGIC-1823-614-95
20. Saperstein A (1995) War and Chaos. American Scientist 83:548–557
21. Tagarev T, Nicholls D (1996) Identification of Chaotic Behavior in Warfare. In: Sulis W, Combs A (eds) Nonlinear Dynamics in Human Behavior. World Scientific, Singapore
22. Maes P (1990) Designing Autonomous Agents. MIT Press, Cambridge, Massachusetts
23. Ferber J (1999) Multi-Agent Systems. Addison-Wesley, Reading, Massachusetts
24. Weiss G (1999) Multiagent Systems. MIT Press, Cambridge, Massachusetts
25. Gilbert N, Troitzsch K G (1999) Simulation for the Social Scientist. Open University Press, Philadelphia, Pennsylvania
26. Gilbert N, Conte R (1995) Artificial Societies. UCL Press, London
27. Conte R, Hegselmann R, Terna P (1997) Simulating Social Phenomena. Springer-Verlag, New York
28. Barrett C (1997) Simulation Science as it Relates to Data/Information Fusion and C2 Systems. Los Alamos
29. Epstein J M, Axtell R (1996) Growing Artificial Societies. MIT Press, Cambridge, Massachusetts
30. Prietula M J, Carley K M, Gasser L (1988) Simulating Organizations. MIT Press, Cambridge, Massachusetts
31. Ilachinski A (2000) EINSTein. Center for Naval Analyses, Alexandria, Virginia
32. Ilachinski A (1999) EINSTein User's Guide. Center for Naval Analyses, Alexandria, Virginia

33. Ilachinski A (1997) Irreducible Semi-Autonomous Adaptive Combat (ISAAC). Center for Naval Analyses, Alexandria, Virginia
34. Ilachinski A (2004) Artificial War. World Scientific, Singapore
35. Ilachinski A (2001) Cellular Automata. World Scientific, Singapore
36. Braitenberg V (1984) Vehicles. MIT Press, Cambridge, Massachusetts
37. Boccara N, Roblin O, Roger M (1994) Automata Network Predator–Prey Model with Pursuit and Evasion. Physical Review E 50:4531–4541
38. Woodcock, A E R, Cobb L, Dockery J T (1988) Cellular Automata: A New Method for Battlefield Simulation. Signal 1:41–50
39. Varela F J, Maturana H, Uribe R (1974) Autopoiesis. Biosystems 5:187–196
40. Lauren M K (2002) Firepower Concentration in Cellular Automata Models. Journal of the Operations Research Society 53:672–679
41. Krantz H, Schreiber T (1997) Nonlinear Time Analysis. Cambridge University Press, Cambridge
42. Lauren M K (1999) Characterizing the Difference Between Complex Adaptive and Conventional Combat Models. Defense Operational Technology Support Establishment, New Zealand
43. Lauren M K, McIntosh G, Moffat J (2007) Fractals: From Physical Science to Military Science. Phalanx 40:8–10
44. Dockery J T, Woodcock A E R (1993) The Military Landscape. Woodhead Publishing Limited, Cambridge
45. Bak P (1996) How Nature Works. Springer-Verlag, New York
46. Roberts D C, Turcotte D L (1988) Fracticality and Self-Organized Criticality of Wars. Fractals 6:351-357
47. Taylor D, Schmal C, Hashim A (2000) Ground Combat Study: Summary of Analysis. Center for Naval Analyses, Alexandria, Virginia
48. Edwards S J A (2000) Swarming on the Battlefield. RAND Corporation, Santa Monica, California
49. Dewar J A, Gillogly J, and Juncosa M (1991) Non-Monotonicty, Chaos, and Combat Models. RAND Corporation
50. Mitchell M (1996) An Introduction to Genetic Algorithms. MIT Press, Cambridge, Massachusetts
51. Thulasiraman K, Swami M (1992) Graphs: Theory and Algorithms. John Wiley and Sons, New York
52. Schroden J, Broyles D (2005) EINSTein Simulations of Squad and Fire Team Operations. Report 11705, Center for Naval Analyses, Alexandria, Virginia
53. Gill A (2001) Impact of Reconnaissance on Mission Success. Defense Science and Technology Organization, Australia, Briefing Slides
54. Baigent D, Lauren M (2000) Exploring the Recce-Counter Recce Scenario with ISAAC. Defense Operational Technology Support Establishment, DOTSE Report 171, NR 1349
55. Schroden J, Broyles D (2005) An EINSTein Simulation of a Conventional Ground Combat Scenario. Report 10873, Center for Naval Analyses, Alexandria, Virginia
56. Broyles D, Trickey W, Schroden J (2005) Ground Combat Modeling in EINSTein. Report 13192, Center for Naval Analyses, Alexandria, Virginia
57. Broyles D, Trickey W (2006) Modeling Dispersed Units in Open and Urban Environments with EINSTein, Volume I. Report 14187, Center for Naval Analyses, Alexandria, Virginia
58. Perla P, Ilachinski A, Hawk C, Markowitz M, Weuve C (2002) Using Gaming and Agent Technology to Explore Joint Command and Control Issues. Report 7164, Center for Naval Analyses, Alexandria, Virginia
59. Woodaman, R (2000) Agent-Based Simulation of Military Operations Other Than War, Small Unit Combat. Master's Thesis, Naval Postgraduate School, Monterey, California

60. Kewley R (2001) Combat Operations Support System for Adaptive Command and Control. United States Military Academy, Briefing Slides
61. Freedman T, Ilachinski A (2004) Riots and Civil unrest: An Agent-Based Model Approach. Center for Naval Analyses Briefing Slides
62. Lauren, M, Stephen R (2000) Modeling Patrol Survivability in a Generic Peacekeeping Setting Using ISAAC. Defense Operational Technology Support Establishment, DOTSE Report 172, NR 1350
63. Cox G (2002) EINSTein Visits the Persian Gulf (via Monterey, CA): A Naval Requirements War "Gamelet" Held at the Naval Postgraduate School, Monterey, California. Annotated Brief, Center for Naval Analyses, Alexandria, Virginia
64. Klingaman, Randall R, Carlton B (2002) EINSTein Model Validation, Military Academy, West Point New York, Department of System Engineering, Operations Research Center of Excellence Technical Report DSE-TR-02-03
65. Axtell R, Axelrod R, Epstein J, Cohen M (1996) Aligning Simulations Models: A Case Study and Results. Computational and Mathematical Organization Theory, 1:123–141
66. Wolfram Mathematica website, http://www.wolfram.com/products/mathematica/index.html
67. Maple website, http://www.maplesoft.com/products/Maple11/professionals/
68. Graphics Server website, http://www.graphicsserver.com/
69. Python developer homepage, http://www.python.org/
70. ISAAC website, http://www.cna.org/isaac/
71. EINSTein website, http://www.cna.org/isaac/PyE_setup.htm
72. EINSTein previous releases website, http://www.cna.org/isaac/einstein_install.htm
73. Apache Software's Xerces XML parser, http://xml.apache.org/xerces-c/index.html
74. wxWidgets developer homepage, http://www.wxwidgets.org/
75. EINSTein's programming guide, http://www.cna.org/isaac/EINSTein/PyEProgRef.pdf
76. Axelrod R (1997) The Dissemination of Culture: A Model with Global Polarization. Journal of Conflict Resolution, 41:203–226
77. Wolfram S (1994) Cellular Automata and Complexity: Collected Papers. Addison-Wesley, Reading, Massachusetts

Part III
Artificial Chemistries

Chapter 10
From Artificial Chemistries to Systems Biology: Software for Chemical Organization Theory

Christoph Kaleta

Artificial Chemistries abstract from real-world chemistries by reducing them to systems of interacting and reacting molecules. They have been used to study phenomena in a wide array of fields like social and ecological modelling, evolution or chemical computing. Artificial Chemistries are inherently difficult to study and, thus, methods have been proposed to analyze their complexity. This chapter outlines how the concept of chemical organization and software dedicated at their analysis can help to ease this task. The chemical organizations of a reaction network correspond to sets of molecules that can coexist over long periods of (simulation-) time. Thus, they can be used to study the full dynamic behavior a system can exhibit without the need to simulate it in every detail. Due to this appealing property, Chemical Organization Theory has been used in the study of a wide array of systems ranging from Artificial Chemistries to real-world chemistries and biological systems. Especially the analysis of biological systems motivated an integration of the tools dedicated to the study of chemical organizations into an application framework from Systems Biology. The benefit of this integration is that tools from Systems Biology can be used without much effort along with the tools for the computation of chemical organizations and vice versa. Thus, software for the analysis of chemical organizations seamlessly integrates into a framework covering almost any aspect of network design and analysis.

10.1 Introduction

The quest of understanding the emergence and development of life on Earth has a long-standing history in research. While the concept of Darwinian Evolution readily explains the appearance of more and more complex organisms through inheritance, mutation and selection, it is still not fully understood how the first organism for selection to act upon came to life. Artificial Life and, in particular, Artificial Chemistry are trying to shed light on this question by taking an approach that is in some parts opposed to the one taken by biology. These fields try to understand real living systems not by considering them in mechanistic detail, but from an abstract point of view. Thus, they study how concepts like self-replication or information processing and storage might have evolved. An abstraction borrowed from real chemistries that is common to all systems in Artificial Chemistries is that of colliding molecules. Thus, [9] defined an Artificial Chemistry as "[..] a man-made system which is similar to a real chemical system." Formally, an Artificial Chemistry is defined by the triple (S, R, A), with S denoting the set of possible species, R denoting the set of collision rules among the species, commonly also referred to as reactions, and A denoting an algorithm describing the reaction vessel and how the collision rules are applied to the species [9]. An example for a simple Artificial Chemistry can be found in Fig. 10.1.

Artificial Chemistries have seen a wide array of applications. Those applications can be aligned on three broad axes: modelling, information processing and optimization [9]. In the first direction they have been used to create and analyze models of ecological and social systems as well as evolution. Prominent fields of Artificial Chemistry in information processing are, for example, chemical computing or DNA computing. Due to their ability to create evolutionary behavior, they have also been used for optimization in the context of evolutionary algorithms. More details on some of these applications in connection to Chemical Organization Theory are outlined in Section 10.3.3. For a comprehensive overview on the field of Artificial Chemistry the reader is referred to [9].

Due to the inherent complexity of systems in Artificial Chemistry, methods to analyze their behavior are needed. One of these is Chemical Organization Theory [8], which allows to study the dynamic behavior of chemical reaction networks, a particular type of Artificial Chemistry. In contrast to concepts that explicitly take into account space (e.g., dissipative particle dynamics, Chapter 11), a chemical reaction network in the context of Chemical Organization Theory is best visualized as a well-stirred reaction tank containing a finite set of species reacting with each other through a finite set of reactions. Thus, the concentrations of the species are assumed to be homogeneous. In consequence, the system can be described by a set of species, reactions among the species and the underlying dynamics of the reactions alone (see Fig. 10.1 for an example). Chemical Organization Theory focuses on the structure of a reaction network (i.e., the set of species and reactions). From this structure it tries to infer constraints on the dynamics of the reaction network. The set of chemical organizations of a reaction network consists of sets of species that are likely to coexist over a long period of simulation-time of the dynamics. In consequence, the dynamics can be mapped to a movement in the set of organizations [8].

The relationship between the structure and the dynamics of a reaction network can be derived by imposing two conditions on a set of species to be an organization. Both serve to guarantee the stability of this set. First, the species set is required to be *closed.* This condition is fulfilled if there is no reaction in the network that could produce a species not within the set. Thus, the dynamic of a reaction network will be restricted to the set of species in an organization if they were present at the beginning of the simulation (with some species possibly vanishing). The second condition, *self-maintenance*, ensures that there exists a flux through the reactions of the network (i.e., an instance of the dynamics, such that no species vanishes). Even though the existence of such a flux does not guarantee that the species of an organization will persist during a simulation, it can be shown in the other direction that if such a species set is encountered during a simulation, it corresponds to an organization. A formal definition of these concepts is given in Section 10.3.1.

Chemical Organization Theory has seen a wide array of applications in both Artificial Chemistry and Systems Biology. In Artificial Chemistry it has been used to study systems for chemical computing [23, 25], to serve as a design principle in chemical computing [7] and to study chemical evolution [24]. In Systems Biology it has been used to study models of HIV infection [22], the diauxic shift in *Escherichia coli* [4, 18], as well as the properties of a genome-scale network of the same organism [5] and effects of gene knock-outs [18]. Some of these applications are presented in more detail in Section 10.3.3.

Due to a close relationship of the concept of a chemical reaction network to the understanding of a metabolic network in Systems Biology, many tools from this field can be applied to the study of such networks. In the other direction, Chemical Organization Theory can be used to complement methods developed in Systems Biology. Thus, the tools for the study of chemical organizations were developed with the aim to allow an easy integration into a whole array of existing tools from Systems Biology.

This integration is achieved on two different levels. First, chemical reaction networks are represented in the *Systems Biology Markup Language* (SBML) format [16]. Even though species and reactions can be stored in a simple text-file, different kinds of analysis of such networks focus on different aspects. Thus, a common representation serves, on the one hand, to circumvent the cumbersome way of interconvert reaction networks between different data formats. On the other hand, an integrated representation of all aspects that might be of interest is achieved. One analysis might focus, for example, on the set of reactions and species, while another analysis necessitates information on the kinetics of the reactions. If a reaction network has been designed with the former kind of analysis in mind, the second kind can be applied by "just" adding information about the kinetics to the already available set of reactions.

Second, an integration on the level of applications is achieved. Using SBML as a representation already makes available a whole array of tools that use this standard (see [32] for a comprehensive list of applications using SBML). Furthermore, the *Systems Biology Workbench* (SBW) [31] offers an even deeper integration of applications that not only encompasses the exchange of reaction networks between two applications but also the possibility to call specific functions offered by each application that interface SBW. Thus, one application can be used to design a reaction network. If one wants to study the dynamic of this network under a particular condition for a certain time span, a function from another application offering this *service* through SBW can be called. Thus, it is, on the one hand, not necessary to develop such a tool anew or integrate the corresponding code into the own project. On the other hand, the cumbersome process of saving the network, initializing the simulation tool, loading the network again and specifying the parameters of the simulation can be circumvented and all functionality is available from the initial application. A more comprehensive overview on SBML and SBW is given in Section 10.2.

This chapter is organized as follows. A short overview on the tools from Systems Biology that can be used with the application for the analysis of chemical organizations is outlined in Section 10.2. A comprehensive introduction into the concept of chemical organizations is given in Section 10.3.1. Details on the algorithms for the computation of chemical organizations and applications of Chemical Organization Theory in the field of Artificial Chemistry are presented in Sections 10.3.2 and 10.3.3. Finally, applications for the computation and analysis of chemical organizations in reaction networks are outlined in Section 10.4.

10.2 Using Tools from Systems Biology: SBML and SBW

Even though Chemical Organization Theory has been introduced in the field of Artificial Chemistries, it has seen a wide array of applications in Systems Biology. Consequently, it makes use of some tools commonly used in this field of research.

10.2.1 Systems Biology Markup Language

Reaction networks can be stored using simple text files. However, this simplicity can result in every research group using proprietary formats that are most appropriate to their needs. In consequence, it is difficult to interconvert reaction networks between different formats. A standardized representation of reaction networks that incorporates the most commonly used aspects of their representation allows a simpler interchange of models and reduces mistakes made during interconversion.

The Systems Biology Markup Language (SBML) [15, 16] is proposed as a standard for the representation of reaction networks in the field of Systems Biology. Through its growing acceptance, SBML is extended to meet the needs of its users. In its current version, *level 2*, it allows the definition of compartments, species, reactions, units, rules and events. Through the definition of compartments and relations between the compartments, some basic spatial differentiation seen in living systems can be integrated into a reaction network. Reactions can be described by rate laws. Since these laws are formulated using *MathML*, arbitrary functions can be described. Using rules, concentrations and parameters can be constrained. Events offer the possibility to perturbate networks at given time-points by, for example, setting the concentration of a species or changing a specific parameter. Additionally it is possible to include user-defined tags that enable, for example, the display of information in HTML format or the incorporation of positional information (e.g., for the display in network editors). Since SBML is used by a growing array of applications virtually covering any aspect of network analysis, it offers a sophisticated base that can also be useful in the field of Artificial Chemistry.

10.2.2 Systems Biology Workbench

Using the Systems Biology Workbench (SBW) [31] the unification of reaction networks based on SBML is extended to the level of applications. SBW allows the integration of some commonly used steps in network design and analysis. SBW basically represents a message passing architecture. Its central element is a broker to which applications, written in a wide variety of programming languages, can interface. Messages can be passed from one application to another allowing applications that interface to SBW to communicate between each other. The aim of SBW is to offer a framework that allows the reuse of software written in heterogeneous programming languages. Each tool interfacing to SBW offers a set of methods that can be called by other applications that interface to SBW. One of the most basic methods is the *doAnalysis()* method that is offered by most SBW-compliant applications. This method is called with a string representing a reaction network in SBML format. This application then loads the network, which can subsequently be examined with the analysis methods offered. In connection with SBW interfacing network editing and design tools like *JDesigner* [31] and *CellDesigner* [20], tools for deterministic and stochastic simulation like *Jarnac* [31] and *Dizzy* [30] as well as analysis tools like *METATOOL* [27] and *Bifurcation Discovery Tool* [6], the applications for the analysis of chemical organizations are integrated into one easily expandable framework.

10.3 Chemical Organization Theory

This section introduces formally the concept of chemical organizations and gives some details about the algorithms that can be used for their computation. This is followed by an overview on recent applications of this concept.

10.3.1 Background

Extending ideas by Fontana and Buss [12], the theory of chemical organizations [8] provides a new method to analyze complex reaction networks. It allows one to predict sets of species that are likely to coexist over long periods of simulation time (i.e, the potential phenotypes of the reaction network). A *reaction network* $\langle \mathcal{M}, \mathcal{R} \rangle$ is described by a set of species $\mathcal{M}$ and a set of reactions $\mathcal{R} \subseteq \mathcal{P}_M(\mathcal{M}) \times \mathcal{P}_M(\mathcal{M})$ among these species, with $\mathcal{P}_M(\mathcal{M})$ denoting the set of all multisets with elements from $\mathcal{M}$. Furthermore, each reaction network $\langle \mathcal{M}, \mathcal{R} \rangle$ implies an $m \times n$ stoichiometric matrix $\mathbf{S} = (s_{i,j})$, where $s_{i,j}$ is the number of molecules of species $i \in \mathcal{M}$ that is produced in reaction $j \in \mathcal{R}$ (i.e., right-hand side minus left-hand side). An example for a reaction network and the associated stoichiometric matrix is given in Fig. 10.1.

To be an organization, a species set has to fulfill two criteria: closure and self-maintenance. The closure condition ensures that no new species appear through the reactions among the species of an organization. We call a set of species M *closed* if there exists no reaction $r = M_1 \times M_2$ with $M_1 \subseteq M$ and $M_2 \nsubseteq M$.

The self-maintenance condition ensures that there exists a flux through the reactions, such that no species in M will vanish over time. The dynamics of the reaction network can be written using the ordinary differential equation

$$\dot{\mathbf{x}}(t) = \mathbf{S}\mathbf{v}(\mathbf{x}(t)),$$

with $\mathbf{x}(t)$ being the concentration values of the species in $\mathcal{M}$ at time t and $\mathbf{v}(\mathbf{x}(t))$ being the flux through the reactions as a function of $\mathbf{x}(t)$. The self-maintenance condition requires the existence of a flux vector $\mathbf{v} \in \mathbb{R}^n$ containing positive entries for all reactions among the species in M and zero entries for the remainder such that the concentration change given as $\mathbf{S}\mathbf{v}$ is nonnegative. Formally, we call a set M *self-maintaining* if:

1. For every reaction $(\mathcal{A} \rightarrow \mathcal{B}) \in \mathcal{R}$ with $\mathcal{A} \in \mathcal{P}_M(M)$, its corresponding flux is $\mathbf{v}_{\mathcal{A}\rightarrow\mathcal{B}} > 0$.
2. For every reaction $(\mathcal{A} \rightarrow \mathcal{B}) \in \mathcal{R}$ with $\mathcal{A} \notin \mathcal{P}_M(M)$, its corresponding flux is $\mathbf{v}_{\mathcal{A}\rightarrow\mathcal{B}} = 0$.
3. For every species $i \in M$, its concentration change is nonnegative: $(\mathbf{S}\mathbf{v})_i \geq 0$.

Please note that this condition also includes the steady state condition that is a commonly used constraint in the stoichiometric analysis of metabolic networks like flux balance analysis [34] or elementary mode analysis [33].

$$
\mathcal{R} = \{
$$

$$
\begin{aligned}
1\,a &\rightarrow 2\,a \\
1\,a + 1\,b &\rightarrow 1\,a + 2\,b \\
1\,d + 1\,c &\rightarrow 1\,a + 2\,c \\
1\,d &\rightarrow 2\,d \\
1\,a &\rightarrow \\
1\,b &\rightarrow \\
1\,c &\rightarrow
\end{aligned}
$$

$$
\}
$$

$$
\mathbf{S} = \begin{pmatrix} 1 & 0 & 1 & 0 & -1 & 0 & 0 \\ 0 & 1 & 0 & 0 & 0 & -1 & 0 \\ 0 & 0 & 1 & 0 & 0 & 0 & -1 \\ 0 & 0 & -1 & 1 & 0 & 0 & 0 \end{pmatrix}
$$

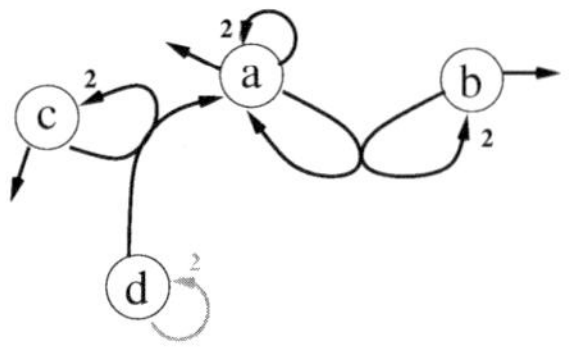

Fig. 10.1 Example for a reaction network. In the upper left, the set of reactions is given. **S** is the stoichiometric matrix of the system. Rows correspond to species and columns correspond to reactions in the same order as in the list of reactions. A schematic drawing of the reaction network is depicted in the lower half of the figure. The light gray reaction is added to the network for the computation of the second Hasse diagram of organizations in Fig. 10.2.

The set of organizations, together with the set inclusion $\subseteq$, forms a partially ordered set that can be visualized as a hierarchical structure in a Hasse diagram (see Figs. 10.2a and 10.2 for the Hasse diagram of the reaction network in Fig. 10.1). Linking to the dynamics, [8] has shown that all fixed-points of the system are instances of organizations, given that the dynamics of the system is described by a set of ordinary differential equations. This represents one of the most important aspects of chemical organization theory, since it allows one to infer statements about the possible qualitative states of a reaction network during simulation from its stoichiometry alone. Furthermore, a simulation of a network can be mapped to a movement in the Hasse diagram of organizations. For a simulation using ordinary differential equations, the floating-point concentrations need to be mapped to a set of present species. This is done by assuming only species present that have a concentrations above a predefined threshold Θ. This threshold could, for example, relate to a concentration equaling one molecule of a species. An example for the mapping of a simulation to the Hasse diagram of organizations can be found in Figs. 10.2b and 10.2c.

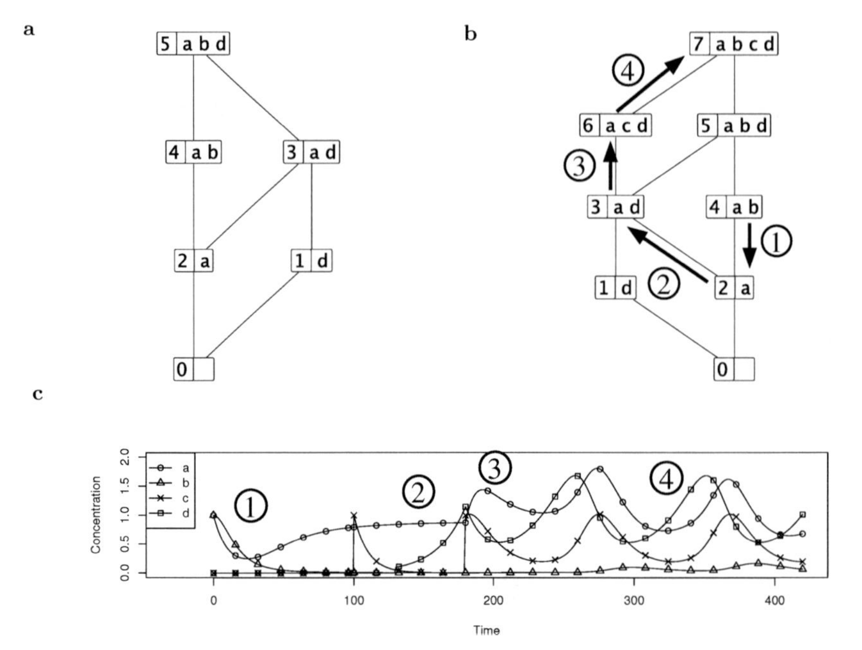

Fig. 10.2 Example for the movement in the Hasse diagram of organizations. Hasse diagrams of the chemical organizations of the reaction network in Fig. 10.1 **a** without and **b** with the self-replicating reaction for species d. **c** Example for the relation between the dynamics of a network and a movement in the Hasse diagram in part b. (1) At $t = 0$, the network is initialized with $[a] = 1$ and $[b] = 1$ corresponding to organization 4. Due to the kinetic laws, species b vanishes (i.e., its concentration falls below the predefined threshold Θ), corresponding to a downward movement into organization 2. There is no organization containing $\{a, c\}$; hence, $[c]$ raised to 1 at $t = 100$ rapidly vanishes (until $t = 130$). (2) At $t = 130$ $[d]$, is raised to 0.1 and rapidly accumulates, since the only reaction consuming d needs c as an additional educt, which is only present in very small amounts. Thus, the system moves to organization $\{a, d\}$, since still $[c] < \Theta$. (3) $[c]$ is raised to 1 to stop the accumulation of d. In consequence, the system moves upward into $\{a, c, d\}$. (4) Until $t = 180$, the decay of b was faster than its production through reaction 2. The peaks of the oscillations of $[a]$ starting in $t = 180$ allow a production of b greater than its decay. Thus, it accumulates, finally leading to an upward movement in the Hasse diagram when $[b] > \Theta$.

10.3.2 Algorithms for the Computation of Chemical Organizations

Two deterministic and one heuristic algorithm for the computation of chemical organizations have been developed [5]. While the deterministic algorithms can be applied to medium-scale networks of up to 100 reactions and molecules, the heuristic algorithm can find a subset of all organizations in larger networks. Thus, even networks with up to 1000 species and reactions have been analyzed [5].

Constructive Approach

The constructive approach for the computation of chemical organizations focuses on the Hasse diagram of organizations. The algorithm starts with the smallest organization of the network. This organization contains all species that can be generated by inflow reactions. If no inflow reaction is present, the smallest organization corresponds to the empty set. Starting from this initial organization, species sets containing this organization are being searched.

In order to find such organizations, the self-maintenance condition is relaxed to the semi-self-maintenance condition. This condition requires that if a species of an organization is consumed by a reaction that it is also produced by another reaction of the organization. It can be shown that each species set that fulfills the self-maintenance condition also fulfills the semi-self-maintenance condition. Given an initial organization, the algorithm iteratively tries to add species to the current organization. Then the closure of the species set is computed. Thus, for all reactions whose educts are present in the species set, the products are added if they are not yet present. This step is iterated until no further product can be added. Subsequently, the algorithm tries to fulfill the semi-self-maintenance condition by adding species such that producing reactions are available to species that are only consumed. Any possible combination of such production reactions are tried. By iteratively using the resulting species sets as initial species sets, all species sets fulfilling the closure and the semi-self-maintenance condition are being found. Using linear programming, each species set is subsequently tested for the self-maintenance condition. If this condition is fulfilled, the species set represents an organization.

An approach to accelerate the search for organizations can be to consider only connected organizations. In this approach, every species that is added to the network has to take part in a reaction that uses at least one species from the current species set. By computing only the connected organizations in the original and a modified version of the reaction network, all organizations can be found.

Flux Based Approach

The second approach for the computation of chemical organizations focuses on the self-maintenance condition. All flux vectors satisfying this condition lie in a convex polyhedral cone $\mathcal{P}$ in the n-dimensional flux space. This cone is defined by its (unique) set of n-dimensional spanning vectors or extreme rays. Thus, each flux vector representing a solution of the self-maintenance condition can be written as a positive linear combination of the extreme rays of $\mathcal{P}$. The inequalities defining $\mathcal{P}$ are given by

$$\mathbf{Sv} \geq 0 \text{ and } \mathbf{v} \geq 0.$$

Various algorithms for the transformation of the representation of $\mathcal{P}$ by inequalities into the set of extreme rays have been proposed. The most prominent is represented by the the double description method [13].

The extreme rays are further simplified by a transformation into the corresponding set of reactions (i.e., the set of nonzero indices). Since all flux vectors satisfying the self-maintenance condition are positive linear combinations of extreme rays, the algorithm subsequently searches for combinations of the extreme rays that satisfy the closure condition. The search for such combinations is split into three steps. In the first step, all elementary organizations of the system are computed. These are organizations that fulfill two properties. First, they are reactive organizations; that is, each species they contain participates in at least one reaction of the organization. Second, there exists no set of other organizations such that their union is equal to any of the elementary organizations. Consequently, each reactive organization can be written as the union of elementary organizations. In the second step, the elementary organizations are combined, yielding all reactive organizations. In a last step, it is iteratively tested for each organization whether there exist species that can be added without changing the set of available reactions. Thus, all organizations are found in the final step.

Heuristic

Empirical results show that the most time-consuming step in the flux-based approach is represented by the computation of the extreme rays of the convex polyhedral cone $\mathcal{P}$ representing the solution space to the self-maintenance condition. The number of extreme rays is growing exponentially with system size and, thus, the time needed for their computation in larger networks can exceed practical limits.

Since the extreme rays are not necessary for the second and third steps of the flux-based approach, a heuristic approximating the set of elementary organizations can be used. A central observation is that since the reactions of each organization fulfill the self-maintenance condition, also the union of two such reaction sets fulfills the self-maintenance condition (i.e., there exists a positive linear combination of both flux vectors satisfying condition (1) of the self-maintenance condition). Consequently, given a set of organizations, we just need to search for combinations of these organizations that fulfill the closure condition to find additional organizations. If this initial set contains all elementary organizations, the full set of organizations of the complete system can be found.

To find such an initial set of organizations, a random walk strategy can be used. This strategy starts with a randomly chosen species that is added to the current species set. Then randomly selected species that are substrate or product to at least one reaction also consuming or producing the last added species are iteratively added. After computing the closure of the final species set, it is tested whether the self-maintenance condition is fulfilled. In this case, an organization has been found.

The random walk strategy is iterated several times, yielding an initial set of organizations. Subsequently, we can search for combinations of those organizations that fulfill the closure condition to find all organizations that can be generated from unions of the initially found organization set. As demonstrated in [5], this approach can be used to find all organizations in reaction networks that contain even 1000 species if the number of organizations is small.

10.3.3 Application of Chemical Organization Theory

Chemical Organization Theory has seen a wide array of applications. In the field of Artificial Chemistry it has been applied to the study of chemical programming and chemical or prebiotic evolution. Although first introduced in the study of Artificial Chemistries, it has also seen many applications in the study of biological networks. One of its appealing properties is the possibility to predict the dynamic behavior of a model under simulation. This is of great importance if the kinetics governing a reaction network are partly or totally unknown, impossible to tackle in current solvers or too complex to analyze their potential behavior.

Application in Chemical Computing

Using chemical reactions as a base for computation has first been proposed by Banâtre and Métayer [1]. Appealing properties of such systems are the massive parallelism were the solution appears as an emergent global behavior [23]. These concepts have been used in *GAMMA* [2], *CHAM* [21], *P-systems* [28] and *MGS* [14]. Such systems are difficult to analyze and, thus, Chemical Organization Theory has been used to tackle some of the problems in their analysis.

One example is the analysis of chemical reaction networks derived from boolean networks. Boolean networks can be transformed into reaction networks using a method described in [18] and [23]. This can be done by transforming boolean variables into two species each: one corresponding to the `true` state of the variable and one corresponding to the `false` state of the variable. System states encompassing both representations of a variable at once can be resolved by the addition of a reaction mimicking mutual destruction (i.e., a reaction taking both reactions as educts and producing nothing).

An example of the transformation of the universal *Nand* operator, presented in [23], is outlined in Fig. 10.3. If the input of the boolean network is simulated by inflow reactions with each input variable being set (i.e., either the species presenting the `true` or the `false` state is supplied), the only chemical organizations of the resulting network contains a species corresponding to the output of the *Nand* operator. Since every boolean function can be constructed from the *Nand* operator, arbitrary boolean networks can be transformed into reaction networks using this approach.

The transformation of boolean networks into reaction networks can also be used to integrate a regulatory network modelled by a boolean formalism into a metabolic network. Doing this in a model of the central metabolism of *Escherichia coli* [18] showed that the chemical organizations of the resulting system corresponded to the growth on several available carbon sources. Additionally, the results of the knock-out of genes could be correctly predicted in most cases. The knock-outs can be simulated by removing the corresponding reactions from the network. The lethality of a knock-out can be assessed through the existence of no organization containing all species that are necessary for the organism to survive.

Design Principles in Chemical Computing

On a more abstract level, Chemical Organization Theory can be used as a guideline in the construction of systems devised for chemical computation [7]. The construction of such a system can be separated into two stages. In the first stage, only the set of species and reactions is constructed. The kinetics governing the reactions are devised in the second step. In the first step, Chemical Organization Theory is iteratively applied to analyze whether the constructed model exhibits the desired behavior based on the set of chemical organizations it contains. After this stage has been completed, kinetic laws can be assigned to the reactions. As argued in [7], such a design principle leads to more robust models. This is due to the fact that chemical organizations confer a structural stability to the system that is independent from the kinetic laws governing the reactions of the system.

A promising prospect of these design principles lies in the emerging field of Synthetic Biology [10]. The research in this area focuses on the combination of components like gene regulatory elements or proteins from different species to construct systems that exhibit a qualitatively new behavior. In [29], genes were used from several species to construct a recombinant strain of *Saccharomyces cerivisae* that could detect trace amounts of explosives. As outlined in [11], principles from engineering can be used to facilitate such developments organized in several layers. One of these layers includes the design of new systems. Since biological systems are often represented as reaction networks, a promising aspect of Chemical Organization Theory is the application in the construction of systems in Synthetic Biology.

a

A, B → C

b

A	B	C
`false`	`false`	`true`
`false`	`true`	`true`
`true`	`false`	`true`
`true`	`true`	`false`

c

$$\begin{aligned} \mathcal{R} = \{ & \\ 1\,\overline{A} + 1\,\overline{B} &\rightarrow 1\,C \quad (1)\\ 1\,\overline{A} + 1\,B &\rightarrow 1\,C \quad (2)\\ 1\,A + 1\,\overline{B} &\rightarrow 1\,C \quad (3)\\ 1\,A + 1\,B &\rightarrow 1\,\overline{C} \quad (4)\\ 1\,A + 1\,\overline{A} &\rightarrow \emptyset \quad (5)\\ 1\,B + 1\,\overline{B} &\rightarrow \emptyset \quad (6)\\ 1\,C + 1\,\overline{C} &\rightarrow \emptyset \quad (7)\\ 1\,C &\rightarrow \emptyset \quad (8)\\ 1\,\overline{C} &\rightarrow \emptyset \quad (9)\\ \} & \end{aligned}$$

d

$\mathcal{R} \cup \{\emptyset \rightarrow 1\,\overline{A} + 1\,\overline{B}\}$	$\{\overline{A},\ \overline{B},\ C\}$
$\mathcal{R} \cup \{\emptyset \rightarrow 1\,\overline{A} + 1\,B\}$	$\{\overline{A},\ B,\ C\}$
$\mathcal{R} \cup \{\emptyset \rightarrow 1\,A + 1\,\overline{B}\}$	$\{A,\ \overline{B},\ C\}$
$\mathcal{R} \cup \{\emptyset \rightarrow 1\,A + 1\,B\}$	$\{A,\ B,\ \overline{C}\}$

e

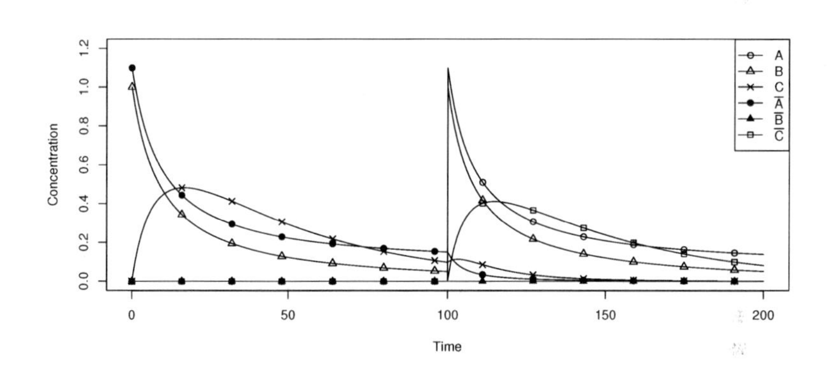

Fig. 10.3 Example for the conversion of a boolean *Nand* function into a reaction network. **a** *Nand* function with input variables A and B and output variable C. **b** Logic table of the *Nand* function. For every possible input of A and B, the output in C is given. **c** Reaction network modelling the *Nand* function. Plain symbol denotes `true` state of a variable and overlined symbol the `false` state. Reactions 5 to 7 are used to avoid inconsistent system states where the two species representing the state of a variable are present at the same time. Reactions 8 and 9 are used to reset the "output" of the network. **d** For every possible state of the input variables, the only organization of the system is given. It contains the input variables as well as a form of the output variable C. **e** Numerical simulation of the reaction network using mass-action kinetics. At $t = 0$, the network is initialized with $[\overline{A}] = 1.1$ and $[B] = 1.0$ corresponding to the input $A =$`false` and $B =$`true`. The output of the network is a rising concentration of the `true` form of variable C (line with crosses). At $t = 100$, another input is given by increasing the concentration of A to 1.1 and B to 1.0. Accordingly, this leads to an increase of the concentration of $\overline{C}$, corresponding to an output of $C =$`false` (line with squares).

Using Chemical Organization Theory for Model-checking

The correspondence of each fixed-point of a reaction network to an organization of the same system can be used to check the consistency of reaction networks. This property can even be extended to states of the system where the concentration of some species grows [19]. Thus, Chemical Organization Theory can be used to check which species of a reaction network can be present in a long-term simulation of the system. Since the analysis does not require an explicit knowledge of the kinetics of the reactions, this approach also extends to networks with partly or completely unknown kinetics.

Analyzing, for example, the reaction network depicted in Fig. 10.1 without reaction (4) and the associated Hasse diagram of organizations in Fig. 10.2a, we find that the largest organization does not contain species c. Thus, neither this species nor reactions using it as a substrate can persist in a long-term simulation of the reaction network.

Study of Chemical Evolution

Chemical Organization Theory can also be used to study chemical evolution, a process thought to have preceded the appearance of the first organism on Earth. Matsumaru et al. [24] analyzed the evolution of a chemistry of binary strings interacting with each other. The interaction between two strings produced new strings and, thus, could be transformed into a set of reaction rules. In a first setting, a system without mutation events was examined. It was found that the diversity of the system (i.e., the number of different strings diminished over time) as did the number of chemical organizations. Finally, only two organizations remained. In a second step, mutation events randomly flipping a position of a small number of strings at given time-points were introduced. In this setting, interesting effects could be observed. These events increased the diversity both in the set of species and organizations even though there is no trivial relation between the numbers of organizations and species.

Events in the evolution of a reaction network can be interpreted as either an upward, downward or sideward movement in the set of organizations of the system. A downward movement corresponds to the disappearance of certain organizations of a set of organizations and an upward movement corresponds to organizations appearing anew. A sideward movement corresponds to a mixture of both types of events. The sideward movement can be caused by species appearing anew, resulting in a upward movement instantaneously followed by a downward movement. While downward movements usually correspond to a decrease in diversity, upward movements correspond to an increase.

10.4 Software for the Computation of Chemical Organizations

The three central aspects of the software for the analysis using Chemical Organization Theory are the computation of chemical organizations, the visualization of the Hasse diagram of organizations and the integration into existing software frameworks for the analysis of complex reaction networks. These aspects are offered by the applications of *OrgTools*. The applications can be downloaded from [17]. A schematic overview on the applications of *OrgTools* and their integration into SBW is given in Fig. 10.4.

Three different tools allow the computation of the chemical organizations of reaction networks. Each of them is dedicated to some commonly appearing setting in which chemical organizations are being computed.

OrgAnalysis aims to offer some basic network editing capabilities and an integration into SBW [31]. Thus, it integrates into a rich framework of commonly used tools for network editing, simulation and analysis.

OrgFinder, in contrast, represents a basic command line tool which offers almost the same options for organization computation as *OrgAnalysis*. Thus, is can be easily integrated into scripts and allows a simple "conversion" of reaction networks into the corresponding set of chemical organizations.

OrgBatch is focused on the batch computation of chemical organizations. This is especially of interest if, for example, the evolution of a reaction network is simulated and the chemical organizations of different time steps of the network should be analyzed.

The central concept in Chemical Organization Theory is represented by the Hasse diagram of organizations. *OrgView* visualizes the Hasse diagram of organizations and thus allows a structured analysis of the chemical organizations of a reaction network. Additionally, it is possible to compare different Hasse diagrams at once and, for example, analyze the evolution of chemical organizations in the course of the evolution of a reaction network.

All of these tools use SBML [16] as a standard for the description of reaction networks. Together with an integration into SBW [31], many commonly used tools for the design and analysis of reaction networks as described in Section 10.2 become easily accessible.

10.4.1 OrgAnalysis

OrgAnalysis offers simple network editing capabilities integrated into an application for organization computation. It can either be started directly or by any other SBW-compliant application through its SBW interface. As it integrates into the Analysis section of SBW, it can be called with a string representing a reaction network in the SBML format. When called, for example,

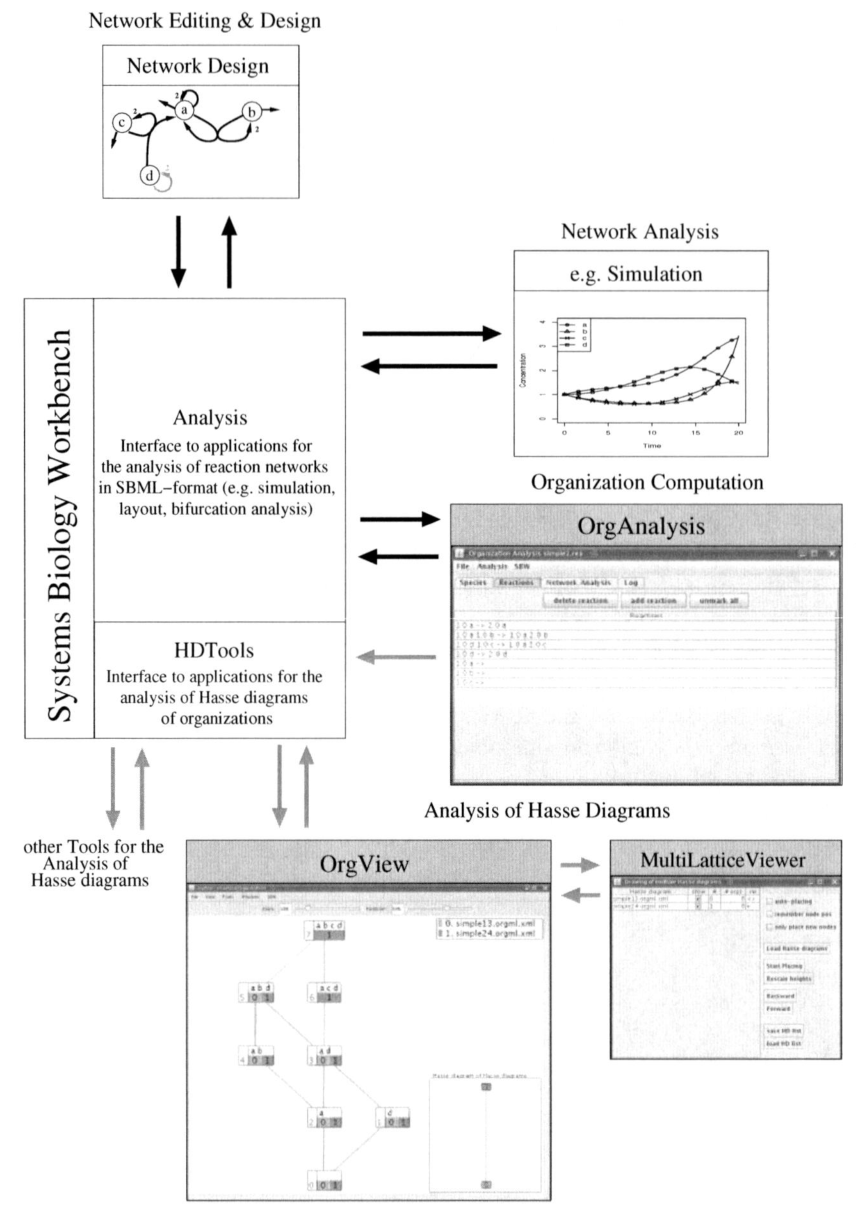

Fig. 10.4 Schematic drawing of the integration of OrgTools into SBW. Applications with gray headings denote *OrgTools*. Black arrows denote reactions networks being passed and gray arrows correspond to Hasse diagrams. First, the network is designed using, for example, *CellDesigner*. Then the network can be passed through the Analysis section of SBW (e.g., for simulation or computation of chemical organizations (*OrgAnalysis*)). The Hasse diagram of organizations can then be passed to *OrgView* for visualization. *OrgView* interfaces to *MultiLatticeViewer* to control the display of several Hasse diagrams at once. If a second Hasse diagram is loaded into *OrgView*, it is automatically forwarded to *MultiLatticeViewer*.

from a network design tool interfacing with SBW, the corresponding reaction network is directly loaded without the prior need of saving the network to a file.

Network Editing

Since there is a whole range of dedicated network editing software available, *OrgAnalysis* only concentrates on editing capabilities that help in the analysis of chemical organizations. Thus, it offers an analysis mechanism that identifies species that are only produced or consumed. Since the self-maintenance condition requires each species of an organization to be producible if it is consumed (cf. Section 10.3.1), the addition of an inflow for a species that is only consumed can lead to new organizations and, thus, to new qualitative behavior of the model during simulation. The metabolite section of *OrgAnalysis* allows one to add outflow or inflow reactions for species. Alternatively, it is possible to set a species to external; that is, it is considered present at any time and buffered by reactions outside of the system. These capabilities are complemented by a simple text-based mechanism for editing reactions.

Computation of Chemical Organizations

All of the presented algorithms for the computation of chemical organizations (Section 10.3.2) are integrated into *OrgAnalysis*. While the flux-based approach has been completely implemented in Java, the heuristic and the constructive approaches use the linear programming library *LpSolve* [3] shipped along with *OrgTools* as a subroutine.

All algorithms offer a range of options to focus the computation of chemical organizations onto certain aspects. Thus, the constructive algorithm by default only computes the set of connected organizations. This option can be of importance if the network contains many organizations that are a superset to a certain reactive organization, without containing any additional reactions [5]. Thus, they contain molecules that do not participate in any reaction within the organization. Additionally, it is possible to calculate regulatory organizations. Following an approach outlined in [18] and [23], this allows the integration of boolean rules governing the availability of reactions in the network. Some reactions can, for example, be constrained to zero flux if certain conditions formulated in a boolean formula are met (see Section 10.3.3 for details on the integration of boolean formalisms into reaction networks).

SBW Interface

The SBW interface allows the integration of *OrgAnalysis* into some commonly used network analysis tools. Additionally, it serves as interface between *OrgAnalysis* and *OrgView* to visualize the computed Hasse diagrams. Several Hasse diagrams can be visualized in the same instance of *OrgView* to allow an easier comparison of the results. Additionally, *OrgAnalysis* can call any application that possesses an interface to the Analysis section of SBW.

Model-checking

As outlined in Section 10.3.3, Chemical Organization Theory can be used to analyze the long-term behavior of reaction networks during simulation. Thus, *OrgAnalysis* offers the capability to directly examine the chemical organizations of a network and identify species and reactions that cannot be present in a long-term simulation. In some cases, only the short-time behavior is of interest. Consequently, if some species or reactions were found absent from any organization, the analysis is repeated by taking into account species whose concentration is either initially nonzero or is set to a positive value at a given time-point. If this second step detects species or reactions that do not appear in any organization, they cannot have a positive concentration, respectively, flux during simulation. This hints to possible modelling inconsistencies [19].

10.4.2 Other Applications for Organization Computation

OrgBatch

In some applications (e.g., in the simulated evolution of a reaction network), it becomes desirable to compute the organizations of a whole list of reaction networks at once. In this case *OrgBatch* can be used. It implements the constructive and the flux-based approach for organization computation, including various options that can be used with both algorithms. Additionally, it is possible to process files by size in order to obtain results for those networks first where computation time is expected to be shortest. For some networks, the computation of organization does not finish in reasonable time. As discussed in [5], this might be due to network size or network topology. In such a case, computation can be aborted either directly or after a user-specified time-span.

OrgFinder

In order to offer an the simplest possible mechanism for the computation of chemical organizations, the commandline tool *OrgFinder* can be used. By just providing a network file in SBML format, the organizations are computed directly and the Hasse diagram is written into a file in the XML-based *OrgML* format. Both deterministic algorithms for organization computation with the same options as in *OrgAnalysis* can be used.

10.4.3 OrgView

The purpose of *OrgView* is the visualization of Hasse diagrams of organizations. The Hasse diagrams can be loaded either through the SBW interface or from a file. In order to analyze a Hasse diagram, an appropriate placing of its nodes is necessary. Thus, different placing algorithms have been implemented. Multiple lattices can be loaded at once to allow an easier comparison of the chemical organizations of different networks.

Loading Hasse Diagrams

Hasse diagrams can be loaded into *OrgView* in two ways. First, *OrgView* can be called with a computed Hasse diagram from *OrgAnalysis* via the *HDTools* section of the SBW menu. Second, a Hasse diagram can be loaded from a file. This file can be either in the XML-based *OrgML* or in the text-based *.ltc* format.

Since information on the network is necessary for the visualization of the Hasse diagram, the corresponding network file in SBML format is additionally required. To facilitate the access to the network file, *OrgView* automatically searches for a network file obeying the following naming convention. For an *OrgML* file with the name *network.orgml.xml* or *network.ltc*, the corresponding network file is expected to be *network.xml* in the SBML format.

Placing of Hasse Diagrams

The nodes of the Hasse diagram of organizations are normally arranged with growing size from bottom to top. To allow a more user-friendly display of lattices, two placing algorithms have been implemented. Both offer fine-tuning of options for user-oriented demands. The simulated annealing-based placer yields the best results in positioning of nodes but has a high requirement in computation time for large Hasse diagrams. The parallel placer offers good results with a reduced requirement in computation time. Additionally, available options include the display of species only in nodes, where they appear anew; that is, each node only displays species that do not appear in any organization that are subset to this organization. Furthermore, the display of species can be completely switched off in order to concentrate on the topology of the Hasse diagram.

Displaying Several Hasse Diagrams at Once

A useful tool in organization analysis is the comparison of the chemical organizations of several reaction networks. Thus, *OrgView* allows the display of several Hasse diagrams at once. The Hasse diagrams of all networks are united and organizations that appear in several networks are indicated. A so-called "Hasse diagram of Hasse diagrams" allows the analysis of the relations of the organization of the networks between each other. Thus, if two networks contain the same set of organizations, they are shown as one node in this meta-diagram. If one network contains a subset of the organizations of another network, the node corresponding to the first Hasse diagram lies below the node corresponding to the second one (see Fig. 10.5 for an example).

The display of several Hasse diagrams at once is facilitated by *MultiLatticeViewer* integrated into *OrgView*. It allows one to load a list of lattices and the specification of the lattices to display. Additionally, it interfaces with SBW. Thus, if the same instance of *OrgView* is called several times with a Hasse diagram from *OrgAnalysis*, the Hasse diagrams are loaded into *MultiLatticeViewer*. Such a functionality is useful when designing a reaction

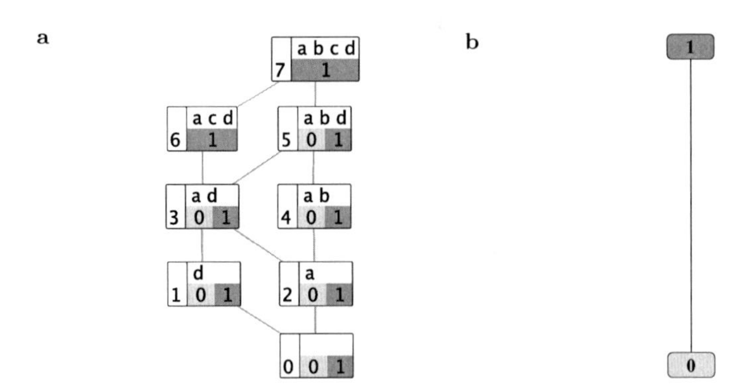

Fig. 10.5 **a** Integrated view of the two Hasse diagrams from Fig. 10.2. **b** Hasse diagram of Hasse diagrams of Fig. 10.2. The Hasse diagrams to which each organization belongs is indicated by a number below the set of species. Hasse diagram 0 contains the chemical organizations of the initial network from Fig. 10.1 and Hasse diagram 1 contains the organizations after the addition of one reaction producing species *d*. All organizations in Hasse diagram 0 appear in Hasse diagram 1. Hence, the Hasse diagram of Hasse diagram consists of a node labeled "0" below the node labeled "1."

network and examining the changes in the set of chemical organization that appear through addition or removal of certain reactions.

10.4.4 Organization Computation on the Web

Since one of the algorithms for organization computation and all of the presented tools have been implemented in Java, an applet that offers a similar functionality is available from [17]. The advantage of an applet is that it can be run without any prior installation on all systems that have a Java-capable web browser. The applet integrates the basic functionalities of *OrgAnalysis* and *OrgView*. Additionally, it contains an interface to a repository of curated biological networks in SBML format, the BioModels database [26] and a list of sample networks that have been studied using chemical organization theory. A tutorial on the use of the applet is linked on [17].

Analysis View

The analysis view allows the loading of network files by pasting a file in SBML format or a URL linking to such a file into the editor. After loading a network, it can be edited either directly in a text editor displaying the SBML or a more human-readable proprietary format, or via the same tools offered in *OrgAnalysis*. Thus, species and reactions can be added, deleted or modified. Additionally, species can be specified as external (i.e., adding an inflow and an outflow reaction). Subsequently, the chemical organizations of the reaction network can be computed and visualized. As in *OrgAnalysis*, it is additionally possible to analyze the long-term behavior of the model during simulation (see Sections 10.3.3 and 10.4.1 for more details). Thus, species and reactions that cannot be present in a long-term simulation of the reaction network or even cannot have any positive concentration or flux during the simulation can be identified.

Interface to the BioModels Database

To enlarge the list of networks available to the applet, an interface to the BioModels database [26] has been implemented. This database contains a whole list of curated biological models. When selecting a model, information available from the corresponding SBML file is displayed. Double-clicking on a model in the list loads it into the applet.

Visualization View

Some basic functionalities of *OrgView* are offered by the applet: Two placing algorithms for Hasse diagrams are available; some of the most commonly used options for the display of Hasse diagrams can also be specified; additionally, the visualized Hasse diagram can be exported into the .eps format.

Web-Integration

Another aspect of the applet is the possibility to directly access some of its capabilities from any webpage. It can be called with several parameters automatizing some of the most commonly used functions. It is, for example, possible to directly load a reaction network from a URL and display the Hasse diagram of organizations in one step. Details on this integration are given in the tutorial in [17].

10.5 Conclusion

In this chapter, *OrgTools*, a set of applications dedicated at the analysis of chemical reaction networks from Artificial Chemistry and Systems Biology, has been presented. These applications integrate into SBW and, thus, come along with a framework that covers almost any aspect in network analysis and design. Chemical Organization Theory is a good example of how work in Artificial Chemistry can benefit the work in Systems Biology. The applications of this concept reach from the study of boolean networks, the evolution of the first life to the study of "real-world" networks from Systems Biology. One of the most appealing properties of Chemical Organization Theory that help the study of such systems is the correspondence of each fixed-point of a reaction network to an organization. Since only information on the algebraic structure of the network (i.e., the set of species and reactions) is needed, even networks where information about them is unavailable can be analyzed. Thus, the species sets in the fixed points of a reaction network can be determined without needing to solve the corresponding kinetic laws analytically. In consequence, even the possible phenotypes of a reaction network (i.e., the set of species that can coexist over a long period of simulation-time) can be determined without an explicit knowledge of the kinetics of the network.

In a simplified setting, the installation of a whole system for the analysis of chemical organizations might be too much of an afford. Tools for the computation of organizations without the necessity of installation have been developed. In consequence, it is possible to compute chemical organizations and perform most of the analysis that has been described in this work only using a Java-capable web browser. Thus, many tools tailored to the specific needs of the user are available.

Acknowledgements I thank Pietro Speroni di Fenizio and Maciej Komosinski for helpful comments on the manuscript.

References

1. Banâtre, J.P., Métayer, D.L.: A new computational model and its discipline of programming. Tech. Rep. RR-0566, INRIA (1986)
2. Banâtre, J.P., Métayer, D.L.: The gamma model and its discipline of programming. Sci. Comput. Program. **15**(1), 55–77 (1990)
3. Berkelaar, M., Eikland, K., Notebaert, P.: lp solve: Open source (mixed-integer) linear programming system, version 5.5 (2005). URL `http://tech.groups.yahoo.com/group/lp_solve/`
4. Centler, F., Fenizio, P., Matsumaru, N., Dittrich, P.: Chemical organizations in the central sugar metabolism of *Escherichia coli.* Math. Model. Biol. Syst. **1**, 105–119 (2007)
5. Centler, F., Kaleta, C., di Fenizio, P.S., Dittrich, P.: Computing chemical organizations in biological networks. Bioinformatics **24**(14), 1611–1618 (2008)
6. Chickarmane, V., Paladugu, S.R., Bergmann, F., Sauro, H.M.: Bifurcation discovery tool. Bioinformatics **21**(18), 3688–3690 (2005)
7. Dittrich, P., Matsumaru, N.: Organization-oriented chemical programming. In: Hybrid Intelligent Systems, 2007. HIS 2007. 7th International Conference on, pp. 18–23 (2007)
8. Dittrich, P., Speroni di Fenizio, P.: Chemical organization theory. Bull. Math. Biol. **69**(4), 1199–1231 (2007)
9. Dittrich, P., Ziegler, J., Banzhaf, W.: Artificial chemistries – a review. Artificial Life **7**, 225–275 (2001)
10. Drubin, D.A., Way, J.C., Silver, P.A.: Designing biological systems. Genes. Dev. **21**(3), 242–254 (2007)
11. Endy, D.: Foundations for engineering biology. Nature **438**(7067), 449–453 (2005)
12. Fontana, W., Buss, L.W.: 'The arrival of the fittest': Towards a theory of biological organization. Bull. Math. Biol. **56**, 1–64 (1994)
13. Fukuda, K., Prodon, A.: Double description method revisited. In: Selected papers from the 8th Franco-Japanese and 4th Franco-Chinese Conference on Combinatorics and Computer Science, pp. 91–111. Springer-Verlag, London, UK (1996)
14. Giavitto, J.L., Michel, O.: MGS: a rule-based programming language for complex objects and collections. In: M. van den Brand, R. Verma (eds.) Electr. Notes Theor. Comput. Sci., vol. 59, pp. 286–304. Elsevier Science Publishers (2001)
15. Hucka, M., Finney, A., Bornstein, B.J., Keating, S.M., Shapiro, B.E., Matthews, J., Kovitz, B.L., Schilstra, M.J., Funahashi, A., Doyle, J.C., Kitano, H.: Evolving a lingua franca and associated software infrastructure for computational systems biology: the Systems Biology Markup Language (SBML) project. Syst. Biol. (Stevenage) **1**(1), 41–53 (2004)
16. Hucka, M., Finney, A., Sauro, H.M., Bolouri, H., Doyle, J.C., Kitano, H., Arkin, A.P., Bornstein, B.J., Bray, D., Cornish-Bowden, A., Cuellar, A.A., Dronov, S., Gilles, E.D., Ginkel, M., Gor, V., Goryanin, I.I., Hedley, W.J., Hodgman, T.C., Hofmeyr, J.H., Hunter, P.J., Juty, N.S., Kasberger, J.L., Kremling, A., Kummer, U., Novère, N.L., Loew, L.M., Lucio, D., Mendes, P., Minch, E., Mjolsness, E.D., Nakayama, Y., Nelson, M.R., Nielsen, P.F., Sakurada, T., Schaff, J.C., Shapiro, B.E., Shimizu, T.S., Spence,

H.D., Stelling, J., Takahashi, K., Tomita, M., Wagner, J., Wang, J., Forum, S.B.M.L.: The Systems Biology Markup Language (SBML): a medium for representation and exchange of biochemical network models. Bioinformatics **19**(4), 524–531 (2003)
17. Kaleta, C.: Organization computation on the web. URL `http://orgtheo.artificialchemistries.org/`
18. Kaleta, C., Centler, F., di Fenizio, P.S., Dittrich, P.: Phenotype prediction in regulated metabolic networks. BMC Syst. Biol. **2**(1), 37 (2008)
19. Kaleta, C., Richter, S., Dittrich, P.: Using chemical organization theory for model-checking (2009). Submitted
20. Kitano, H., Funahashi, A., Matsuoka, Y., Oda, K.: Using process diagrams for the graphical representation of biological networks. Nat. Biotechnol. **23**(8), 961–966 (2005)
21. Ma, W., Johnson, C., Brent, R.: Implementation of the Chemical Abstract Machine on a MIMD computer. Tech. rep., CS Lab, ANU, Australia (1996)
22. Matsumaru, N., Centler, F., di Fenizio, P.S., Dittrich, P.: Chemical organization theory applied to virus dynamics (Theorie chemischer organisationen angewendet auf infektionsmodelle). Inform. Technol. **48**(3), 154–160 (2006)
23. Matsumaru, N., Centler, F., Speroni di Fenizio, P., Dittrich, P.: Chemical organization theory as a theoretical base for chemical computing. Int. J. Unconv. Comp. **3**(4), 285–309 (2007)
24. Matsumaru, N., di Fenizio, P.S., Centler, F., Dittrich, P.: On the evolution of chemical organizations. In: S. Artmann, P. Dittrich (eds.) Proc. 7th German Workshop on Artificial Life, pp. 135–146. IOS Press, Amsterdam, NL (2006)
25. Matsumaru, N., Lenser, T., Hinze, T., Dittrich, P.: Toward organization-oriented chemical programming: A case study with the maximal independent set problem. In: F. Dressler, I. Carreras (eds.) Advances in Biologically Inspired Information Systems, *Studies in Computational Intelligence (SCI)*, vol. 69, pp. 147–163. Springer-Verlag, Berlin, Heidelberg, New York (2007)
26. Novére, N.L., Bornstein, B., Broicher, A., Courtot, M., Donizelli, M., Dharuri, H., Li, L., Sauro, H., Schilstra, M., Shapiro, B., Snoep, J.L., Hucka, M.: BioModels database: A free, centralized database of curated, published, quantitative kinetic models of biochemical and cellular systems. Nucleic Acids Res. **34**(Database issue), D689–D691 (2006)
27. Pfeiffer, T., Sánchez-Valdenebro, I., no, J.C.N., Montero, F., Schuster, S.: METATOOL: For studying metabolic networks. Bioinformatics **15**(3), 251–257 (1999)
28. Păun, G.: From cells to computers: computing with membranes (P systems). Biosystems **59**(3), 139–158 (2001)
29. Radhika, V., Proikas-Cezanne, T., Jayaraman, M., Onesime, D., Ha, J.H., Dhanasekaran, D.N.: Chemical sensing of DNT by engineered olfactory yeast strain. Nat. Chem. Biol. **3**(6), 325–330 (2007)
30. Ramsey, S., Orrell, D., Bolouri, H.: Dizzy: Stochastic simulation of large-scale genetic regulatory networks. J. Bioinform. Comput. Biol. **3**(2), 415–436 (2005)
31. Sauro, H.M., Hucka, M., Finney, A., Wellock, C., Bolouri, H., Doyle, J., Kitano, H.: Next generation simulation tools: the systems biology workbench and biospice integration. OMICS **7**(4), 355–372 (2003)
32. SBML Community: Systems Biology Markup Language (SBML) website. URL `http://www.sbml.org/`
33. Schuster, S., Fell, D.A., Dandekar, T.: A general definition of metabolic pathways useful for systematic organization and analysis of complex metabolic networks. Nat. Biotechnol. **18**(3), 326–332 (2000)
34. Varma, A., Palsson, B.: Metabolic flux balancing: Basic concepts, scientific and practical use. Bio/Technology **12**, 994–998 (1994)

Chapter 11
Spatially Resolved Artificial Chemistry

Harold Fellermann

Although spatial structures can play a crucial role in chemical systems and can drastically alter the outcome of reactions, the traditional framework of artificial chemistry is a well-stirred tank reactor with no spatial representation in mind. Advanced method development in physical chemistry has made a class of models accessible to the realms of artificial chemistry that represent reacting molecules in a coarse-grained fashion in continuous space. This chapter introduces the mathematical models of Brownian dynamics (BD) and dissipative particle dynamics (DPD) for molecular motion and reaction. It reviews calibration procedures, outlines the computational algorithms, and summarizes examplary applications. Four different platforms for BD and DPD simulations are presented that differ in their focus, features, and complexity.

11.1 Introduction

The traditional mindset of artificial chemistry (AC) is a well-stirred tank reactor with possible inflow and outflow of substrates (see Dittrich et al. [8] for a review of AC). In this framework, spatial heterogeneities within a chemical solution are usually not taken into account: The focus lies on the mere presence, absence, or concentration of chemicals rather than their spatial organization in the reaction vessel.

It is well known, however, that spatial structures can play a crucial role in chemical systems and can drastically alter the outcome of reactions. Such structures can be either imposed from the outside or they can be the result of the chemical reactions themselves. An example for imposed heterogeneities are upheld concentration gradients along a reaction vessel. An example for self-organized structures are self-assembled lipid aggregates which play a prominent role in molecular biology and its primordial origins and are thus of primary interest for the field of Artificial Life (AL) and AC.

The surplus of complexity that results from the presence of lipid structures is manifold: Closed impermeable or semipermeable membranes (liposomes) compartmentalize the reaction space thereby allowing for a variety of chemical regimes at once; proteins embedded in such membranes allow for specific transport of substances (under the use of energy even against a concentration gradient); and lipid structures can increase the effective concentration of hydrophobic chemicals and alter rate constants of reactions therein (milieu effect).

Notable AL and AC models that explicitly incorporate space are the pioneering works by Varela and Maturana [46], the lattice molecular automaton (LMA) by Mayer and colleagues [29, 30], Hutton's model of self-replicating DNA strands [24], and a model of self-reproducing abstract cells by Ono and Ikegami [36]. The common ground of all these models is that space is treated by means of cellular automata: Molecules or parts of molecules are modeled as point particles that occupy sites on a two-dimensional (2D) or three-dimensional (3D) lattice. Motion of these particles and reactions between them are defined by transition rules and an (possibly stochastic) updating algorithm is used to follow a system state through time.

Advanced method development in computational physical chemistry paired with the dramatic increase in computational power made a new class of models accessible to the realm of AC. While being similar to the above-mentioned cellular automata in that space is modeled explicitly, these models operate off-lattice; that is, particle positions are not confined to sites on a grid. Instead of a fixed set of transition rules, particle motion is determined by means of differential equations. Among these new methods, Brownian dynamics (BD) and dissipative particle dynamics (DPD) are the most prominent ones.

Off-lattice simulation techniques have several advantages compared to cellular automata: (i) Not being constrained to fixed lattice sites and having translational invariant interactions results in a significantly smaller set of required transition rules, (ii) continuous particle coordinates allow one to closely connect to mechanistic physical theories, and (iii) unconstrained motion of the particles avoids some of the artifacts found in lattice models (see, e.g., [3]). However, the price one has to pay for these advantages is generally a higher computational effort both in implementation and runtime.

The remainder of the chapter is organized as follows: Section 11.2 presents the theoretical concepts of BD and DPD. First, the general principle behind coarse-grained off-lattice simulation methods are outlined in Section 11.2.1. This overview is followed by detailed discussions of individual aspects of the physics (Sects. 11.2.2 through 11.2.4) and implementation (Sect. 11.2.5). The theoretical section is concluded by a summary of recent applications in Section 11.2.6. Section 11.3 is dedicated to the presentation of software to perform off-lattice simulations with the possible capacity for chemical reactions. The software packages – ESPresSo (Sect. 11.3.1), Spartacus (Sect.

11.3.2), Smoldyn (Sect. 11.3.3), and LAMMPS (Sect. 11.3.4) – are summarized in one section each that list respective features and shows examplary simulation setups where appropriate.

11.2 Concepts

11.2.1 Basic Principles of Coarse-Grained, Off-Lattice Simulation Techniques

Both BD and DPD are instances of coarse-grained modeling techniques in which the spatial structure of molecules is represented explicitly, although not in full atomistic detail. Instead, groups of atoms within a molecule are lumped together into point particles, usually called *beads*. These beads are then connected by elastic springs to form the whole molecule. Small molecules such as water are even considered to be lumped together into a single bead by groups of 3 to 5 molecules (see Fig. 11.2.1). The number of solvent molecules per bead is referred to as the *coarse-graining* parameter. While it is possible to relate coarse-grained representations to physical molecules [20], qualitative studies often content themselves with simply specifying functional groups like hydrophobic or charged parts of a molecule without a particular reactant in mind.

Concerning the treatment of solvent molecules, one can either explicitly represent them by beads or implicitly account for their effect on the interactions of other beads. If one is not interested in the dynamics of the solvent (e.g., hydrodynamic modes of the system), its implicit treatment can save significant computational effort, since most of the calculations are typically spent on solvent-solvent interactions.

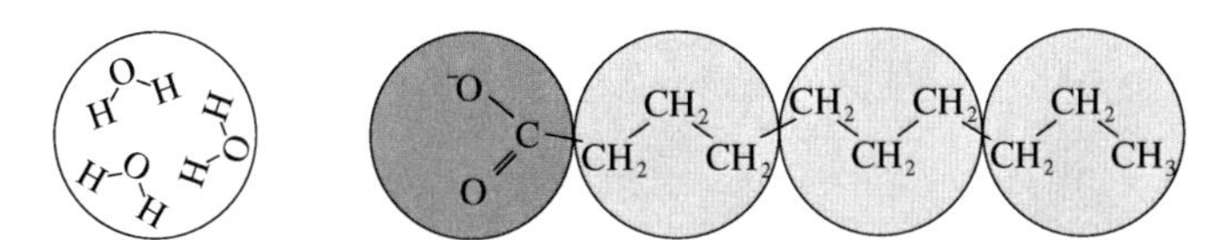

Fig. 11.1 Example coarse-grained representation of decanoic acid (a fatty acid surfactant) and water. The molecules are modeled by three types of beads: one representing the hydrophobic tail of the surfactant (3 hydrocarbon groups), one representing the hydrophilic carboxyle group, and one representing water molecules of approximately the same molecular volume.

An off-lattice simulation consists of a set of N beads in a 2D or 3D box. In explicit simulations, the space is considered to be densely filled, whereas in implicit simulations, the bead density will depend on the concentration of solved molecules. Since only a finite volume is simulated, boundary conditions need to be defined to determine the outcome of collisions with the wall. Most common are periodic boundary conditions, but closed, reflecting boundaries or a combination of the two can also be found in the literature. Each bead has a position, velocity, and a type (for an example configuration, see Fig. 11.2). The type of the bead determines its mass and its interactions with other beads. The motion of the beads follows *Newton's Second Law of Motion* according to which a particle accelerates proportional to the force that acts on it and inversely proportional to its mass. Written as an ordinary differential equation (ODE) in bead position and velocity, the law reads:

$$\begin{aligned} \dot{\mathbf{r}}_i(t) &= \mathbf{v}_i(t), \\ \dot{\mathbf{v}}_i(t) &= \frac{1}{m_i}\mathbf{F}_i(t), \end{aligned} \tag{11.1}$$

where $\mathbf{r}_i$ is the position, $\mathbf{v}_i$ is the velocity, m_i is the mass of bead i, and dots denote time derivatives (instantaneous changes in time). $\mathbf{F}_i$, the force that acts on bead i, collects all bead interactions.

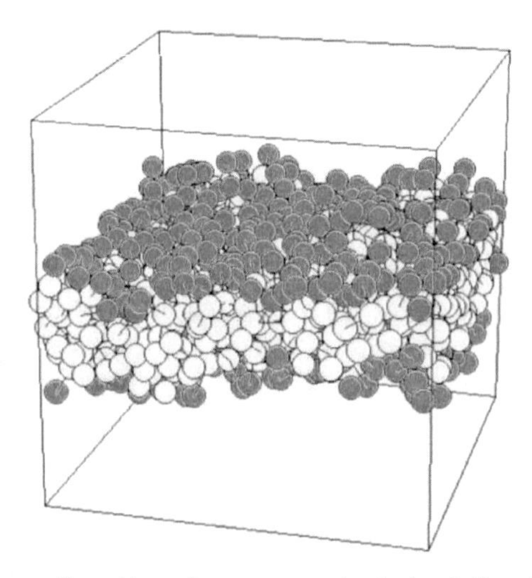

Fig. 11.2 Example configuration of a coarse-grained simulation: The image shows a bilayer membrane composed of **HTT** trimers (green-yellow surfactants) in water (not shown).

If all atoms were simulated individually (as is done in fully atomistic Molecular Dynamics (MD)), the force could be expressed as the negative gradient of some potential ϕ_i such that Eq. 11.1 constantly transforms kinetic into potential energy and vice versa. However, when individual atoms are lumped together into coarse-grained beads, energy is also exchanged with the internal energy of these beads. Instead of explicitly accounting for the internal energy of each bead, coarse-grained methods employ a *Langevin* formalism to express the energy exchange with internal degrees of freedom: The force $\mathbf{F}_i$ is expressed by three additive components – a conservative one ($\mathbf{F}_i^{\mathrm{C}}$), a dissipative one ($\mathbf{F}_i^{\mathrm{D}}$), and a random one ($\mathbf{F}_i^{\mathrm{R}}$):

$$\mathbf{F}_i = \mathbf{F}_i^{\mathrm{C}} + \mathbf{F}_i^{\mathrm{D}} + \mathbf{F}_i^{R}. \tag{11.2}$$

The dissipative force models friction by which kinetic energy is dissipated into successively less coherent motion, thereby removing energy from the explicitly represented mesoscale into the assumed underlying microscale (internal energy of the beads). This energy flow is counteracted by a random force that models the effect of microscopic thermal noise on the mesoscale (Brownian motion). As the name suggests, $\mathbf{F}_i^{\mathrm{R}}$ involves random variables, turning Eq. 11.1 into a stochastic differential equation, such that the change in system energy can be captured by an Uhlenbeck-Ornstein process. Combined, the dissipative and random forces act as a thermostat to regulate the temperature (i.e., the kinetic energy of the explicitly modeled mesoscale).

Several proposals have been made to combine off-lattice simulations with chemical reaction kinetics. To enable this, a set of chemical reaction rules is added to the system description that defines allowed transitions between bead types. To account for these reactions, the numerical integrator is intertwined with a stochastic process that applies bead transformations to the reaction vessel: Within each time step, first the positions and velocities of all beads are updated, and, second, reaction rules are applied subsequently to all beads in the system.

To summarize, the coarse-grained models described in this chapter draw on a set of (possibly interconnected) beads whose motion is governed by bead-bead interactions expressed by potential functions and by the action of a thermostat. Additionally, transition rules between beads can be defined to describe chemical reactions. The following sections will detail each of these aspects.

A simple example of the overall setup is given in Fellermann and Solé [14] where the system consists of beads of three molecules: water, fatty acid surfactants, and oily fatty acid esters. Water is represented explicitly by beads of type **W**, esters are represented as dimers of two interconnected **T** beads, and surfactants are modeled as a **T** bead with a surfactant head **H** bead attached to it. The potential interactions are chosen such that the model qualitatively reproduces the phases of these binary and ternary systems. A catalytic reaction is defined by which ester molecules are transformed into surfactant molecules in the vicinity of other surfactant molecules:

$$\mathbf{TT} + \mathbf{HT} \longrightarrow 2\mathbf{TT}. \tag{11.3}$$

This reaction models the hydrolysis of the ester bond under the assumption that the produced alcohol is small enough to be neglected.

In this setup, the ester forms oil droplets in water which are coated by surfactants. The size distribution of these droplets is determined by the water/oil/surfactant ratio. The relatively slow reaction constantly transforms ester into new surfactants, thereby changing the oil/surfactant ratio of the system. The change in concentrations is accompanied by a shift in surface-to-volume ratio of the droplets which respond by a shape change from spherical to elongated, rod-like aggregates. Once a critical threshold ratio is reached, the oil core is not sufficiently large to stabilize the aggregate which divides into two spherical aggregates. When ester is supplied constantly, the growth and division process continues and leads to exponential growth of the aggregates.

11.2.2 Interaction Potentials

In the picture of Newtonian mechanics, each bead has a potential energy that results from its interactions with other beads. This energy corresponds, for example, to pressure in thermally fluctuating fluids, but could also have other (e.g., electrostatic origins). Alternatively, one might say that the bead "feels" the potential energy of its neighborhood and responds to it by moving away from high-energy states – known as the principle of least constraints.

As commonly done in particle-based simulations, BD and DPD assume that the potential of a bead i can be expressed as the sum of pairwise interactions with neighboring beads:

$$V_i = \sum_{j \neq i} V_{ij}(\mathbf{r}_i, \mathbf{r}_j).$$

Since the space is assumed to be isotropic, the potential between particles i and j can only depend on their distance rather than their absolute position. Using

$$V_{ij}(\mathbf{r}_i, \mathbf{r}_j) = \phi_{ij}(r_{ij}),$$

where $r_{ij} = |\mathbf{r}_i - \mathbf{r}_j|$, the conservative force in Eq. 11.2 becomes

$$\mathbf{F}_i^{\mathrm{C}} = -\boldsymbol{\nabla} \sum_{i \neq j}^{N} \phi_{ij}(r_{ij}). \tag{11.4}$$

For further computational convenience, ϕ_{ij} is usually truncated after a certain distance r_{c}, such that $\phi_{ij}(r) = 0$ for $r \geq r_{\mathrm{c}}$.

It has to be pointed out that the potential functions used in coarse-grained simulation techniques must not be taken for mere mechanical energies as in atomistic (MD) simulations. Instead, the potential functions used in coarse-grained simulations are rather to be understood as a free energy that captures systemic properties of the system such as heat, pressure, pH, or the concentration of cosolvents [37]. This is particularly important for the study of entropy-driven processes underlying lipid systems, since entropy is included in the free energy. Although there is no closed theory to derive the functional form of the potential energy employed in coarse-grained simulations from first principles, it is still possible to relate the potential functions to structure properties of the system under consideration. For example, there is a one-to-one mapping between the potential function and the radial distribution function which expresses the average probability for two beads to be separated by a certain distance. Measuring the radial distribution function of detailed atomistic MD simulations allows one to construct energy functions of a coarse-grained representation thereof (see Lyubartsev et al. [27] and references therein). A further difficulty arises from the fact that the dynamics in coarse-grained simulations should represent a time average of the assumed underlying atomistic motion, since the fast degrees of motion, like most vibrational modes of covalent bonds, are meant to be removed in the coarse-graining process and instead comprised in the noise term of the thermostat.

In general, the spatial and temporal averaging of the coarse-graining process motivates the use of much smoother interaction potentials than the ones commonly applied in atomistic MD simulations. In particular, they typically do not possess a singularity at $r = 0$, meaning that two beads are allowed to sit right on top of each other (although under an energy "penalty" that may render this event practically impossible). This can be justified by the observation that (i) the bead position merely identifies the center of mass of a collection of molecules and (ii) even if the bead represents a single molecule or part thereof, it denotes the average position of the molecule during some short time interval; for both of these mean positions, it is perfectly legal for two positions to coincide. Whichever justification one might assent to, the resulting *soft core* potential allows one to run the numerical integrator of Eq. 11.1 with a significantly wider time step compared to functional forms that include a singularity.

Given these difficulties, many studies (that treat the solvent explicitly) content themselves with simple potentials of the form

$$\phi_{ij}(r) = \begin{cases} \frac{1}{2} a_{ij} \left(1 - \frac{r}{r_c}\right)^2 & \text{if } r < r_c \\ 0 & \text{otherwise,} \end{cases} \tag{11.5}$$

where $a_{ij} > 0$ denotes the mutual repulsion strength between two beads of type i and j. Note that there is no theoretical foundation for this function other than being the most simple confined function with continuous derivative. It has to be pointed out, however, that Eq. 11.5 relies on the explicit treatment of solvent, since the absence of an energy minimum would not allow for the formation of structures.

Having fixed the functional form of the potential, the matrix (a_{ij}) is the premier place for model calibration. A standard calibration procedure to mesoscopic observables has been suggested by Groot and Rabone [20, 21, 28] (see, however, Füchslin et al. for issues on its scaling behavior [15]). Generally, the procedure starts by matching the diagonal interaction parameters a_{ii} to the compressibility of the physical fluid. The off-diagonal elements a_{ij} $(i \neq j)$ express the mutual solubility of substances and are calibrated in a later step to parameters obtained from mixing theories, in particular the Flory-Huggins theory of polymer mixing where mutual interaction parameters are derived from free-energy considerations in a lattice model of polymers. For specific systems, other mesoscopic observables have been suggested for calibration (e.g., [9, 13, 18, 28]).

Additional terms can be added to the "ground" potential if needed. Most prominently, mechanical potentials between bonded beads have been use to model extended molecules. Venturoli and Smit [48] were the first to introduce harmonic spring potentials into DPD simulations, where bonded beads feel the additional potential

$$\phi_{ij}^{\mathrm{S}} = k_{\mathrm{S}} \left(r_{ij} - r_{\mathrm{S}} \right)^2 ,$$

with r_{S} being the optimal bond distance and k_{S} the spring constant. Shillcock and Lipowsky [43] have added a three-body angular potential to model stiffness in extended polymer chains. Higher-order potentials used in MD simulations (e.g., torsion potentials) are barely needed in coarse-grained simulations due to the simplistic representation of molecules. Other notable add-ons are Coulomb forces for electrostatic interactions [19], gravity [50], and shear forces [6, 44].

11.2.3 Thermostats

As pointed out in Section 11.2.1, thermostats in coarse-grained simulations are an integral part of the system description: Whereas MD simulations utilize thermostats merely to keep the system at constant temperature and generally try to minimize their impact on simulation results, coarse-grained simulations interpret the energy exchange with a heat bath as physically meaningful. As a consequence, coarse-grained models employ entirely different thermostats than the ones found in MD simulations.

The theoretical foundation of thermostats for coarse-grained simulations is the Langevin formalism for diffusive motion [16]: a spherical particle in a steady medium will experience a viscous drag proportional to its velocity and stochastic kicks from other particles randomly hitting it. In the presence of a possible external force $\mathbf{F}$, the equation of motion reads

$$\dot{\mathbf{v}}(t) = \frac{1}{m} \left(\mathbf{F}(t) - \gamma \mathbf{v}(t) + \boldsymbol{\xi}(t) \right) , \tag{11.6}$$

where $\gamma > 0$ is a friction coefficient and $\boldsymbol{\xi}$ is an uncorrelated random vector with zero mean and finite variance: $\langle \boldsymbol{\xi}_i(t), \boldsymbol{\xi}_j(t') \rangle = 2\sigma\delta_{ij}\delta(t - t')$. In the context of Eq. 11.2, the first term of the sum represents $\mathbf{F}^{\mathrm{C}}$, the second one represents $\mathbf{F}^{\mathrm{D}}$, and the last one represents $\mathbf{F}^{\mathrm{R}}$. The equilibrium temperature is given by γ and σ and resolves to σ/γ. Off-lattice models that employ this thermostat are commonly referred to as Brownian dynamics (BD). They are particularly suited for implicit simulations, since the effect of the solvent is already accounted for in the thermostat. It has to be emphasized, however, that Eq. 11.6 models motion in a steady medium, as the friction is proportional to the absolute velocity of the bead, rather than the velocity relative to the surrounding medium. Thus, the dynamics of the system is purely diffusive and neglects hydrodynamic modes.

To incorporate hydrodynamics into coarse-grained simulations, Hoogerbrugge and Koelman introduced the method of dissipative particle dynamics (DPD) [22] whose thermostat has become increasingly popular. The DPD thermostat decomposes all forces into pairwise contributions:

$$F_i = \sum_{j \neq i} \mathbf{F}_{ij} = \sum_{j \neq i} \mathbf{F}_{ij}^{\mathrm{C}} + \mathbf{F}_{ij}^{\mathrm{D}} + \mathbf{F}_{ij}^{\mathrm{R}}, \tag{11.7}$$

which are required to be central,

$$\mathbf{F}_{ij} = -\mathbf{F}_{ji} \quad \text{and} \quad \mathbf{F}_{ij} \propto \hat{\mathbf{r}}_{ij}, \tag{11.8}$$

where $\hat{\mathbf{r}}_{ij}$ is the unit vector pointing from bead j to i. The centrality asserts that the linear and angular momentum of the system is preserved, therefore allowing one to analyze hydrodynamic flows. In fact, it can be shown that the DPD thermostat implements a numerical solver for the Navier-Stokes equations [11]. With these requirements, the only (local and memory-free) way [12] to achieve an Uhlenbeck-Ornstein process whose equilibrium distribution is a Gibbs ensemble is to set

$$\begin{aligned}
\mathbf{F}_{ij}^{\mathrm{D}} &= -\frac{\sigma^2}{2k_bT}\omega^2(r_{ij})\,(\mathbf{v}_{ij} \cdot \hat{\mathbf{r}}_{ij})\,\hat{\mathbf{r}}_{ij}, \\
\mathbf{F}_{ij}^{\mathrm{R}} &= \sigma\omega(r_{ij})\zeta_{ij}\hat{\mathbf{r}}_{ij},
\end{aligned}$$

where r_{ij} is the Euclidean distance between beads i and j, and $\mathbf{v}_{ij}$ is the relative velocity between bead j and i. ζ_{ij} is an uncorrelated random variable with zero mean, Gaussian statistics, and a variance of $1/\Delta t$ for the numerical time step Δt. In order to preserve linear momentum, it has to hold that $\zeta_{ij} = \zeta_{ji}$. σ is the friction coefficient of the medium (related to its Reynolds number), k_bT defines the energy unit where T denotes the temperature in Kelvin, and k_b is the Boltzmann constant. ω is a dimensionless weighing function which is not specified by the general formalism. Most studies employ a weight function similar to the soft core interaction potential:

$$\omega = \begin{cases} \left(1 - \frac{r}{r_c}\right)^2 & \text{if } r < r_c \\ 0 & \text{otherwise.} \end{cases}$$

While it is known that ω has impact on the overall system dynamics, like viscosity [10] and temperature conservation [39], no proposal has yet been made on how to derive the weight function from first principles.

The scalar product in Eq. 11.9 ensures that $\mathbf{F}^{\mathrm{D}}$ is maximal when two beads approach each other and zero when the particles move parallel (see Fig. 11.3). This leads to the alignment and collaborative motion of nearby beads. As a result, the pairwise coupling of the dissipative and random forces in DPD results in faster dynamics than the ones observed in BD simulations. However, the accelerated dynamics of the DPD thermostat are actually an artifact of the method: In DPD, mass diffusion is too high when compared to travel of momentum. The Schmidt number (the ratio of kinematic viscosity vs. self-diffusion coefficient) is about three orders of magnitude lower in the DPD simulation of water than in the real system [10, 21].

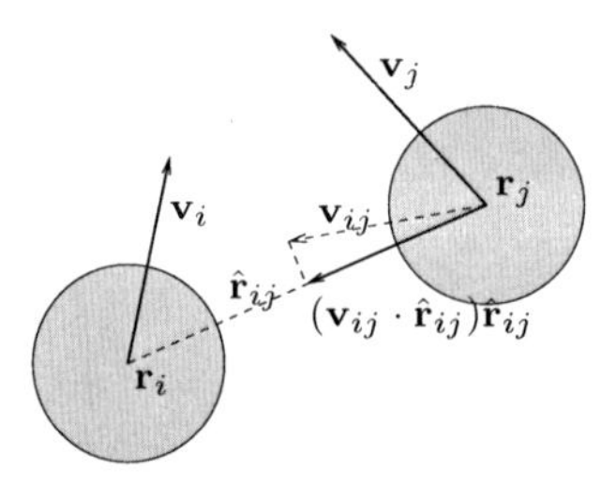

Fig. 11.3 Schematic of the dissipative force (friction) in DPD. The force acts central on the line given by the bead positions $\mathbf{r}_i$ and $\mathbf{r}_j$ (parallel to $\mathbf{r}_{ij}$). Its magnitude depends on the relative velocity $\mathbf{v}_{ij}$: Friction is maximal when beads approach each other directly and zero when they move parallel.

11.2.4 Chemical Reactions

Probably the first approach to extend spatially resolved reactor vessels with chemical reactions was the work by Ono [36], who incorporated a stochastic process for reactions into a BD simulation. This approach has later been put on a more rigorous theoretical foundation (the Smoluchowski model for diffusion-limited reactions) by Andrews and Bray [1] and Monine and Haugh [31]. Similar approaches are the work of Buchanan et al. [5, 17], Fellermann et al. [13, 14], and Liu et al. [26], who extended DPD by similar means to incorporate reactions.

Most chemical reactions can be classified as unimolecular or bimolecular reactions – depending on the number of molecules that participate as reactants. Higher-order reactions barely occur in nature due to the unlikelihood of three molecules hitting each other concurrently. Following this formalism, inflow of chemicals can be regarded as a zeroth-order reaction that has no educt, whereas outflow of chemicals can be regarded as a first-order reaction that has no product. Reactions can be further classified as syntheses or analyses, depending on whether covalent bonds are formed or broken by the reaction. The formation and braking of covalent bonds can result in a change of physico-chemical properties of the molecular species. In the coarse-grained representation of BD and DPD, this is expressed as a type change of the reacting beads. Depending on the coarse-graining level and representation of molecules, bond braking and formation might even happen completely below the resolution of the model such that chemical reactions are mere type transformations that do not affect explicitly modeled covalent bonds. In the formalism presented here, chemical reactions only occur between beads that represent a single molecule or part of a molecule – no attempt has been made yet to define chemical reactions between beads that group together several molecules. Examples of possibles reactions are

$$
\begin{array}{rll}
 & \xrightarrow{k} X & \text{inflow (zeroth-order)} \\
A & \xrightarrow{k} & \text{outflow (first-order)} \\
A & \xrightarrow{k} X & \text{molecular reconfiguration (first-order)} \\
AB & \xrightarrow{k} X + Y & \text{unimolecular analysis (first-order)} \\
A + B & \xrightarrow{k} XY & \text{bimolecular synthesis (second-order)}
\end{array}
$$

Here, A, B, X, and Y denote beads that may be part of an extended molecule. Concatenated symbols refer to bonded beads. For simplicity, all reactions are written as irreversible reactions. Reversible reactions can be represented as pairs of irreversible reactions, where each member represents one of the two directions.

Evidently, bimolecular reactions occur only when the two reagents are in close vicinity. For diffusion-limited reactions (i.e., in the absence of an activation energy barrier) this distance is given by the sum of the molecular radii of the two reagents. If there is an activation barrier, one possibility is to define an effective binding radius smaller than the sum of molecular radii. The size of this binding radius can be related to the effective reaction rate by the Smoluchowski equation

$$
r_{\mathrm{b}} = \frac{k}{4\pi(D_{\mathrm{A}} + D_{\mathrm{B}})}, \tag{11.9}
$$

where D_A and D_B are the diffusion coefficients of the two reagents and k is the effective (measured) reaction rate [1]. Following Collins and Kimball [7], one can alternatively define a probability for reactions to occur within the unaltered reaction radius. Likewise, unimolecular reactions either occur with a certain probability

$$1 - e^{-k\Delta t},$$

or – if explicit bond breaking is involved in the reaction – once the reacting beads separate by more than a certain unbinding radius $r_{\mathrm{u}} > r_{\mathrm{b}}$. r_{u} can be related to r_{b} by the equation

$$r_{\mathrm{u}} = \frac{r_{\mathrm{b}}}{1 - k/k_{\mathrm{r}}}, \tag{11.10}$$

where k_{r} is the rate of geminate recombination [1]. When a pair of bond forming and breaking reactions is used to model a reversible reaction, care must be taken that the reactants are separated by more than r_{b} after bond breaking. Note that Eqs. 11.9 and 11.10 are only valid for numerical time steps Δt smaller than the root mean square displacement of the reacting particles. Andrews and Bray [1] give corrected equations for larger simulation time steps, as well as a comparison of the deterministic and stochastic method. Zeroth-order reactions do not depend on any educt concentration, and their occurrence is simply determined by comparing the reaction rate to a Poisson distributed random number with average $k\Delta t$ for the simulation time step Δt.

Ono [35] and Fellermann et al. [13, 14] use an extended variant of the Collins and Kimball scheme that accounts for the presence of nearby catalysts. Other than modeling the exact reaction mechanism by which the catalyst participates in the reaction scheme, it is assumed that the mere presence of catalytic molecules enhances the effective reaction rate. The equation to calculate this effective rate is

$$k = k_i + \sum_C k_C \left(1 - \frac{r}{r_{\mathrm{cut}}}\right), \tag{11.11}$$

where r is the distance between the catalyst and any of the reactants and the sum runs over all catalysts C with $r < r_{\mathrm{cut}}$.

All of the stochastic approaches assume that reactions occur independently, which is violated when a single bead might react with one of several nearby reagents. A proper mathematical framework to deal with this stochastic independence would be provided by the Master equation [16]. However, its computation is too expensive to be redone in each updating step, which is why current studies commonly ignore the problem of interdependence and merely try to avoid it by using small reaction and catalyst rates and constantly reshuffling the list of reactions and reagents to avoid biases.

Having defined the set of possible reaction rules, the integrator for the equations of motion is intertwined with a stochastic process such that in each time step, particle motion and chemical reactions are taken care of sequentially.

Further method development needs to be done in connecting the potential, kinetic, and chemical energy reservoirs: The change in internal (chemical) energy associated with bond formation and breaking and change of potential energy due to bead type transformation would need to be accounted for in a velocity change of the reacting beads leading to local heating. This would, for example, allow one to capture effects that rely on activation energy barriers (a first step in this direction has been taken by Yingling and coworkers [54]).

11.2.5 Updating Schemes and Spatial Organization

At the heart of all BD and DPD simulators lies an integrator for the stochastic differential equation (11.1). Owing to the special form of the Newtonian dynamics, an inexpensive high-order integrator can be derived from two Taylor expansions:

$$\mathbf{r}_i(t+\Delta t) = \mathbf{r}_i(t) + \dot{\mathbf{r}}_i(t)\Delta t + \frac{1}{2}\ddot{\mathbf{r}}_i(t)\Delta t^2 + \frac{1}{6}\dddot{\mathbf{r}}_i(t)\Delta t^3 + \mathcal{O}(\Delta t^4),$$

$$\mathbf{r}_i(t-\Delta t) = \mathbf{r}_i(t) - \dot{\mathbf{r}}_i(t)\Delta t + \frac{1}{2}\ddot{\mathbf{r}}_i(t)\Delta t^2 - \frac{1}{6}\dddot{\mathbf{r}}_i(t)\Delta t^3 + \mathcal{O}(\Delta t^4).$$

Adding these two equations and substituting $\ddot{\mathbf{r}}_i(t) = \mathbf{a}_i(t)$ leads to a very simple forth-order integrator known as *Verlet* integrator [49]:

$$\mathbf{r}_i(t+\Delta t) = 2\mathbf{r}_i(t) - \mathbf{r}_i(t-\Delta t) + \mathbf{a}_i(t)\Delta t^2 + \mathcal{O}(\Delta t^4).$$

Verlet-based integrators outperform standard procedures like Runge-Kutta not only because of their computational inexpensiveness but also because they preserve phase space volume in conservative systems, which reduces artifacts such as gradual temperature increase.

If velocities are to be known explicitly (e.g., to compute the kinetic energy of the system) the substitution $\mathbf{v}_i(t) = \left(\mathbf{r}_i(t) - \mathbf{r}_i(t-\Delta t)\right)/\Delta t + \frac{1}{2}\mathbf{a}_i(t)\Delta t + \mathcal{O}(\Delta t^2)$ leads to a variant called the *Velocity Verlet* algorithm:

$$\mathbf{r}_i(t+\Delta t) = \mathbf{r}_i(t) + \mathbf{v}_i(t)\Delta t + \frac{1}{2}\mathbf{a}_i(t)\Delta t^2, \tag{11.12}$$

$$\mathbf{v}_i(t+\Delta t) = \mathbf{v}_i(t) + \frac{1}{2}\left(\mathbf{a}_i(t) + \mathbf{a}_i(t+\Delta t)\right)\Delta t, \tag{11.13}$$

which is fourth order in positions $\mathbf{r}_i$ and second order in velocities $\mathbf{v}_i$. This is a commonly used integrator for BD simulations. There is a subtlety, however, when accelerations $\mathbf{a}_i$ do not only depend on positions but also on velocities, as is the case for $\mathbf{F}^{\mathrm{D}}$ in the DPD equations. Equation 11.13 then becomes

$$\mathbf{v}_i(t+\Delta t) = \mathbf{v}_i(t) + \frac{1}{2m_i}\left(\mathbf{F}_i(\mathbf{r}(t),\mathbf{v}(t)) + \mathbf{F}_i(\mathbf{r}(t+\Delta t),\mathbf{v}(t+\Delta t))\right),$$

where the term $\mathbf{v}_i(t+\Delta t)$ appears in both sides of the equation. To overcome this problem, Groot and Warren [21] have suggested a predictor-corrector-like integrator which has become the de facto standard in DPD simulations. In their integrator, the positional update is unchanged (Eq. 11.12). To compute the velocities, the algorithm first makes a prediction (Eq. 11.14) followed by a correction based on the force field of the predicted state (Eq. 11.15):

$$\tilde{\mathbf{v}}_i(t+\Delta t) = \mathbf{v}_i(t) + \frac{\lambda}{m_i}\mathbf{F}_i(\mathbf{r}(t),\mathbf{v}(t)) \qquad \lambda \in [0,1], \tag{11.14}$$

$$\begin{aligned}\mathbf{v}_i(t+\Delta t) &= \mathbf{v}_i(t) \\ &+ \frac{1}{2m_i}\left(\mathbf{F}_i(\mathbf{r}(t),\mathbf{v}(t)) + \mathbf{F}_i(\mathbf{r}(t+\Delta t),\tilde{\mathbf{v}}(t+\Delta t))\right).\end{aligned} \tag{11.15}$$

If $\mathbf{F}_i$ does not depend on $\mathbf{v}$, the original Velocity Verlet integrator is recovered for $\lambda = 0.5$. Due to the stochastic nature of the force field, the order of the integrator is unclear. There is no imperative on how λ should optimally be chosen. Reported values differ between $\lambda = 0.5$ [34] and $\lambda = 0.65$ [21]. Thus, λ clearly depends on other systems parameters and needs to be fine-tuned to the system under consideration.

Subsequent studies [4, 34, 47] have identified artifacts of the Verlet-based algorithms, predominantly in the radial distribution function of the system [47]. Building upon work by Pagonabarraga et al. [38], Vattulainen et al. [47] have proposed a more elaborate so-called *self-consistent* integrator in which the system temperature is constantly measured and compared to its target value. The deviance is used to fine-tune the dissipation rate of the system. In general, the performance of DPD integrators is still an active area of research due to the stochastic nature of the interactions.

In addition to well-suited integrators, simulators for coarse-grained simulations need to provide efficient means to access neighbors of beads. Fast neighbor look-up is essential for the computation of $\mathbf{F}^{\mathrm{C}}$ in the case of BD (Eq. 11.4) and $\mathbf{F}^{\mathrm{C}}$, $\mathbf{F}^{\mathrm{D}}$, and $\mathbf{F}^{\mathrm{R}}$ in the case of DPD (Eq. 11.7). If all particles were naively held in a simple list or array, look-up time would scale quadratic with the number of particles. By taking advantage of the limited cutoff range r_{c} for all forces, careful bead management can reduce the number of look-ups to scale linear with the number of beads. To achieve this, an algorithm called *domain decomposition* or *linked lists* is commonly used: The space is partitioned in cells with a side length of at least the cutoff radius r_{c}. Each cell holds a list of particles it contains, as well as information about neighboring cells (dashed grid in Fig. 11.4). For each bead, neighbor look-up can now be restricted to the 3×3-Moore neighborhood of the cell that holds the current bead (solid square in Fig. 11.4). If bead interactions are symmetric (e.g., due to Eqs. 11.4 and 11.8), the number of cells to check can even be reduced to a half. With this management, an estimated 62.8% of all tested pairs will be within the cutoff radius (solid circle in Fig. 11.4) and 29.9% in three dimensions. It should be emphasized that the grid representation of the space is only an algorithmic organization principle and does not confine the actual position of beads within each cell. Implementation can either use one linked list for all beads or separate lists for each bead type, which can improve the performance when bead interactions are highly type-specific (as in case of chemical reactions).

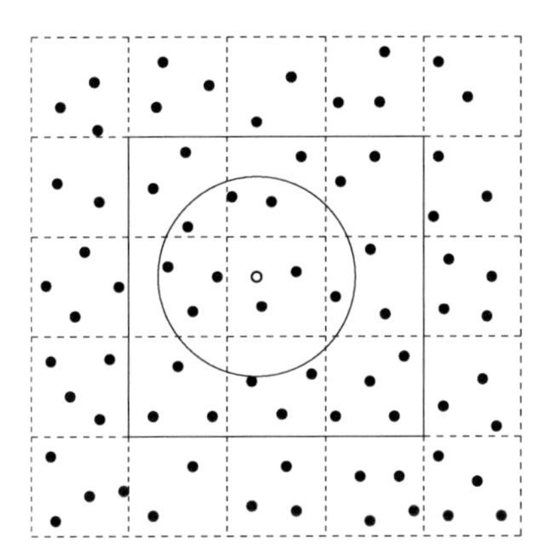

Fig. 11.4 Internal partitioning of a 2D space (dashed grid) filled with beads (black dots). The size of the cells is at least the cutoff radius of all involved interactions (radius of the solid circle). To find all interaction partners for a given bead (white dot), it is sufficient to consider only beads within a 3×3-Moore neighborhood around the cell that contains the bead (solid square). Note that the grid representation is only internal and does not restrict the actual position of beads within each cell.

11.2.6 Applications

Pure BD and DPD models that do not incorporate chemical reactions have been used extensively in areas as diverse as rheology, material sciences, soft matter studies, and molecular biology. Concerning the latter, the study of lipid strctures such as membranes and vesicles, their self-assembly, fission, fusion, rapture, and dissolution has arrested particular attention [20, 32, 33, 48, 52, 53]. The recent incorporation of chemical reactions into these models has initiated several simulations in the area of artificial chemistry and artificial life.

Gazzola et al. [17] have used reactive DPD to study the catalytic effect of self-assembled lipid structures due to effective upconcentration of hydrophobic reagents. Simulating abstract chemistries, they observed that the affinity of reagents to reside either in the solvent or in the interior of self-assembled lipid structures (micelles) alters the effective reaction networks that emerge in the reaction vessel.

Ono [35] and Fellermann and Solé [14] have studied the self-assembly, growth, and replication of minimal life-like entities: lipid aggregates equipped with a minimal artificial metabolism. Ono presents 2D BD simulations of vesicular structures driven by a catalytic turnover of precursor molecules; Fellermann and Solé performed 3D DPD simulations of surfactant-coated oil droplets that replicate by means of an auto-catalytic nutrient turnover (described in Sect. 11.2.1).

Complementary to the above work are qualitative simulations on non-enzymatic replication of short genetic polymer sequences. Bedau et al. [2] have analyzed the ability of complementary polymer strands to spontaneously replicate in solution, whereas Fellermann et al. [13] have studied a similar process at the oil-water interface of the aforementioned lipid aggregates. Taken together, these works present integrated simulations of the spontaneous emergence of artificial life-like aggregates, or "protocells" [42].

11.3 Available Software and Tools

Several BD and DPD simulators exist both in commercial as well as open-source applications. The following list is not meant to be complete, but tries to identify those programs that stand out for maturity, extensibility, or interoperability. The programs are similar in that they use configuration files to set up a simulation. The configuration file is sent to a non-interactive simulator that calculates the system trajectory and writes out the result of various analyzers.

11.3.1 ESPresSo

ESPResSo is a fairly developed and feature-rich open source simulation package for soft matter research (BD and DPD among others) developed by Limbach et al. [25] and licensed under the GNU General Publishing License (GPL). The software is available for Windows, Unix/Linux, and Mac OS. ESPresSo is written in C and can operate in parallel mode via MPI and can be controlled via Tcl scripts [45]. Online visualization is delegated to VMD, an independent open-source software for molecular visualization [23]. Out of the box, ESPresSo does not include chemical reactions.

In ESPresSo, simulations are set up via configuration scripts that define system size, boundary conditions, bead types, interactions, an initial condition, analyzers, and so on. These configuration files are actual Tcl scripts that offer powerful means for arbitrarily complex simulation setups. An example script to set up and run a simulation is shown in Fig. 11.5. The system consists of 1780 **W** beads (water), 600 **TT** dimers (oil), and 10 **HT** dimers (surfactants) in a box of size 10^3. Bead interaction parameters are $a_{\mathbf{W},\mathbf{W}} = 25, a_{\mathbf{W},\mathbf{H}} = 15, a_{\mathbf{W},\mathbf{T}} = 80, a_{\mathbf{H},\mathbf{H}} = 35, a_{\mathbf{H},\mathbf{T}} = 80$, and $a_{\mathbf{T},\mathbf{T}} = 15$. The DPD thermostat is chosen with a temperature of 1 and friction parameter of 4.5. Integration step size and number of steps is set to 0.01 and 1000, respectively, and the system is finally integrated. Various analyzers could further be incorporated.

The package offers support for various boundary conditions, a variety of potentials for bonded and non-bonded short-range interactions, electrostatics (Coulomb forces), rigid bonds, as well as bond-angle and dihedral interactions. ESPresSo also provides tabulated interactions to incorporate interaction potentials measured, for example from more detailed interactions (see Sect. 11.2.2).

ESPresSo comes with a variety of predefined analyzers for common observables. Among the most common ones are the following:

- statistical physics
  - energies (total, kinetic, Coulomb, non-bonded, bonded)
  - pressure (total, ideal, Coulomb, non-bonded, bonded)
  - stress tensor (total, ideal, Coulomb, non-bonded, bonded)
  - mean square displacement
- distribution and correlation functions
  - radial distribution function
  - structure factor
  - van Hove autocorrelation function
  - aggregate size distribution
  - free-volume distribution
- analyzers for polymer chains
  - end-to-end distance
  - radius of gyration

In addition, custom analyzers can be defined with ease. Analyzers are ordinary functions that can be used anywhere in the configuration script. ESPresSo provides a few built-in commands for statistics (averaging, errors, etc.) and plotting, the latter by delegation to gnuplot [51].

Owing to the flexibility of the Tcl-based design, it is easy to incorporate chemical reactions into ESPreSso. Figure 11.6 shows an exemplary algorithm that implements the reaction (11.3) based on a simplified version of Eq. 11.11. The implementation is only meant for the purpose of illustration: For a productive setup, the algorithm should be written in a compilable computer language and wrapped into Tcl.

```
# System parameters
set box_l     10.            ;# box size
set n_W      1780            ;# number of W beads
set n_TT      600            ;# number of TT dimers
set n_HT       10            ;# number of HT dimers
set n_total 3000             ;# total number of particles

# Integration parameters
set cut 1.0                  ;# cut off range r_c
set int_steps     100        ;# micro-integration steps (between analysis)
set int_n_times  1000        ;# macro-integration steps

setmd time_step 0.01
setmd skin      0.4
setmd box_l $box_l $box_l $box_l
thermostat dpd 1.0 4.5 $cut

# Interaction setup
inter 0 harmonic 100 0.75

inter 0 0 soft-sphere 25 2 $cut 0
inter 0 1 soft-sphere 80 2 $cut 0
inter 0 2 soft-sphere 15 2 $cut 0
inter 1 1 soft-sphere 15 2 $cut 0
inter 1 2 soft-sphere 80 2 $cut 0
inter 2 2 soft-sphere 35 2 $cut 0

# Particle setup

# water beads
for {set i 0} { $i < $n_W } {incr i} {
    set posx [expr $box_l*[t_random]]
    set posy [expr $box_l*[t_random]]
    set posz [expr $box_l*[t_random]]
    part $i pos $posx $posy $posz type 0
}

# TT dimers
for {} { $i < [expr $n_W+$n_TT] } {incr i 2} {
    set posx [expr $box_l*[t_random]]
    set posy [expr $box_l*[t_random]]
    set posz [expr $box_l*[t_random]]
    part $i pos $posx $posy $posz type 1
    part [expr $i+1] pos $posx $posy [expr $posz+0.8] type 1 bond 0 $i
    part $i bond 0 [expr $i+1]
}

# HT dimers
for {} { $i < [expr $n_W+$n_TT+$n_HT] } {incr i 2} {
    set posx [expr $box_l*[t_random]]
    set posy [expr $box_l*[t_random]]
    set posz [expr $box_l*[t_random]]

    part $i pos $posx $posy $posz type 1
    part [expr $i+1] pos $posx $posy [expr $posz+0.8] type 2 bond 0 $i
    part $i bond 0 [expr $i+1]
}

# Integration
set j 0
for {set i 0} { $i < $int_n_times } { incr i} {
    integrate $int_steps
    # [...] do some analysis here
}
```

Fig. 11.5 Example Tcl script to control ESPresSo for the example system of Section 11.2.1. See text the for explanations.

```
proc react {reaction}
{
    # reaction is { educt product rate catalyst cat_rate cat_range}
    # educt and product are list of 2 elements, specifying bead types


    set educt     [lindex $reaction 0]
    set product   [lindex $reaction 1]
    set rate      [lindex $reaction 2]
    set catalyst  [lindex $reaction 3]
    set cat_rate  [lindex $reaction 4]
    set cat_range [lindex $reaction 5]


    foreach {part_a} [part]
    {
        # iterate over all possible educts
        set a_id    [lindex $part_a 0]
        set a_type  [lindex $part_a 6]
        set a_bonds [lindex $part_a 33]
        if { $a_type != [lindex $educt 0] } { continue }
        set b_id    [lindex [lindex $a_bonds 0] 1]
        set b_type  [part $b_id print type]
        if { "$a_type $b_type" != $educt } { continue }


        # compute effective reaction rate
        set k $rate
        foreach {cat} [analyze nbhood $a_id $cat_range]
        {
            set c_type [part $cat print type]
            if { $c_type != $catalyst } { continue }
            set k [expr $k+$cat_rate]
        }


        # exchange bead types if a reaction occurs
        if { [expr $k*[setmd time_step]] > [t_random] }
        {
            part $a_id type [lindex $product 0]
            part $b_id type [lindex $product 1]
            continue
        }
    }
}



# [...]
# This part replaces 'integrate $int_steps' from the example in Fig. 11.5

    for {set t 0} { $t < $int_steps } { incr t}
    {
        integrate 1
        react "{1 1} {1 2} 0 2 0.5 $cut"
    }
# [...]
```

Fig. 11.6 Algorithm for reconfiguration reactions in ESPresSo that implements a simplified version of Eq. 11.11 (not implementing the linear decrease with catalyst distance). The function `react` is called with 6 arguments: the reactant, product, spontaneous reaction rate, catalyst bead type, the catalyst's rate enhancement, and range. Reactant and product are assumed to be Tcl lists of 2 elements. Together with the system setup in Fig. 11.5, this implements the example given in Section 11.2.1.

11.3.2 Spartacus

Spartacus is an experimental open source framework for BD and DPD simulations developed by the author and licensed under the GNU GPL. The software is available for Mac Os, Linux. Spartacus is similar in function to EsPresSo, but focuses on chemistry rather than physics. It incorporates all algorithms for chemical reactions that have been described in Section 11.2.4. The framework consists of a core simulation engine written in C and Python [41] and can be controlled and extended via Python scripts, and inspected at runtime by a graphical interface. Figure 11.7 shows the example configuration of Figure 11.5 implemented for Spartacus.

Spartacus offers predefined analyzers comparable to ESPresSo but with emphasis on chemistry instead of physics:

- statistical physics
  - energies (kinetic, potential)
  - pressure (virial and excess pressure)
  - mean square displacement
- distribution and correlation functions
  - velocity correlation and autocorrelation
  - radial distribution function
  - aggregate size distribution
- system chemistry
  - bead numbers
  - reactivity
  - compositional entropy (of aggregates)

Additional analyzers can be defined in the configuration file. Spartacus prints out the result of analyzers for further processing. Alternatively, the system state or screen shots can be saved along a trajectory. Scripts are provided to assemble movies from this data.

11.3.3 Smoldyn

Smoldyn is a spatially resolved simulator for chemical reaction networks with focus on molecular biology. It has been implemented by S. Andrews and licensed under the GNU Lesser Publishing License (LGPL). Smoldyn does not consider bead interactions but merely diffusive motion (implemented as momentumless random walk). Consequently, molecules can only be represented by single beads and their internal structure cannot be modeled – the framework is therefore most appropriate for higher coarse-graining parameters.

```
# set system size, bead number, and density
from simulation.grid import Space3D
size = 10
space = Space3D


# define bead types
from simulation.chemistry import Particle
class W(Particle) :
    pass


class H(Particle) :
    color = "green"


class T(Particle) :
    color = "yellow"


H.binds = [T]
T.binds = [T,H]
particles = [W,H,T]


# reactions
from simulation.chemistry import Reaction
reactions = [
    Reaction(
        [T,T], [H,T],          # educt, product
        1.0, 0.0,              # range (unused), spontaneous rate
        H, 1.0, 0.5            # catalyst, range, and rate
    )
]


# random initial condition
initial_condition = RandomInit(n_W=1780, n_T=600, n_H=10)


# define interactions
import simulation.physics.dpd as dpd


cut = 1
class Physics(dpd.Physics) :
    matrix = {
        (W,W) : (25,cut),
        (W,H) : (15,cut),
        (W,T) : (80,cut),
        (H,H) : (35,cut),
        (H,T) : (80,cut),
        (T,T) : (15,cut),
    }


# thermostat variables and integration time step
gamma = 4.5
sigma = 3.0
time_step = 0.01
physics = Physics(cut, gamma, sigma, time_step)
```

Fig. 11.7 The same system setup as in Fig. 11.5 implemented for Spartacus in the Python programming language. Unlike ESPresSo, Spartacus uses configuration files only for the system setup but not for the actual integration.

The main focus of Smoldyn is its accurate incorporation of chemical reactions [1]. Reactions can be zeroth-, first-, and second-order (type change) reactions. Smoldyn implements reactions based on the Smoluchowski formalism (see Sect. 11.2.4). Smoldyn further allows one to define surfaces and compartments for membranes and vesicles by geometrical shapes. These structures can reflect, absorb, or bind beads and may have an influence on their reactivity and diffusion.

Simulations in Smoldyn are set up by plain text configuration files, implying that the software is not extensible by scripting. The time evolution of a system can be inspected at runtime via a graphical interface and can be captured as a movie. Additionally, analyzer results can be printed out for further analysis. Analyzers focus on bead numbers, either in total, or located at specific surfaces or compartments.

11.3.4 LAMMPS

LAMMPS (Large-scale Atomic/Molecular Massively Parallel Simulator) is a fast and feature-rich molecular (MD) dynamics software that includes code for coarse-grained DPD simulations. The software has been developed by S. J. Plimpton, P. Crozier, and A. Thompson [40] and licensed under the GNU GPL. LAMMPS is implemented in C++ and runs on single-processor desktop machines but is designed with massively parallel architectures in mind (employing MPI). LAMMPS does not support chemical reactions and is not extensible by scripting but defines a clear interface for C++ extensions.

LAMMPS consists of a core engine that is controlled by a textual configuration file. It can print out the system's time evolution. Alternatively, LAMMPS allows one to print out the results of custom analyzers that can be defined in the configuration file. Since the configuration file does not define a full-featured programming language, the range of possible analyzers is limited (e.g., to time and ensemble averages of per-atom quantities). More complex observables need to be calculated from the system's trajectory files. The package includes scripts to generate movies and can produce output for VMD [23] and other visualization softwares.

11.4 Conclusion

This chapter has introduced modeling techniques and software for spatially resolved, coarse-grained artificial chemistry – primarily Brownian dynamics (BD) and dissipative particle dynamics (DPD) with incorporated chemical reactions. The chapter has derived the underlying mathematical models and has reviewed implementation principles, means of calibration. The presented methods allow one to simulate molecular motion, reaction, and spatial organization in an integrated framework, which makes them valuable tools in the study of emergent physico-chemical structures (e.g., lipid aggregates) and processes that influence or are influenced by those structures, as the existing examples from the areas of artificial chemistry, artificial life, and systems chemistry indicate.

Four existing software packages (ESPresSo, Spartacus, Smoldyn, and LAMMPS) have been presented. These applications differ in focus and features and vary in their requirements on hardware and user-knowledge from pedagogical tools to high-performance computing applications. Most of the presented programs are scriptable/extensible, so that flexible simulation setup and incorporation into existing simulation environments can be achieved easily.

References

1. Andrews, S.S., Bray, D.: Stochastic simulation of chemical reactions with spatial resolution and single molecule detail. Phys. Biol. **1**, 137–151 (2004)
2. Bedau, M., Buchanan, A., Gozzala, G., Hanczyc, M., Maeke, T., McCaskill, J., Poli, I., Packard, N.: Evolutionary design of a DDPD model of ligation. In: Proceedings of the 7th International Conference on Artificial Evolution EA'05, pp. 201–212. Springer, Berlin (2006)
3. Bedrov, D., Smith, G.D., Freed, K.F., Dudowicz, J.: A comparison of self-assembly in lattice and off-lattice model amphiphile solutions. J. Chem. Phys. **116**(12), 4765–4768 (2002)
4. Besold, G., Vattulainen, I., Karttunen, M., Polson, J.M.: Towards better integrators for dissipative particle dynamics simulations. Phys. Rev. E **62**, 7611–7614 (2000)
5. Buchanan, A., Gazzola, G., Bedau, M.A.: Evolutionary design of a model of self-assembling chemical structures. In: Systems Self-Assembly: multidisciplinary snapshots. Elsevier, London (2006)
6. Chen, S., Phan-Thien, N., Fan, X.J., Khoo, B.C.: Dissipative particle dynamics simulation of polymer drops in a periodic shear flow. J. Non-Newtonian Fluid Mech. **118**(1), 65–81 (2004)
7. Collins, F.C., Kimball, G.E.: Diffusion-controlled reaction rates. J. Colloid Sci **4**(425) (1949)
8. Dittrich, P., Ziegler, J., Banzhaf, W.: Artificial chemistries – a review. Artif. Life **7**, 225–275 (2001)
9. Dzwinel, W., Yuen, D.A.: Matching macroscopic properties of binary fluids to the interactions of dissipative particle dynamics. J. Mod. Phys. C **11**(1), 1–25 (2000)
10. Eriksson, A., Jacobi, M.N., Nyström, J., Tunstrøm, K.: Effective thermostat induced by coarse-graining of SPC water. J. Chem. Phys. **129**(2), 024,106 (2008)
11. Español, P.: Hydrodynamics from dissipative particle dynamics. Phys. Rev. E **52**(2), 1734–1742 (1995)
12. Español, P., Warren, P.: Statistical mechanics of dissipative particle dynamics. Europhys. Lett. **30**, 191–196 (1995)
13. Fellermann, H., Rasmussen, S., Ziock, H.J., Solé, R.: Life-cycle of a minimal protocell: a dissipative particle dynamics (DPD) study. Artif. Life **13**(4), 319–345 (2007)
14. Fellermann, H., Solé, R.: Minimal model of self-replicating nanocells: a physically embodied, information-free scenario. Phil. Trans. R. Soc. Lond. Ser. B **362**(1486), 1803–1811 (2007)

15. Füchslin, R.M., Fellermann, H., Eriksson, A., Ziock, H.J.: Coarse-graining and scaling in dissipative particle dynamics. J. Phys. Chem. (2008). Submitted
16. Gardiner, C.: Handbook of Stochastic Methods for Physics, Chemistry, and Natural Sciences, *Springer Series in Synergetics*, vol. 13, 3rd edn. Springer, New York (2004)
17. Gazzola, G., Buchanan, A., Packard, N., Bedau, M.: Catalysis by self-assembled structures in emergent reaction networks. In: M. Capcarrere, A.A. Freitas, P.J. Bentley, C.G. Johnson, J. Timmis (eds.) Advances in Artificial Life, *Lecture Notes in Computer Science*, vol. 4648, pp. 876–885. Springer (2007)
18. Groot, R.D.: Mesoscopic simulation of polymer-surfactant aggregation. Langmuir **16**, 7493–7502 (2000)
19. Groot, R.D.: Electrostatic interactions in dissipative particle dynamics-simulation of polyelectrolytes and anionic surfactants. J. Chem. Phys. **118**(24), 11,265–11,277 (2003)
20. Groot, R.D., Rabone, K.L.: Mesoscopic simulation of cell membrane damage, morphology change and rupture by nonionic sufactants. Biophys. J. **81**, 725–736 (2001)
21. Groot, R.D., Warren, P.B.: Dissipative particle dynamics: bridging the gap between atomistic and mesoscale simulation. J. Chem. Phys. **107**(11), 4423–4435 (1997)
22. Hoogerbrugge, P., Koelman, J.: Simulating microscopic hydrodynamic phenomena with dissipative particle dynamics. Europhys. Lett. **19**, 155–160 (1992)
23. Humphrey, W., Dalke, A., Schulten, K.: VMD – visual molecular dynamics. J. Molec. Graphics **14**, 33–38 (1996). URL `http://www.ks.uiuc.edu/Research/vmd/`
24. Hutton, T.J.: Evolvable self-replicating molecules in an artificial chemistry. Artif. Life **8**(4), 341–356 (2002)
25. Limbach, H.J., Arnold, A., Mann, B.A., Holm, C.: ESPResSo – an extensible simulation package for research on soft matter systems. Comput. Phys. Commun. **174**(9), 704–727 (2006)
26. Liu, H., Qian, H.J., Zhao, Y., Lua, Z.Y.: Dissipative particle dynamics simulation study on the binary mixture phase separation coupled with polymerization. J. Chem. Phys. **127**, 144,903.1–144,903.8 (2007)
27. Lyubartsev, A.P., Laaksonen, A.: On the reduction of molecular degrees of freedom in computer simulations. Lect. Notes Phys. **640**, 219–244 (2004)
28. Maiti, A., McGrother, S.: Bead-bead interaction parameters in dissipative particle dynamics: relation to bead-size, solubility, and surface tension. J. Chem. Phys. **120**(3), 1594–1601 (2003)
29. Mayer, B., Köhler, G., Rasmussen, S.: Simulation and dynamics of entropy-driven, molecular self-assembly process. Phys. Rev. E **55**(4), 4489–4500 (1997)
30. Mayer, B., Rasmussen, S.: The lattice molecular automaton (LMA): a simulation system for constructive molecular dynamics. Int. J. Mod. Phys. C **9**(1), 157–177 (1998)
31. Monine, M.I., Haugh, J.M.: Reactions on cell membranes: comparison of continuum theory and Brownian dynamics simulations. J. Chem. Phys. **123**(7), 074,908 (2005)
32. Noguchi, H., Gompper, G.: Dynamics of vesicle self-assembly and dissolution. J. Chem. Phys. **125**, 164,908 (2006)
33. Noguchi, H., Takasu, M.: Adhesion of nanoparticles to vesicles: a Brownian dynamics simulation. Biophys. J. **83**, 299–308 (2002)
34. Novik, K.E., Coveney, P.V.: Finite-difference methods for simulation models incorporating nonconservative forces. J. Chem. Phys. **109**(18), 7667–7677 (1998)
35. Ono, N.: Artificial chemistry: computational studies on the emergence of self-reproducing units. Ph.D. thesis, University of Tokyo, Tokyo (2001)
36. Ono, N., Ikegami, T.: Self-maintenance and self-reproduction in an abstract cell model. J. Theor. Biol. **206**(2), 243–253 (2000)
37. Pagonabarraga, I., Frenkel, D.: Dissipative particle dynamics for interacting systems. J. Chem. Phys. **115**, 5015–5026 (2001)
38. Pagonabarraga, I., Hagen, M.H.J., Frenkel, D.: Self-consistent dissipative particle dynamics algorithm. Europhys. Lett. **377**(42) (1998)

39. Pastorino, C., Kreer, T., Müller, M., Binder, K.: Comparison of dissipative particle dynamics and langevin thermostats for out-of-equilibrium simulations of polymeric systems. Phys. Rev. E **76**, 026,706 (2007)
40. Plimpton, S.J.: Fast parallel algorithms for short-range molecular dynamics. J. Comp. Phys. **117**, 1–19 (1995)
41. Python Software Foundation: the Python programming language. URL `http://www.python.org/`
42. Rasmussen, S., Bedau, M., Chen, L., Deamer, D., Krakauer, D., Packard, N., Stadler, P. (eds.): Protocells: Bridging Nonliving and Living Matter. MIT Press, Cambridge, MA (2008)
43. Shillcock, J.C., Lipowsky, R.: Equilibrium structure and lateral stress distribution of amphiphilic bilayers from dissipative particle dynamics simulations. J. Chem. Phys. **117**(10), 5048–5061 (2002)
44. Sims, J.S., Martys, N.S.: Simulation of sheared suspensions with a parallel implementation of QDPD. J. Res. Natl. Inst. Stand. **109**(2), 267–277 (2004)
45. Tcl Core Team: The tool command language (Tcl). URL `http://www.tcl.tk/`
46. Varela, F.J., Maturana, H.R., Uribe, R.: Autopoiesis: the organization of living systems. BioSystems **5**(4), 187–196 (1974)
47. Vattulainen, I., Karttunen, M., Besold, G., Polson, J.M.: Integration schemes for dissipative particle dynamics simulations: from softly interacting systems towards hybrid models. J. Chem. Phys. **116**, 3967–3979 (2002)
48. Venturoli, M., Smit, B.: Simulating self-assembly of model membranes. PhysChemComm **10** (1999)
49. Verlet, L.: Computer "experiments" on classical fluids. I. Thermodynamical properties of Lennard-Jones molecules. Phys. Rev. **159**, 98–103 (1967)
50. Warren, P.B.: Vapor-liquid coexistence in many-body dissipative particle dynamics. Phys. Rev. E **68**, 066,702 (2003)
51. Williams, T., Kelley, C.: gnuplot – an interactive plotting program. URL `http://www.gnuplot.info/`
52. Yamamoto, S., Hyodo, S.: Budding and fission dynamics of two-component vesicles. J. Chem. Phys. **118**(17), 7937–7943 (2003)
53. Yamamoto, S., Maruyama, Y., Hyodo, S.: Dissipative particle dynamics study of spontaneous vesicle formation. J. Chem. Phys. **116**(13), 5842–5849 (2002)
54. Yingling, Y.G., Garrison, B.J.: Coarse-grained chemical reaction model. J. Phys. Chem. B **1008**, 1815–1821 (2004)

Part IV
Artificial Life Arts

Chapter 12
Simulated Breeding: A Framework of Breeding Artifacts on the Computer

Tatsuo Unemi

This chapter describes a basic framework of simulated breeding, a type of interactive evolutionary computing to breed artifacts, whose origin is *Blind Watchmaker* by Dawkins. These methods make it easy for humans to design a complex object adapted to his/her subjective criteria, just similarly to agricultural products we have been developing over thousands of years. Starting from randomly initialized genome, the solution candidates are improved through several generations with artificial selection. The graphical user interface helps the process of breeding with techniques of multifield user interface and partial breeding. The former improves the diversity of individuals that prevents being trapped at local optimum. The latter makes it possible for the user to fix features he/she already satisfied. These methods were examined through artistic applications by the author: SBART for graphics art and SBEAT for music. Combining with a direct genome editor and exportation to another graphical or musical tool on the computer, they can be powerful tools for artistic creation. These systems may contribute to the creation of a type of new culture.

12.1 Introduction

For over a thousand years, mankind has been utilizing the technique of breeding to obtain useful plants and animals to help human life. Almost all agricultural products and domesticated animals we see now are the results of these processes of many generations. Even in the new century of highly improved genetic engineering, it is still impossible to build such a complex living system by human hands in a similar manner with mechanical and electric machines.

It is also a fact that we already have highly sophisticated technologies in our hands to make complex machines and systems, such as robots, transportation control systems, hi-tech aircrafts, space shuttles, and so on. Many researchers and developers are struggling and competing with each other to design and implement amazing gadgets day by day in laboratories and private companies. Recent improvement of information technologies accelerates this movement. However, it is hard to say that we are receiving enough results on these technologies because of the bottleneck of the developmental cost. We need a longer time to design and implement more complex products.

One method to overcome this bottleneck may be to introduce a method of breeding we have been using for living organisms. Some of the currently developed machines are complex enough for objects of breeding. Features that enable us to breed organisms are reproduction with changes, the essential functions for evolvability. All living systems on the Earth intrinsically have this mechanism supported by genetic inheritance and mutation. Artifacts have no ability to reproduce itself in nature, but the computer can help us to realize it. If we build an interactive software to simulate breeding process, it might enable us to obtain our desired design in a reasonable time. In addition, there is the possibility to give us a quite new products that we have unconsciously given up because of the developmental cost.

This chapter presents a framework of *simulated breeding* that realizes breeding the artifacts on the computer through an overview of two sorts of applications. One is SBART for drawing abstract computer graphics, and another is SBEAT for composing short musical pieces.

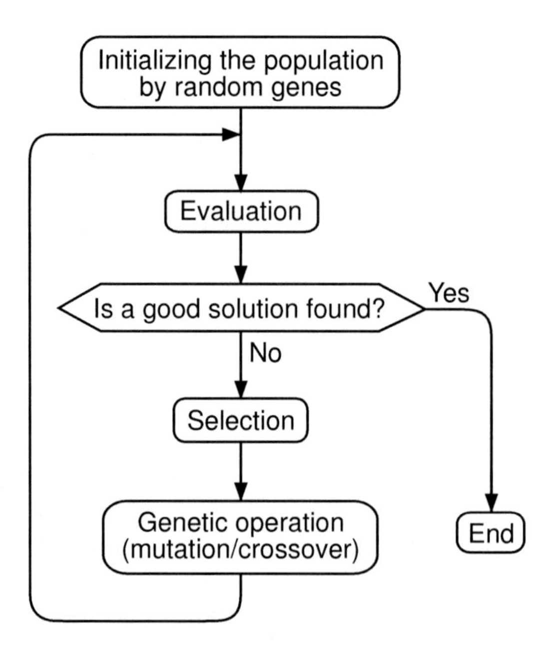

Fig. 12.1 A flowchart of a simple genetic algorithm.

12.2 Basic Framework of IEC

Research on computational intelligence and artificial life has brought some useful algorithms to produce complex systems inspired from biological mechanisms such as morphology, evolution, learning, herding, and so on. One of the most successful technique in terms of engineering applications may be evolutionary computing for optimization in various types of problems that had been thought difficult to solve. This method was originally developed from a type of optimization of morphological design of living organisms to gain a high rate of reproductive success in the physical and ecological environment. Figure 12.1 shows a flowchart of a simple genetic algorithm, a typical evolutionary computing scheme, consisting of a loop of evaluation, selection, crossover, and mutation.

An ordinary type of evolutionary algorithm uses a predefined fitness function to give a criterion of optimization. It just corresponds to the condition of natural selection. This framework works well if the human designer can draw an appropriate procedure to compute fitness values to evaluate each individual. However, we often go through difficulties in figuring it out explicitly by some reasons, such as multiobjectivity, subjective criteria, dynamic environment, and so on. In a design of room arrangement of a house, for example, each member of the resident family has different preference criteria from others. They are, of course, subjective and often dynamically changing with the natural and social environment.

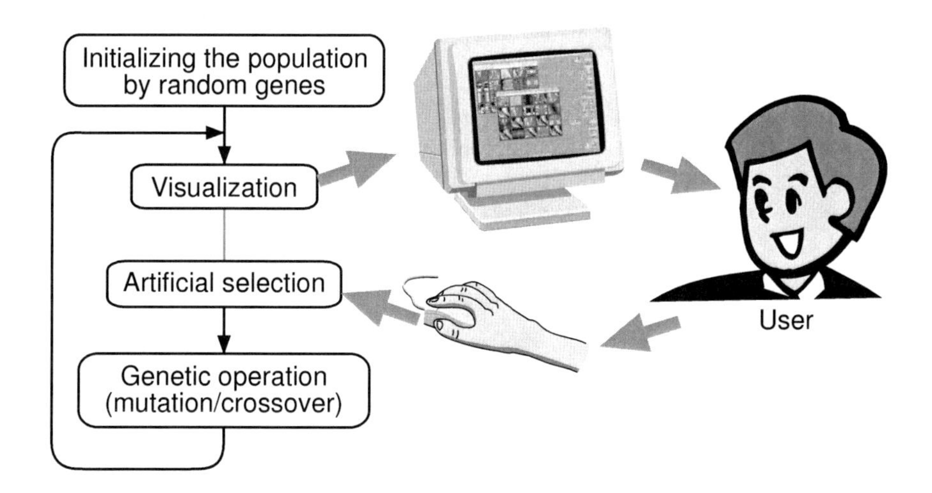

Fig. 12.2 A schematic framework of interactive evolutionary computing (IEC).

Interactive evolutionary computing (IEC) [16] is a promising technique to find better solutions in the domains for optimization by user's subjective criteria, whose root can be found in *Blind Watchmaker* by Dawkins [3]. Differently from ordinary methods of evolutionary computing, fitness values are not calculated automatically by the predefined evaluation function but are given by the user for each individual in some manner. Some researchers are calling this method the *interactive genetic algorithm* (IGA) [15], because it can be seen as a modified version of the *genetic algorithm* (GA) [5]. Usually, the application systems have a method for the user to rate each individual, typically a graphical user interface using a slider or a set of radio buttons. In the method called *simulated breeding*, the user directly picks up his/her favorite individuals as parents for the next generation. This means that the fitness values can take only one (selected) or zero (not selected). It disables stochastic selection, but has an advantage to reduce the number of user's operations to assign the fitness values. Figure 12.2 shows a schematic framework of IEC, where the phases of evaluation and selection in simple GA are replaced with visualization and artificial selection.

Another source of advantage of IEC compared to the other types of design support framework of *generate and test* is explicit separation between genotype and phenotype. Genotype is information on a genome that changes through genetic operation. Phenotype is the object of evaluation that performs in the environment. In natural living organisms, genotype is genetic information on DNA, and phenotype is the body and its performance for survival and reproduction. This separation makes it flexible to design the search space for an application domain.

12.3 SBART and SBEAT

The author developed several types of applications of simulated breeding over the years. One is named SBART for computer graphics and another is named SBEAT for computer music. These artistic domains are mostly suitable for IEC because they are strongly dependent on subject criteria of evaluation.

12.3.1 SBART

SBART [23] is an application of simulated breeding to breed an abstract image of two-dimensional (2D) computer graphics. The basic mechanism is based on the idea by Sims [14], who implemented his system on a combination of a supercomputer and graphic workstations in 1991. Some researchers who wanted to experience this innovative system developed their small system on personal computers. SBART is one of these systems that was developed in 1993 on a Unix workstation and was exported to MacOS in 1998. The basic idea is the same between Sims's system and SBART, but some features described here were uniquely modified and added in SBART. This software

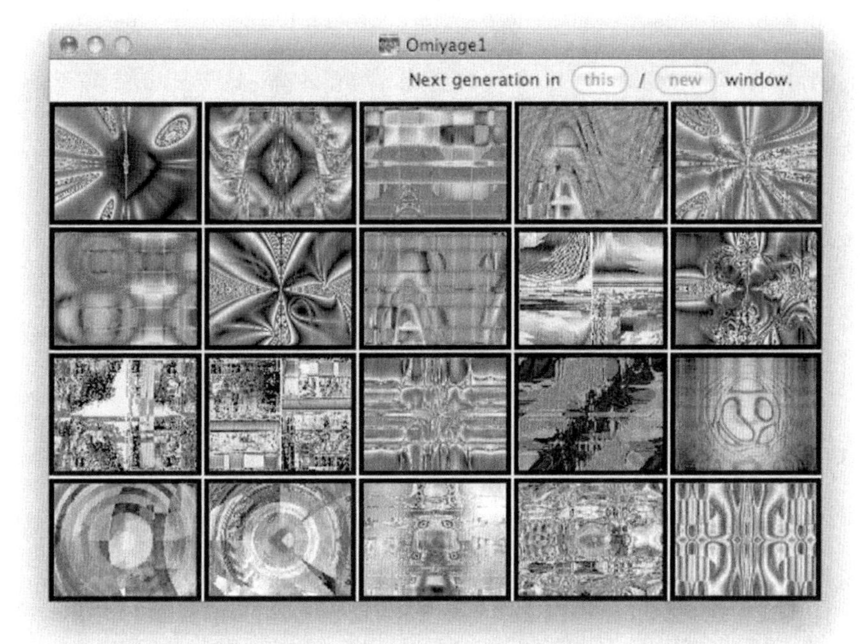

Fig. 12.3 A field window of SBART containing 20 individuals.

has been widely distributed through the Internet and CD-ROMs attached with some books and journals.

Figure 12.3 shows an example of a field window that contains 20 individual drawings, each of which is a candidate of parents for the next generation. The user selects one or more of his/her favorite individuals to reproduce offspring. Through some iterations of generation changes, the user can obtain satisfactory results.

The structure of the genotype is a mathematical expression that calculates color value for each pixel from the 2D coordinates x and y. Each of the intermediate values in the calculation is a three-dimensional (3D) vector that is interpreted as a color in hue, saturation, and brightness at the result. For example, the genotype of the bottom left individual in Fig. 12.3 is

$$\mathrm{and}(-(1.180), \mathrm{and}(\mathrm{0XY}, \mathrm{XY0}) + \mathrm{XY0} - \mathrm{hypot}(\mathrm{hypot}(\mathrm{Y0X}, \mathrm{0XY}), 1.680)),$$

where four types of binary functions, $+$, $-$, "and," and "hypot," and one unary function, $-$, are used. 0XY, XY0, and Y0X are variable vectors that mean $\langle 0, x, y \rangle$, $\langle x, y, 0 \rangle$, and $\langle y, 0, x \rangle$ respectively. SBART has six types of vector variables in all permutations of x, y, and 0. Constant scalar values are expanded into a vector containing three same values; that is, 1.180 in the above expression is expanded into $\langle 1.180, 1.180, 1.180\rangle$. Almost all functions calculate each element values in a vector independently; that is, $\langle x, y, 0\rangle + \langle 1.180, 1.180, 1.180\rangle = \langle x+1.180, y+1.180, 1.180\rangle$. Some exceptional functions calculate the result by combination of elements of argument vectors. In the case of the function named "max," for example, the value is defined as

$$\max(\langle x_1, x_2, x_3\rangle, \langle y_1, y_2, y_3\rangle) = \begin{cases} \langle x_1, x_2, x_3\rangle & \text{if } x_1 > y_1 \\ \langle y_1, y_2, y_3\rangle & \text{otherwise.} \end{cases}$$

When more than one individual is selected, crossover operation is applied. When only one individual is selected, the mutants are produced as the population of the next generation. These genetic operations are done in a style of *genetic programming* [6]. Crossover is an exchange of subtrees between parents. Mutation is replacement of a node: function, variable, or constant with a randomly selected one. The target subtree and node are selected randomly.

SBART has not only a breeding mechanism but also some utilities to import some image files and a single movie file to produce a type of collage, to draw an arbitrary size of product image to export to another graphics tool, and to create a movie allocating time variable into the 0 element of variable vectors. Some visual jockeys (VJs) are using this system to produce a material of their own video clips for playing at a dance club.

12.3.2 SBEAT

Sound and music are also attractive targets for the application of IEC techniques because they also strongly depend on subjective criteria for artistic production. One of the key issues for building a successful system in this domain is how the user checks and selects suitable individuals from a population.

Biles [1] has proposed an alternative method to solve this problem and implemented it in his system named GenJam. This system helps the user to create improvisational phrases in Jazz music. In GenJam, the user listens to endless phrases generated from individual genotypes in turn. The user pushes the "g" or "b" key according to his/her good or bad feelings about the phrase. It is not necessary for the user to assign fitness values explicitly or to know the correspondence between the individual and the phrase.

Another type of implementation of musical application of IEC has been developed as the Sonomorph system by Nelson [8, 9]. It shows nine candidates on the screen by drawing the scores in the form of a collection of horizontal line segments just like piano roll paper. SBEAT [21] was developed by the author in 2000 independently with Sonomorph, which means the author had no knowledge about Sonomorph when considering the basic idea. SBEAT shows nine individuals on a field window shown in Fig. 12.4. A phenotype of

Fig. 12.4 A typical field window of SBEAT containing nine initial individuals.

each individual is a score of 16 beats and several parts, maximally 23 parts in version 3. Four of the whole parts chosen by the user are shown in the subwindow. The first version of SBEAT treated only three parts: guitar, bass, and drums. It was extended to eight parts: five solos, piano, drums, and percussion in the second version SBEAT2 [22]. The next version, SBEAT3 [25], handles 13 solo parts, 2 piano (or chord) parts, and 8 drums and percussion parts.

The genotype of SBEAT3 is a set of three 2D arrays for pitches, rhythm, and velocity. Each array contains 16 by 23 elements, each of which corresponds to beat and part. It uses a type of recursive algorithm to produce a basic melody from genotype to guarantee a natural similarity between parents and children. It is also useful to produce individuals in the initial population acceptable for the user. Refer to [26] for details.

SBEAT2 and SBEAT3 have some utilities to set tempo and scale, to change the instruments set for each part, to integrate the result scores into a longer tune, and to save the score into a file of standard MIDI format (SMF). It is possible to compose a complete tune only by these systems, but the users may export the result in an SMF file to another application of music sequencer or digital audio workstation to mix with another kind of composition and to add various types of effects.

12.4 Breeding in a Field Window

Visualized phenotypes of a population are displayed together in a *field window* as shown in Figs. 12.3 and 12.4. We assume that it is possible to visualize each phenotype without expensive cost for the computation and that the population size is about from 9 to 30. A field window is divided into from 3×3 to 5×6 grids, showing each individual in each rectangle of grid. These 9 to 30 individuals are the members of the current population and there is no other member of the population. The user explicitly selects one or more individuals from the field as parents for the next generation. When only one individual is selected, all individuals except the selected one are replaced by mutants of the selected individual in the next generation. When more than one individual is selected, all individuals are replaced by the children of selected parents produced with a crossover operation.

From the results of various settings of experimental implementation of SBART, the appropriate number of individuals simultaneously shown on the screen should be in the range from 16 to 30. The user often had to produce alternative children to obtain improved candidate if the number was less than 16. Some good individuals were ignored if the number was greater than 30.

There is another issue in the domain of sound and music. It is possible to select favorite graphical drawings in seconds, but the user needs a longer time to evaluate sound and music. It is possible to compare candidates simultaneously in graphics domains, but individuals must be examined independently in sound and musical domains. So a system for sound and music has to have two types of selection methods for each individual. One is for selection as a parent, and the other one is for selection for playing. In the case of SBEAT, three types of buttons are attached to each individual subwindow as show in Fig. 12.4: a play button with speaker icon, a protection button with lock icon, and a selection checkbox. The role of protection is described in the Section 12.5.2.

The collection of genotypes in a field population can be recorded in a disk file similarly to any other document files of a word processor, a drawing tool, and so on. It can be opened again as a field window later if the user saved it. Menu items entitled “New,” “Open,” “Save,” “Save as . . . ,” and “Reset” are available with functions that can be easily guessed from an analogy with other documentation tools. A field is always filled with a fixed number of individuals, so the "New" field is not empty but initialized with individuals of random genes.

12.5 Multifield User Interface

In [20], we propose a design of a graphical user interface using the *multifield* for simulated breeding method. The term *field* is used here as a population of visualized individuals that are candidates of selection from an analogy with fields in an experimental farm. A multifield interface enables the user to breed his/her favorite phenotypes by selection independently in each field, and he/she can copy arbitrary individuals into another field. As known among the researchers of evolutionary computing, a small population likely leads to premature convergence trapped by a local optimum, and migration among plural populations is useful to escape from this trap. For IEC applications, it is impossible to take a large size of population because the user has to observe all phenotypes in the population as possible to evaluate them instead of predefined fitness function in the ordinary framework of evolutionary computation. The multifield user interface is a suitable method for IEC to provide easy implementation of migration and wider diversity.

We often suffer a complicated multimodal landscape in a structural optimization problem because the search space constructed by solution candidates is usually high dimensional. It is necessary to examine as many candidates as possible to find the best solution because each candidate has a lot of neighbors in a high-dimensional space. One of the key techniques for successful search is a method to keep a diversity of individuals in the population. *Island model* [11] is one of the methods to keep diversity.

After geographical separation of a continent into islands, a population of species that lived in a continent is divided into subpopulations in each island, and they usually reach distinct organisms through independent evolutionary process since mating between different islands is prohibited and there are many suboptimal points for the structure of an organism. As the scheme theory [5] in the genetic algorithm indicates, a crossover operation, a combination of parts of different genomes, has the possibility to spawn a better individual by combination of good genes from different individuals. We can expect to get better solutions by migration of individuals among different islands after reaching some convergence in each island. The multifield user interface is a method to bring a similar effect with the island model.

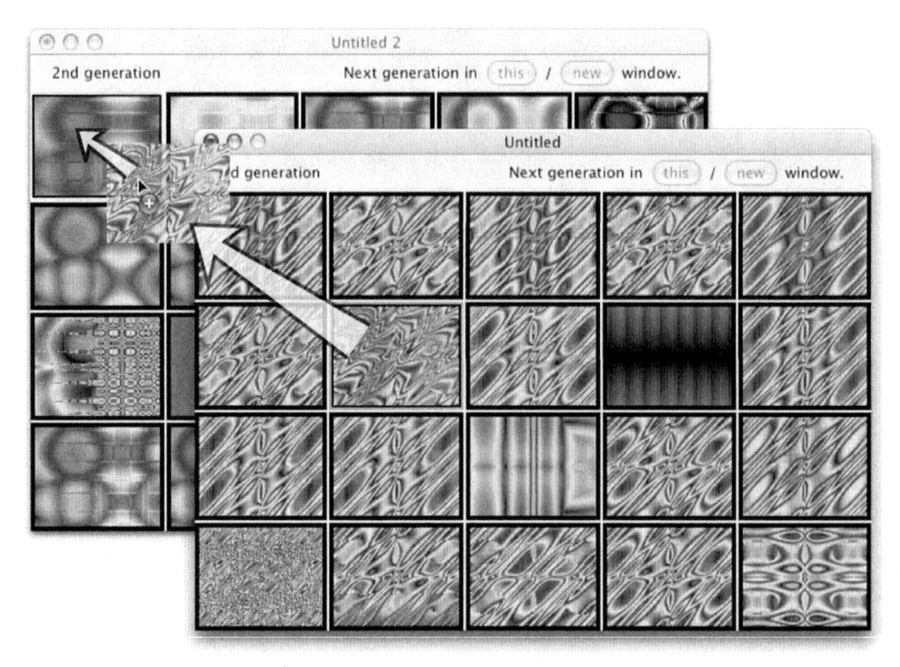

Fig. 12.5 Migration of an individual by *drag & drop*.

12.5.1 Migration Among Fields

The user can open an arbitrary number of field windows by choosing the "New" and "Open" items from the File menu if the memory capacity is adequate. In addition to these operations, a new field window spawns when the user clicks the "new" button at the top right of each field window instead of "this" to generate a population of the next generation. The new field window is filled with the children of selected parents.

SBART and SBEAT have two types of methods to migrate an individual between fields. The first one is to move it by *copy & paste*. The operation includes four steps:

1. Select the individual that the user wishes to move on the field.
2. Copy it into the copy buffer.
3. Select the individual that should be discarded by overwriting.
4. Paste it.

Selection of an individual on the field is done simply by clicking the subwindow. A red border frame of a rectangle area indicates that the individual is selected. It takes six steps if the user uses the Edit menu to both copy and paste because of invoking the Edit menu from the menu bar. Shortcut keys can reduce it by two steps.

The other way is *drag & drop*. As Fig. 12.5 shows, the user can copy any individual by pointing to it with the mouse cursor, pressing the left button, moving the mouse pointer keeping the mouse button pressed, and releasing the button at the destination rectangle. A small individual image moves following the mouse's move. This method needs a fewer number of operations than the first one. It is easier for the user who knows well another application with similar operation such as a file manager.

12.5.2 Protection of Individual

Not only to save genes in a disk file, the user sometimes wish to keep some interesting individuals without modification temporally. It is realized by *protection* of individual. The user can protect his/her selected individual by pushing the "Protect" button in the "Edit menu" and the "Context menu"[1] in SBART. This button is attached in each subarea for individuals in SBEAT. Protected individuals can neither be overwritten by migration nor replaced by offspring.

An alternative design of individual protection is to facilitate the other type of window that keeps the arbitrary number of genome as a profile. This method may lead to a filing system including library files; however, we have not tried to implement it yet. Some method for efficient retrieval for the desired genome from a large-scale database will be needed for this type of filing system.

12.5.3 Effects of Multifiled Interface

Populations processed through independent evolution usually reach unique features for each. In these cases, we can expect that a new one will be produced by migration among independent fields. Figure 12.6 shows offspring produced by crossover between individuals that came from different fields. Some part of features of both parents pass to the children, but it is unknown which feature remains because the crossover points are determined randomly. If the gene coding allows redundant representation or noneffective part, it may produce children of more unexpected features.

[1] "Context menu" is a term in MacOS that is called "pop-up menu" in Motif.

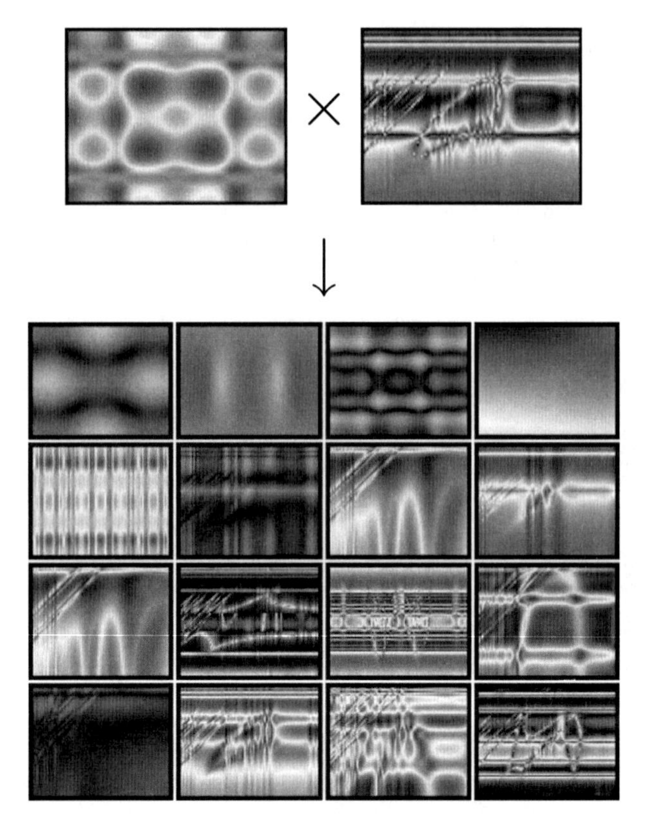

Fig. 12.6 An example of offspring spawned by migration and crossover between two individuals bred independently in the different fields.

12.6 Partial Breeding

It is a good method to divide a complicated problem into independent subproblems to solve it efficiently. However, it is often difficult because of dependency among parts of the problem features. A musical tune consists also of complex information including melody, rhythm, tempo, timbre, and so on. We can easily divide information into these functional parts, but evaluation of the tune usually depends on the combination of them. The combination of good melody and good timbre is not always good.

One of the advantages of evolutionary computation compared with the other optimization technique by search is that we can design different structure of search space than the structure of solution space using mapping between genotype and phenotype. Phenotype corresponds to the solution candidate, and genotype corresponds to the search point. A musical tune can be divided into sections, parts, bars, and so on. It is also helpful to build each of the parts or sections independently and to combine them later. In addition, we can independently breed each of the functional features of music, rhythm, melody, timbre, and so on by encoding them in separate parts of a genotype.

Because of dependency among features in terms of the effect for the quality of the solution, it is difficult to obtain a good solution by optimizing each feature step by step independently. We often need to revise a feature previously optimized during optimization of another feature. Even in the ordinary style of evolutionary computation, it is still a research theme how we should apply evolutionary computation to multiobjective optimization problems. Fortunately in IEC, the user can control the evolutionary process by indicating which parts and features should be fixed and which parts should be modified if the system has a user interface of partial protection.

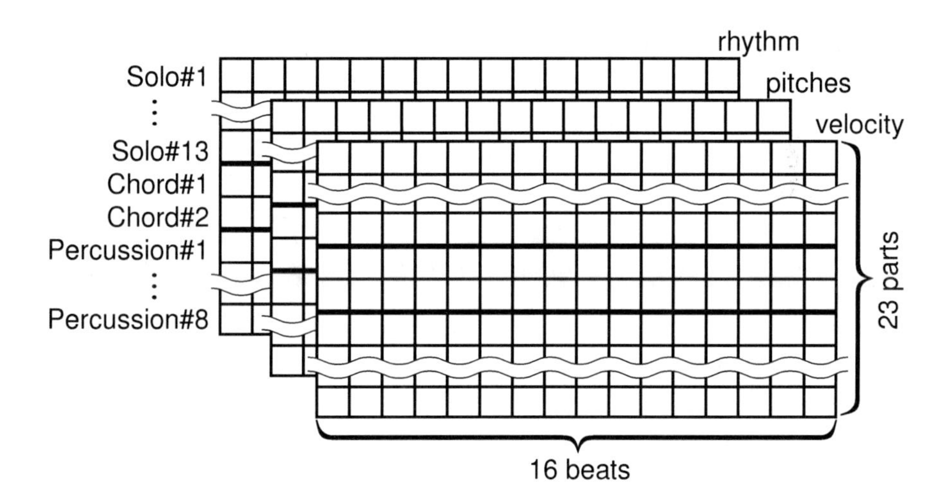

Fig. 12.7 Structure of genotype in SBEAT3.

We designed the structure of genotype for SBEAT as consisting of three types of chromosomes for rhythm, pitches, and velocity, as shown in Fig. 12.7. Velocity, which means loudness of the sound of note as a technical term in music, is an important factor for tune to sound natural for humans' ears. The individual in SBEAT population is a bar of 16 beats and 23 parts. Each of the chromosomes is a 2D array of 16 by 23 to include information for each possible note. Some loci in the chromosomes of pitches and velocity are ignored if the corresponding loci of rhythm chromosome indicate rest or continuation.

We designed a graphical user interface to make it possible for the user to indicate which part should be the objects of genetic operation. In ordinary cases, mutation is applied to all features. We introduced check buttons to indicate protection against mutation. Figure 12.8 shows the dialog window of SBEAT, which includes many buttons, sliders, and menus. The short slider allocated for each part is for setting the correspondence between musical parts

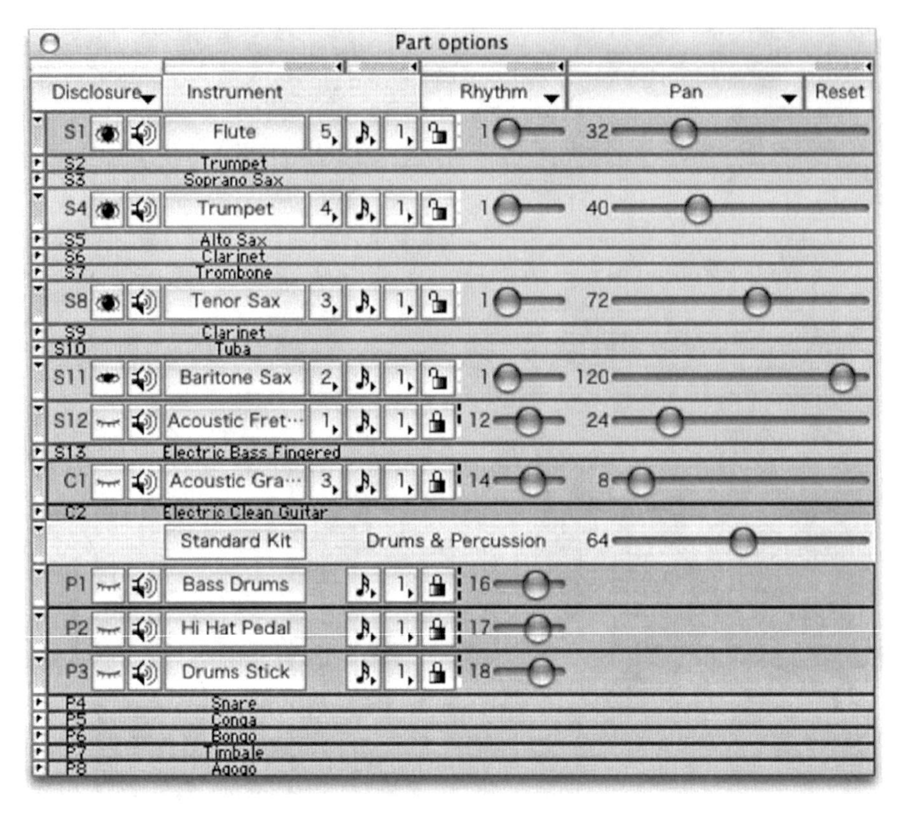

Fig. 12.8 Part option dialog of SBEAT3.

and genotypical parts. A button with a lock icon is for indicating protection. These operations are applied to one of three chromosomes indicated by the pull-down menu at the top row of these sliders and buttons.

It is possible for the user to indicate protection of arbitrary parts of a genotype that corresponds to functional or physical part of phenotype at any time he/she wants. To reduce the number of operations for pressing arbitrary number of buttons at once, we implemented a method to choose them by the *press-drag-release* operation of the mouse. The buttons in the rectangular area indicated by mouse operation act as being clicked. It makes it easy to revise any part again and again by breeding independently until any acceptable solution is found.

This method was invented through an application for music, but it will be useful for another domain where the solution candidates are complex but well structured [24]. However, it is difficult to apply this method to SBART, because the user hardly understands which part of a genotype affects which feature of a phenotype.

12.7 Direct Genome Operation

Breeding is a good method for producing novelty, but it is mostly redundant when we know what type of direct modification brings a better result. The genome editor was designed to answer this requirement by allowing the user to edit chromosomes directly.

The user is allowed to edit the genome of any individual on the field. Figure 12.9 shows an example screen shot of genome editor in SBART. As shown in the right window of the figure, the user selects a node from the tree structure of a genome to cut, copy, paste, and swap with the subtree in the copy buffer and to replace it with another symbol. The new function symbol, variable, and constant can be indicated using the dialog at the left side of the figure.

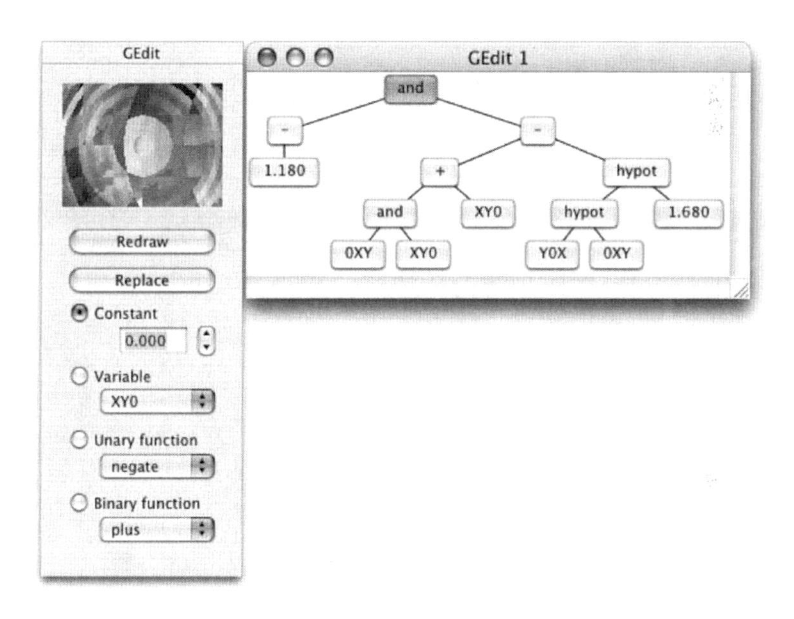

Fig. 12.9 Genome editor of SBART.

Figure 12.10 shows an example screen shot of genome editor in SBEAT. It has two windows, a score of an individual and an editing panel. The user selects a part to be edited using a pop-up menu[2] and then operates buttons allocated to each beat of each chromosome. The button of each beat has a different type of function among chromosomes because the data types of the allele are different among the types of chromosomes.

In the case of SBART, this functionality of direct editing is not so useful because it is usually difficult for the user to understand the concrete corre-

[2] "Pop-up menu" is a term in MacOS that is called "option menu" in Motif.

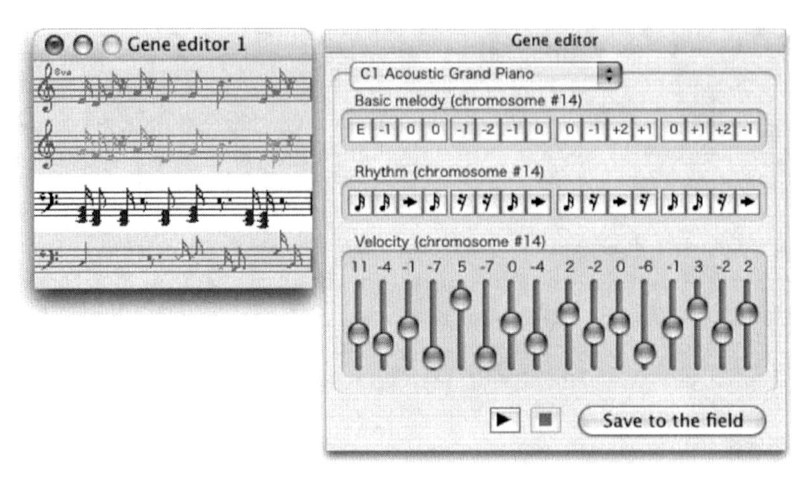

Fig. 12.10 Genome editor of SBEAT.

spondence between genotype and phenotype. However, in the case of SBEAT, it is relatively clear which loci affect which feature and part in a phenotype even though loci and features do not always have a one-to-one correspondence. It can play a different role that allows the user to input the score already known if possible.[3] Collection of the mutant of a known melody can be useful for making an arrangement of a tune.

12.8 Production Samples

The readers have already seen some examples of images produced with SBART in Fig. 12.3, but to understand the potential possibility of this system, this section shows other examples produced using augmented functions to embed an external video data.

Kamei, a VJ in Tokyo, suggested that it would be nice if SBART could import a movie file to embed and to deform it, similarly to importing image files to make a collage. It needs a fast and sophisticated smoothing algorithm among related pixels in video frames that wastes much of computation resources. It seemed to be impossible to process it within a reasonable time by the personal computer some years ago. However, now GHz CPUs and some gigabytes of memory on board are available at reasonable prices. The newest version of SBART includes this feature, and we are investigating some technical issues and developing new algorithms and user interfaces for achieving feasible usability [27].

[3] In the current implementation of SBEAT, some types of scores cannot be mapped backward to genotype due to the algorithm of morphology.

Two new nonterminals were added to realize this functionality. One is named "movief" that extracts pixels' HSB color values from one frame of time t at the position indexed by referring to the first and second elements of the argument vector as x- and y-coordinates. The other one, named "moviec," similarly extracts color values, but it assumes the movie data as the 3D volume and picks a boxel up by referring to three elements.

a. Original external movie.

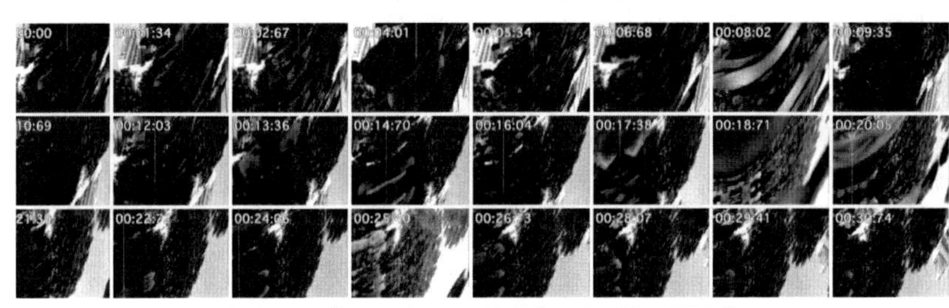

b. movief(hypot(0.648, XTY) + XTY + YXT)

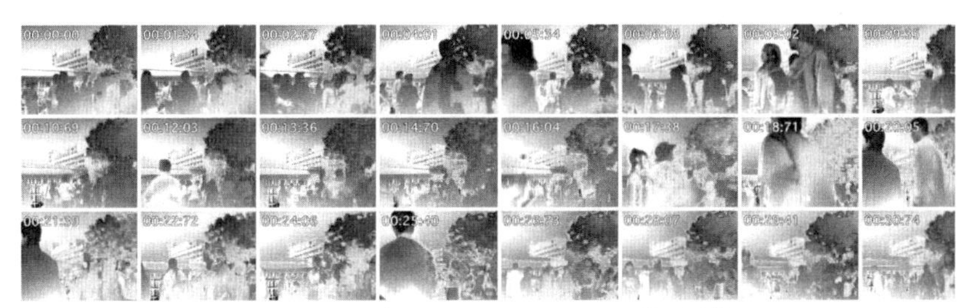

c. −0.203 + (0.648 + YXT + movief(XYT))

Fig. 12.11 Sample frame sequences produced with the frame image method.

The upper sequence of movie frames in Fig. 12.11 is external movie data in half-size of NTSC DV format, 320×240 pixels of frame size, 14.98 frames per second, and 30 seconds duration. It includes $14.98 \times 30 = 449$ frames, then, totally, $320 \times 240 \times 449 = 34{,}483{,}200$ boxels. It was taken by a handy camcorder in the campus during a campus festival in the fall of 2003.

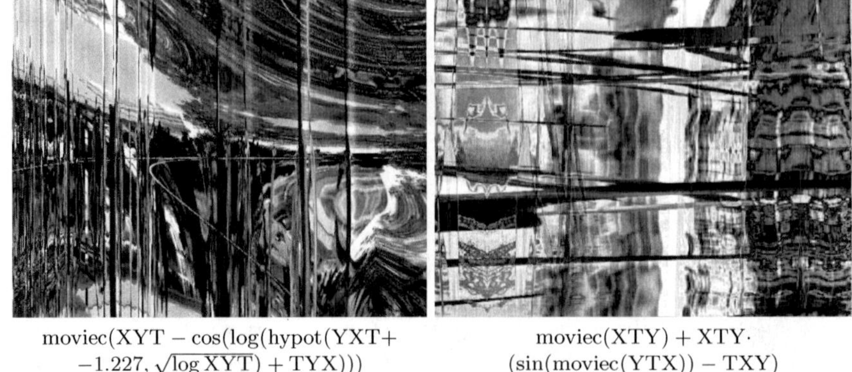

moviec(XYT − cos(log(hypot(YXT+ −1.227, $\sqrt{\log \text{XYT}}$) + TYX)))   moviec(XTY) + XTY· (sin(moviec(YTX)) − TXY)

Fig. 12.12 Sample images produced from the external movie in Fig. 12.11 by the boxel method.

Figure 12.12 shows two typical productions using the boxel method. The genotype of the left-side image has the function "moviec" at the root of the tree structure. This makes nonlinear transformation of the shape of original pattern. The time axis was expanded along the horizontal axis. Some vertical lines included in the resultant image indicate rapid reaction of automatic exposure adjustment by the camcorder. Small vertical vibrations along the horizontal axis means it was difficult to keep the handy gadget steady in the operator's hand.

The right-hand image has a more complex genotype that includes "moviec" at the intermediate nodes. It produces not only deformation of the shape but also modification of colors. As shown in both images, the boxel method brings us a new effect to produce more complicated and unpredictable results from some types of external movies by inheriting the complexity of a 3D pattern involved in the movie data.

The middle- and lower-frame sequences in Fig. 12.11 are sample movies produced from genotypes that includes "movief." The middle one includes the function at the root, and the lower one includes it at just above a leaf of variable XYT. The former one produces shape deformation, and the latter one produces color modification, similar to the two examples for the boxel method.

We also examined another film taken at a concert of a student big band. It can produce a very interesting and effective movie clip that seems useful to create a music video or so on.

Figure 12.13 is a score of example production bred using SBEAT that consists of six parts. The target of breeding in SBEAT is a bar of 16 beats and 23 parts, as described before. This tune was composed by integration of nine bars, each of which restricted only five solo parts and one chord part. A set of sample tunes composed by the author in both SMF and MP3 format is available from the web page of SBEAT.

Fig. 12.13 Sample score composed with SBEAT.

12.9 Future Works

Of course, there are many types of further works to be tried in the future concerning the simulated breeding method. From the viewpoint of engineering, it would be important to examine various types of alternative methods usable for similar purpose and to compare them to each other in several aspects such as efficiency, usability, flexibility, and so on. For the technical evaluation, it must be important to execute experiments on the human-machine interface employing a number of subjects as typical users. One feature that should be examined is the method of artificial selection. In some application domain or for some types of users, stochastic selection by fitness-value rating might be essential to obtain satisfactory results efficiently.

Some researchers are struggling to build similar systems using their original ideas of musical application of IEC technique such as [18, 19]. Including the previous pioneers' approaches including sound filtering [10] and sound generation [2], it may be fruitful to combine several different ideas to build up a more practical integrated system to support human activities of music composition.

SBEAT was designed under the assumption of off-line composition; that is, the processes of composition and performance are separated. However, it is also potentially usable for a live performance on time. The system needs some extension to have two separated channels for sound output for a breeder and listeners, because mutation may sometimes spawn less interesting sounds.

The combination of SBART and SBEAT should be considered as an integrated system to make a computer graphics animation with music. The variety of result products should be enhanced to realize this plan because the products from current systems seem unsuitable to combine with each other. Using a video editing tool on the personal computer currently available, it is possible to combine the products of SBART and SBEAT. However, interactive breeding between these two systems seems important to improve the final results.

It is possible to apply the framework of IEC to a variety of fields other than graphics and music. Robot motion [7] and group behavior of people [4] are also interesting targets from a viewpoint of artistic application. The author is conducting a media art project called Project 7^2 [28, 29] with Bisig from 2003. It is a challenge to make a variety of artworks by a combination of audio visual feedback and swarm simulation. We have released four artworks and have been distributing two softwares on the Internet – DT1: Flocking Orchestra and DT4: Identity SA. We have not applied IEC to the project yet, but it would be valuable to try it to breed more interesting behavior of reactive swarm in a virtual space.

12.10 Conclusion

The framework of simulated breeding has been described. Through the experience with SBART and SBEAT, we may expect future expansion of application of these systems and their framework. In addition to the basic framework of IEC, multifield user interface, partial breeding, and direct genome editing extend efficiency and usefulness of breeding.

Simulated breeding provides not only an alternative method to support human's design activity but also quite a new style of production based on cooperation between human and machine. The only thing the user has to do is selection according to his/her preference. The user needs to know a little about the operation of breeding but does not need to know how the target system can be built up. Although more than 22 years have passed since the original idea of *Blind Watchmaker*, and some artists have employed this idea for their productions, such as Latham [17] and Rooke [13], it is still not popular enough compared with the potential power. In these years, some researchers organized workshops on artistic applications of evolutionary computing and published paper collections and books, such as [12]. These ideas will contribute to create a type of new culture in the near future.

Acknowledgements The author appreciates those who concern our two projects, SBART and SBEAT. Masafumi Nishikawa, Shinichi Sakata, Takanori Yanagisawa, Eiichi Nakada, Manabu Senda, Nobuyuki Mizuno, Megumi Soda, Kayoko Mizuno, and Seiji Takaki contributed to these projects as graduate and undergraduate students at Soka University. Harley Davis suggested that the author introduce new functions "and" and "mix" to SBART in 1993. These functions are implemented in the current version of SBART. Karl Sims, who developed the original breeding system of 2D Computer Graphics, gave the author permission to distribute the source and binary code of SBART through the Internet in 1993. Taro Kamei, a VJ in Tokyo, introduced SBART in his article in a magazine as a tool to make materials for a video clip in 2000.

References

1. Biles JA (1994) GenJam: A Genetic Algorithm for Generating Jazz Solos. In: Proceedings of International Computer Music Conference. International Computer Music Association, San Francisco, CA, pp 131–137.
2. Dahlstedt P (2001) Creating and Exploring Huge Parameter Spaces: Interactive Evolution as a Tool for Sound Generation. In: Proceedings of International Computer Music Conference, Instituto Cubano de la Musica and International Computer Music Association, Havana, pp 235–242.
3. Dawkins R (1986) Blind Watchmaker, Longman, Essex, UK.
4. Funes P (2003) The Interactive Evolution of Configurations for "The Games". Workshop on User-centric Evolutionary Computation in GECCO, Chicago, IL.
5. Goldberg DE (1989) Genetic Algorithms in Search, Optimization and Machine Learning. Addison-Wesley, Reading, MA.
6. Koza JR (1992) Genetic Programming: On The Programming of Computers by Means of Natural Selection, MIT Press, Cambridge, MA.
7. Lund HH, Miglino O, Pagliarini L, Billard A, Ijspeert A (1998) Evolutionary Robotics: A Children's Game. In: Proceedings of IEEE Fifth International Conference on Evolutionary Computation, pp 154–158.

8. Nelson GL (1993) Sonomorphs: An Application of Genetic Algorithms to Growth and Development of Musical Organisms. In: Proceedings of the Fourth Biennial Art & Technology Symposium, Connecticut College, pp 155–169.
9. Nelson GL (1995) Further Adventures of the Sonomorphs. In: Proceedings of the Fifth Biennial Art & Technology Symposium, Connecticut College, pp 51–64.
10. Ohsaki M, Takagi H (1998) Improvement of Presenting Interface by Predicting the Evaluation Order to Reduce the Burden of Human Interactive EC Operators. In: Proceedings of IEEE Conference on Systems, Man and Cybernetics, San Diego, CA, pp 1284–1289.
11. Pettey CB, Leuze MR, Grefenstette JJ (1987) A Parallel Genetic Algorithm. In: Proceedings of the Second Int. Conf. on Genetic Algorithms, Cambridge, MA, pp 155–161.
12. Romero J, Machado P (eds.) (2008) The Art of Artificial Evolution: A Handbook on Evolutionary Art and Music, Springer-Verlag, New York.
13. Rooke S (1999) Artist Talk. In: Dorin A and McCormack J (eds.) Proceedings of First Iteration, CD-ROM. Melbourne, Victoria, Australia.
14. Sims K (1991) Artificial Evolution for Computer Graphics. Computer Graphics 25:319–328.
15. Smith JR (1991) Designing Biomorphs with an Interactive Genetic Algorithm. In: Proceedings of the Fourth International Conference on Genetic Algorithms, San Diego, CA, pp 535–538.
16. Takagi H (2001) Interactive Evolutionary Computation: Fusion of the Capacities of EC Optimization and Human Evaluation. Proceedings of the IEEE 89:1275–1296.
17. Todd S, Latham W (1992) Evolutionary Art and Computers. Academic Press, New York.
18. Tokui N, Iba H (2000) Music Composition with Interactive Evolutionary Computation. In: Proceedings of the Third International Conference on Generative Art, Milan, Italy.
19. Unehara M, Onisawa T (2001) Composition of Music Using Human Evaluation. In: Proceedings of 2001 IEEE International Conference on Fuzzy Systems. Melbourne, Victoria, Australia.
20. Unemi T (1998) A Design of Multi-Field User Interface for Simulated Breeding. In: Proceedings of the third Asian Fuzzy Systems Symposium. Masan, Korea, pp 489–494.
21. Unemi T, Nakada E (2001) A Tool for Composing Short Music Pieces by Means of Breeding. In: Proceedings of the IEEE Conference on Systems, Man and Cybernetics 2001. Tucson, AZ, pp 3458–3463.
22. Unemi T, Senda M (2001) A New Musical Tool for Composition and Play Based on Simulated Breeding. In: Dorin A (ed), Proceedings of Second Iteration. Melbourne, Victoria, Australia, pp 100–109.
23. Unemi T (2002) SBART 2.4: An IEC Tool for Creating Two-Dimensional Images, Movies and Collages. Leonardo 35:189–191.
24. Unemi T (2002) Partial Breeding – A Method of IEC for Well-structured Large Scale Target Domains. In: Proceedings of the IEEE Conference on Systems, Man and Cybernetics 2002, CD-ROM Proceedings, TP1D, Hammamet, Tunisia.
25. Unemi T (2002) A Tool for Multi-part Music Composition by Simulated Breeding. In: Proceedings of the Eighth International Conference on Artificial Life, Sydney, Australia, pp 410–413.
26. Unemi T (2002) A Design of Genetic Encoding for Breeding Short Musical Pieces. Bilotta A, Gross D, et al (eds), ALife VIII Workshops, ALMMA II, Sydney, Australia, pp 25–29.
27. Unemi T (2004) Embedding Movie into SBART – Breeding Deformed Movies. In: Proceedings of the IEEE Conference on Systems, Man and Cybernetics 2004, The Hague, Netherlands.
28. Unemi T, Bisig D (2003) Project 7^2. http://www.intlab.soka.ac.jp/~unemi/1/DT/
29. Unemi T, Bisig D (2004) Playing Music by Conducting BOID Agents – A Style of Interaction in the Life with A-Life. In: Proceedings of the Ninth International Conference on the Simulation and Synthesis of Living Systems, Boston, MA, pp 546–550.

Chapter 13
The Evolution of Sonic Ecosystems

Jon McCormack

This chapter describes a novel type of artistic artificial life software environment. Agents that have the ability to make and listen to sound populate a synthetic world. An evolvable, rule-based classifier system drives agent behavior. Agents compete for limited resources in a virtual environment that is influenced by the presence and movement of people observing the system. Electronic sensors create a link between the real and virtual spaces, virtual agents evolve implicitly to try to maintain the interest of the human audience, whose presence provides them with life-sustaining food.

13.1 Introduction

One thing that foreigners, computers and poets have in common is that they make unexpected linguistic associations.

Jasia Reichardt [26]

Music and art are undoubtedly fundamental qualities that help define the human condition. While many different discourses contribute to our understanding of art making and art interpretation, two implicit themes connect all artworks. The first is the act of creation. Even the most abstract or conceptual artworks cannot escape the fact that, as ideas, objects, or configurations, they must be made. Second, the importance of novelty, either perceived or real, is a fundamental driving force behind any creative impetus or gesture. Artists do not seek to create works that are identical to their previous creations or the previous work of others.

Artificial life (AL) methodologies can play an important role in developing new modes of artistic enquiry and musical composition. For artists, AL can offer new methodologies for the creative arts. For the first time in the history of art, AL suggests that, in theory at least, it may be possible to create artificial organisms that develop their own autonomous creative practices –

to paraphrase the terminology of Langton [16], *life-as-it-could-be* creating *art-as-it-could-be*.

In addition, AL has important contributions to make in our understanding of genuine novelty,[1] often referred to under the generalized term *emergence* [6, 11, 21].

13.1.1 Artificial Life Art

Techniques from cybernetics and artificial life have found numerous applications in the creative arts. General contemporary overviews can be found in [2, 3, 32, 39], for example.

Cybernetics has a rich and often overlooked history in terms of computing and the arts. The seminal ICA exhibition *Cybernetic Serendipity*, held in London in the summer of 1968, was one of the first major exhibitions to look at connections between creativity and technology [25]. Even the title suggests notions of novelty and discovery, a key theme for many works and critics in the decades that have followed the exhibition. Interestingly, the curators shunned distinctions between art and science and instead focused on ideas and methodologies that connected the two.

One particularly relevant concept from cybernetics is that of *open-ended behavior*, what Ashby referred to as *Descartes dictum:* How can a designer build a device that outperforms the designer's specifications [1]? Cybernetist Gordon Pask built an "ear" that developed, not through direct design, but by establishing certain electrochemical processes whereby the ear formed and developed in response to external stimuli [7].

The goal of the work described here is to create an open-ended artistic system that is *reactive* to its environment. In order to address this goal, two important problems were explored during the design and development of the work: first, how we can create a virtual AL world that evolves toward some subjective criteria of the audience experiencing it, without the audience needing to explicitly perform fitness selection and, second, how the relationship between real and virtual spaces can be realised in a way that integrates those spaces phenomenologically. The resultant artwork developed by the author is titled *Eden*.

---

[1] The concept of novelty is a vexed one with many different interpretations in the literature and could easily occupy an entire chapter in itself. Some authors argue that novelty and emergence have no relation [23], whereas others see them as fundamentally the same [6]. In the sense that the term is used in this chapter, novelty suggests that which has never existed before, hence the issues surrounding novelty are connected with determinism [11]. For art, almost every new artwork is in some sense novel; however, we may at least be able to apply criteria that suggest a degree of novelty, such as descriptive causality and explainable causality. Moreover, in an AL sense, we require not only the artwork to be novel, but the behavior of the virtual agents to be novel as well.

In terms of software, *Eden* is an AL environment that operates over a cellular lattice, inhabited by agents who have, among other capabilities, the ability to make and "listen" to sound. Agents use an internal, evolvable, rule-based system to control their behavior in the world. The virtual environment that the agents inhabit develops in response to the presence and movement of people experiencing the system as an artwork. The work is conceptualised and designed as an *artificial ecosystem* where virtual and real species interact with their biotic and a-biotic environment.

This software system will be described more fully in Sections 13.2 and 13.3 of the chapter. Interaction with the work is detailed in Section 13.4, with a summary of results and brief conclusion in Sections 13.5 and 13.6.

13.1.2 Related Work

The software system described in this chapter draws its technical inspiration from John Holland's *Echo* [15], particularly in the use of classifier systems for the internal decision-making system of agents. Many others have used evolutionary systems as a basis for musical composition, but in the main for compositional *simulation* [33, 38], rather than as a new form of creative tool for the artist and audience.

The *Living Melodies* system [9] uses a genetic programming framework to evolve an ecosystem of musical creatures that communicate using sound. *Living Melodies* assumes that all agents have an innate "listening pleasure" that encourages them to make noise to increase their survival chances. The system described in this chapter, *Eden,* contains no such inducement, beyond the fact that some sonic communication strategies that creatures discover should offer a survival or mating advantage. This results in the observation that only some instances of evolution in *Eden* result in the use of sonic communication, whereas in *Living Melodies*, *every* instance evolves sonic communication. *Living Melodies* restricts its focus to music composition, whereas *Eden* is both a sonic and visual experience.

13.2 *Eden*: An Artificial Life Artwork

Eden is a "reactive" AL artwork developed by the author. The artwork is typically experienced in an art gallery setting, but in contrast to more traditional artworks, it is designed as an *experiential* environment, whereby viewers participation and activity within the physical space have important consequences over the development of the virtual environment.

The artwork is exhibited as an installation and can be experienced by any number of people simultaneously. It consists of multiple screens, video projec-

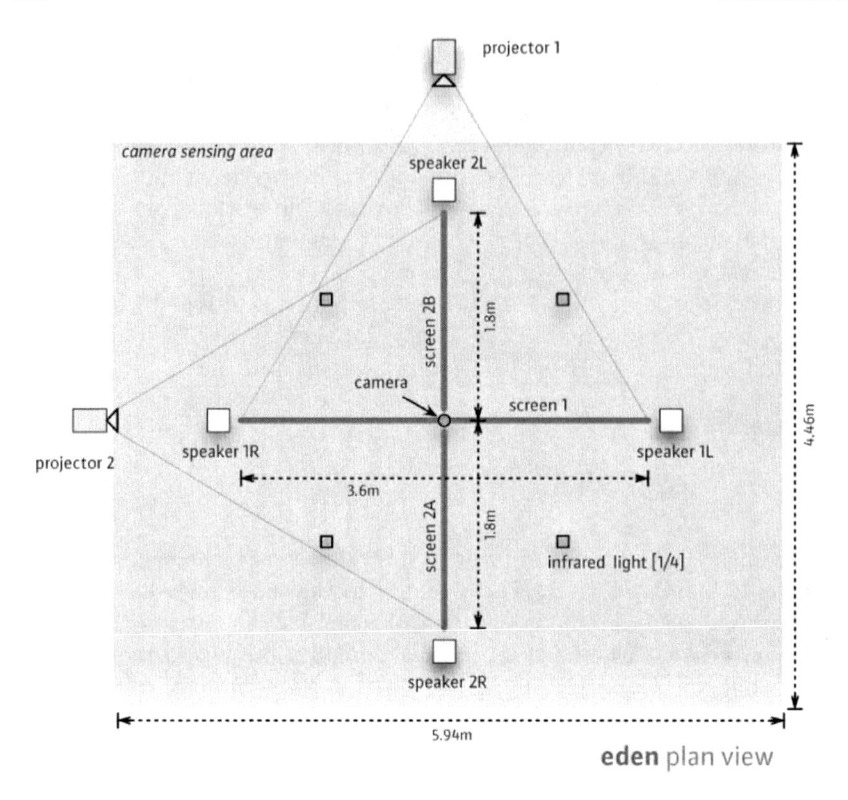

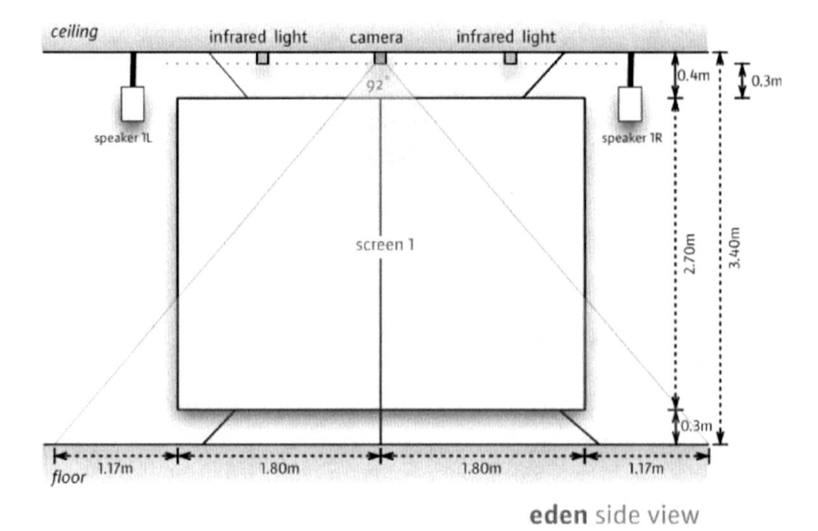

Fig. 13.1 Plan and side views of *Eden* showing the layout screens, speakers, projectors, and camera sensor area. The active sensing area extends approximately one metre past the edge of the screens.

tors, audio speakers, an infrared camera and lighting system, computers, and custom electronics. Figure 13.1 shows layout plans and Figure 13.2 shows a simulated visualisation of the work. As shown in these figures, physically the work consists of two semitransparent screens suspended from the ceiling of the exhibition space. The screens are positioned at 90° to each other, forming an X shape when viewed in plan. The ambient lighting is minimal – making the screens and the light they reflect and transmit the predominant source of visual interest in the space. The screens' transparency enables them to be viewed from either side and creates a layered visual effect that merges the real and virtual boundaries. Multichannel audio is provided by a number of speakers placed on the periphery of the main screen area.

In addition to this audio-visual infrastructure, an infrared digital video camera is placed above the screens, looking down at the space surrounding the projection screens. This area is illuminated by an infrared lighting system, which is invisible to the naked eye and so does not affect the perceptual properties of the work. The purpose of this camera sensor system is to measure the position and movement of people experiencing the work in its immediate vicinity. It is not necessary that the audience has any direct knowledge of this sensing. The purpose of sensing the real space of the artwork is as an environmental stimulus for the virtual agents' world and ultimately to con-

Fig. 13.2 Simulation of *Eden* running in a typical gallery environment, illustrating the effect of using transparent screens to visualise the work.

tribute to selective pressures that aim to encourage a symbiotic relationship between people experiencing the work and the agents populating the virtual world. The role of sensing and its effect on the development of the virtual environment portrayed in the work are detailed in Section 13.4.

13.3 Agents and Environments

This section gives technical details on the major software components of the system, with particular emphasis on the mechanisms that facilitate development of sonic agents within the system. Further details, particularly the *payoff* and *bidding* processes for rule selection, may be found in [18].

13.3.1 The Eden World

The environment projected onto the screens is known as the Eden *world.* In implementation terms, the world consists of a two-dimensional toroidal cellular lattice that develops using a global, discrete, time-step model – a popular AL model based on the theory of cellular automata [8, 35]. Each cell in the lattice may be populated by one of the following entities:

- *Rock:* inert matter that is impervious to other entities and opaque to sound and light. Rock is placed in cells at initialisation time using a variation of the *diffusion limited aggregation* (DLA) model [40]. Rocks provide refuge and contribute to more interesting spatial environmental behavior of the agents.
- *Biomass:* a food source for evolving entities in the world. Biomass grows in yearly[2] cycles based on a simple feedback model, similar to that of *Daisyworld* [36]. Radiant energy (in "infinite" supply) drives the growth of biomass. The amount of radiant energy falling on a particular cell is dependent on a number of factors, including the local absorption rate of the biomass and global seasonal variation. Probabilistic parameters can be specified at initialisation time to control these rates and variations. The efficiency at which the biomass converts radiant energy into more biomass is also dependent on the presence of people in the real space of the artwork. This dependency is detailed in Section 13.4.
- *Sonic Agents:* mobile agents with an internal, evolvable *performance system.* Agents get energy by eating biomass or by killing and eating other agents. More than one agent may occupy a single cell. Since these agents are the most complex and interesting entity in the world, they are described in detail in Section 13.3.2.

[2] An *Eden* year lasts 600 *Eden* days, but passes by in about 10 minutes of real time.

A real-time visualisation of the world produces images that are projected onto the screens, as illustrated in Fig. 13.1 (in this case, there are two worlds, each running on a separate computer, but connected as a single logical world running over two computers). The visualisation process is described more fully in Section 13.3.3. The sound the agents make as they move about the world is played with approximate spatial correspondence by a series of loudspeakers.

13.3.2 Agent Implementation

Sonic agents are the main evolving entity in the world. Essentially, the agent system uses a learning classifier system (LCS) similar to that of Holland's *Echo* system [15]. An agent consists of a set of *sensors*, a rule-based *performance system*, and a set of *actuators*. This configuration is illustrated in Figure 13.3. Sensors provide measurement of the environment and internal introspection of an individual agent's status. The performance system relates input messages from the sensors to desired actions. The actuators are used to show intent to carry out actions in the world. The success or failure of an intended action will be dependent on the physical constraints in operation at the time and place the intent is instigated. Actuators and actions are detailed later in this section.

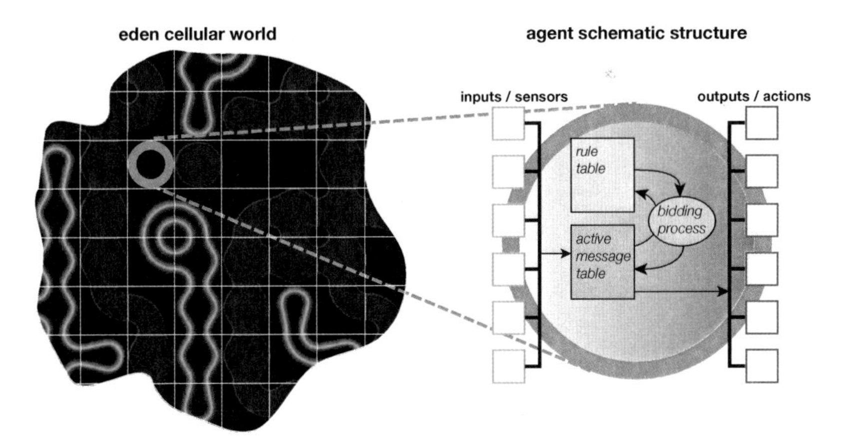

Fig. 13.3 A section of the *Eden* cellular lattice in visual form (left). To emphasize the lattice structure, grid lines have been layered over the image. The image shows rocks (solid), biomass (outline), and an agent (thick circle). The diagram (right) shows the agent's internal schematic structure, consisting of a number of sensors, a performance system that evolves, and a set of actuators.

At initialisation of the world, a number of agents are seeded into the population. Each agent maintains a collection of internal data, which includes the following:

- *Current age*, an integer measured in time steps since birth. Agents live up to 100 years and cannot mate in their first year of life.
- *Health index:* an integer value indicating the overall health of the agent. A value of 100 indicates perfect health; if the health index falls to 0, the agent dies. An agent can lose health via a sustained negative *energy level* differential (explained next) by bumping into solid objects, such as rocks, or being hit by other agents. In addition, the loss in health from being hit by another agent depends on both its mass and health index.
- *Energy level:* a measure of the amount of energy the agent currently has. Agents gain energy by eating biomass or other agents. Energy is expended attempting to perform actions (regardless of their success); a small quantity of energy is expended even if no action is performed at a given time step. If an agent's energy level falls to zero, the agent dies and its body is converted to new biomass in the cell in which it died.
- *Mass*: an agent's mass is linearly proportional to its energy level, plus an initial "birth mass" that is normally distributed over the population.

13.3.2.1 Sensors

Sensors provide a way for an agent to measure itself and its environment [24]. Sensor data are presented as bit strings constructed from local environmental conditions and from the internal data structures held by the agent. Sensor data are updated once every time step. An agent can use a range of sensor types, but the sensors themselves do not undergo any evolution and are fixed in function, sensitivity, and morphology. It is up to an individual agent's performance system to make use of a particular sensor, so data from a particular sensor will only be used in the long term if it provides useful information that assists the agent's survival or mating prospects. Sensor use does not incur any cost to the agent. Sensor information available to an agent consists of:

- A simple local vision system that detects the "colour" of objects on facing and neighboring cells (the range is limited to a single cell). Rocks, biomass, and agents all have different "colours," which enables an agent to distinguish between them.
- A sensor to detect the local cell nutritional value. Cells that contain biomass or dead agents have a high nutritional value; rocks and empty cells do not.
- A sound sensor that detects sound pressure levels over a range of frequency bands. Sound can be detected over a much larger spatial range than vision and also with greater fidelity.

- An introspection of *pain*. Pain corresponds to a negative health index differential and would usually indicate attack by another agent or that the agent is bumping into rocks.
- An introspection of the current energy level.

13.3.2.2 Actuators

Actuators are used to signal an agent's intent to carry out an action in the world. The physical laws of the world will determine whether the intended action can be carried out or not. For example, the agent may intend to "walk forward one cell," but if that cell contains a rock, the action will not be possible. Furthermore, all actions cost energy, the amount dependent on the type of action and its context (e.g., attempting to walk into a rock will cost more energy than walking into an empty cell).

As with the sensors, the number and function of actuators are fixed and do not change as the performance system evolves. Actions will only be used in the long term if they benefit the agent. Analysis of actions used by agents who are successful in surviving shows that not all agents make use of the full set of actuators.

Agents may perform any of the following actions:

- *Move* forward in the current direction.
- *Turn* left or right.
- *Hit* whatever else is in the cell occupied by the agent. Hitting another agent reduces that agent's health level using a non-linear combination of the mass, health, and energy level of the agent performing the hit. Hitting other objects or being hit will cause pain and a loss of health.
- *Mate* with whatever is currently occupying the current cell. Obviously, this is only useful if another agent is in the same cell. In addition, mating is only possible if the age of both agents is greater than 1 year.
- *Eat* whatever is currently occupying the current cell. Agent's can only eat biomass or dead agents (which turn into biomass shortly after death).
- *Sing*: make a sound that can be heard by other agents. Sound is detailed more fully in Section 13.3.4.

Performing an action costs energy, so agents quickly learn not to perform certain actions without benefit. For example, attempting to eat when your nutritional sensor is not activated has a cost but no benefit. Attempting to move into a rock has a cost greater than moving into an empty cell.

Agents may also choose not to perform any action at a given time step (a "do nothing" action), but even this costs energy (although less than any other action).

13.3.2.3 Performance System

The performance system connects an agent's sensors to its actuators (Fig. 13.3). It is based on the classification system of [15]. Sensory data arrive from the sensors in the form of a *message*, a binary string of fixed length.[3] Messages are placed in an *active message table*, a first-in, first-out (FIFO) list of messages currently undergoing processing. Each agent maintains a collection of *rules*, stored in a database or *rule table*. Rules consist of three components: a *condition string*, an *output message*, and a *credit*. Condition strings are composed from an alphabet of three possible symbols: {1,0,#}. At each time step, the message at the head of the active message table is processed by checking for a match with the condition string of each rule in the rule table. A 1 or 0 in the condition string matches the corresponding value in the message at the same index. A # matches either symbol (0 or 1). So, for example, the message 10010111 is matched by any of the condition strings 10010111, 10010##1, and ########. The condition string #######0, however, would not match.

Rules whose condition strings match the current message bid for their output message (also a bit string of the same length as sensor messages) to be placed in the active message table. This bid is achieved by calculating the rule's *strength*. Strength is the product of the rule's credit (detailed shortly) and its *specificity*. Specificity is a unit normalized value, equal to 1− the normalised number of # symbols in the condition string. So, for example, a condition string consisting entirely of # symbols has a specificity of 0; a string with 75% # symbols has a specificity of 0.25; and so on.

For each rule that matches the current message under consideration, its strength is calculated. The rule with the highest strength is selected and then places its output message into the active message table. If more than one rule has the highest strength, then a uniform random selection is made from the winning rules. The selected rule places its output message into the active message table. Most output messages are *action messages*[4], that is, they trigger an actuator. Action messages are removed from the table once they have been translated into actuator instructions.

The process outlined in this section is illustrated in Figure 13.4.

13.3.2.4 Credits and Payoffs

Each rule maintains a credit, essentially a measure of how useful this particular rule has been in the past. Rules begin with a default credit value and

[3] A message length of 32 bits is used, but the actual length does not concern the processes described. Larger message lengths allow more bandwidth in sensor messages, but require more storage.

[4] Action messages are distinguished from other messages by a marker bit in the string being set – all other message types are guaranteed not to set this bit.

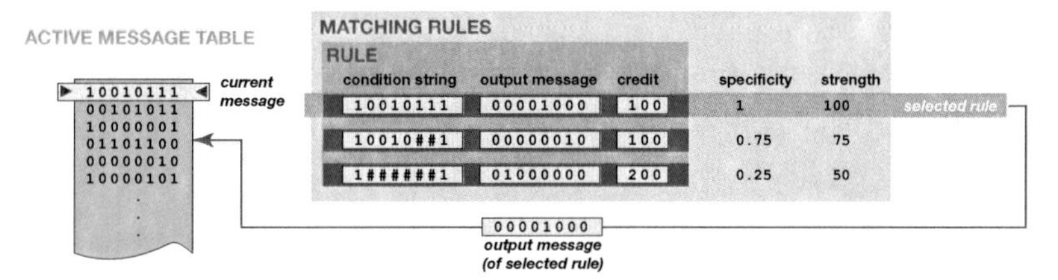

Fig. 13.4 The rule matching and bidding process. The top message from the active message table is selected and becomes the current message. Rules whose condition string matches the current message have a strength calculated as the product of their credit and specificity. The rule with the highest strength then becomes the selected rule and its output message is added to the active message table. The current message is then discarded and the process repeats. Some messages are action messages and trigger actions.

earn or lose credit based on how useful the rule is in assisting the agent to live and mate in the world. As described earlier in this section, agents maintain an energy level and health index. The differentials of these quantities are monitored, and when they reach a certain threshold, a *credit payoff* is performed. The credit payoff rewards or punishes rules that have been used since the last payoff (held in a separate list), by altering their credit according to frequency of use and the magnitude of the change in energy since the last payoff. Further details regarding this process may be found in [18].

The credit payoff system enables rules that, over time, assist in increasing health and energy to be rewarded; those that decrease health and energy will decrease in credit. The rationale being that the next time rules have a chance to bid, if they have been useful in the past, they will probably be useful in the current situation.

The number of time steps between successive payoffs is dependent on the rate of change in the agent's health (i.e., the magnitude of the differential). For example, if a creature is being attacked and losing health quickly, payoffs will be more frequent. The rules involved in letting the agent get hit will also decrease in credit quickly (hopefully soon being outbid by other rules that may prove more successful if the agent is to survive).

Maintaining a list of rules that have been used since the previous payoff allows rules that indirectly increase health to receive appropriate credit. For example, while the rule to "eat when you find food" is a good one, you may need to walk around and look for food first to find it. The rules for walking and turning, although they decrease health in the short term, may result in finding food. This increases health in the longer term. Overall, if such rules are helpful in increasing health, their credit will increase. A rule whose strength falls to zero will be automatically removed from the agent's rule table, since it is on longer able to bid anything for use.

As specified in Section 13.3.2.3, a rule's strength is the product of its credit and specificity. This is necessary, since rules that are more specific will be used less often, as they match fewer messages. Rules that are more specific will have less chance to receive credit payoffs but still may be useful. When two or more rules with the same credit match a message, the more specific rule will have greater strength and thus will be selected over the more general one.

13.3.2.5 Agent Evolution

The credit payoff system allows rules that have contributed to the agent's survival to be used more often. However, this will only select the best rules from the currently available set. The problem remains as to how the agent can discover better rules than those it currently uses.

Genetic algorithms follow a Darwinian metaphor in that they operate as a search method over the phase space of possible phenotypes in a given system, searching over the *fitness landscape* for individuals of higher *fitness* [12, 22]. In the *Eden* system, a rule functions as the genetic unit of selection and new rules are brought into an agent's genome via the standard operations of *crossover* and *mutation* (see the references for explanations of these terms).

Recall from Section 13.3.2 that mating is a possible action an agent can perform. If two agents successfully mate, they produce a new agent whose rule table is a combination of the parents' tables. A proportion of rules from each parent selected based on their strength – the rules of highest strength from each parent being favoured for selection. These selected rules undergo crossover and mutation operations, as per the schema system of Holland [14], resulting in the creation of new rules. Mutation rates vary according to the behavior of people experiencing the artwork in the exhibition space.

Since rules of highest strength are selected from each parent and those rules may have been discovered during the parents' lifetime, the evolutionary process is Lamarckian [4]. This design decision was used to allow more rapid adaptation to changing environmental conditions: a necessary feature if the agents' in the artificial ecosystem are to adapt to the behavior of people experiencing the work in real time. Another way to consider this approach is that parents teach their offspring all the good things they have learnt so far in their lifetime: a kind of social learning.

13.3.3 Image

Representation of the entities of *Eden* is achieved using tiling patterns, loosely based on Islamic ornamental patterns [13]. Only the representation of biomass will be considered here. The visual representation of the biomass is based on a 16-tile set. A tile is chosen for a particular cell based on the neighbor relationships of adjacent cells. For the purposes of tile selection, tiles are selected based on the binary occupancy condition of the cell's neighbors, considering only cells of the same type. For the 16-tile set, only immediate orthogonal cells are considered – thus there are 16 possible configurations of neighboring cells. Figure 13.5 shows the individual tiles and the neighbor relation necessary for the tile to be used. The resultant images formed by a grid of cells (illustrated in Figure 13.6) form a continuous mass of substance, as opposed to squares containing individual entities. These minimalist geometric textures suggest abstract landscapes rather than the iconic or literal individual representations that are common in many artificial life simulations. This design decision forms an integral aesthetic component of the work.

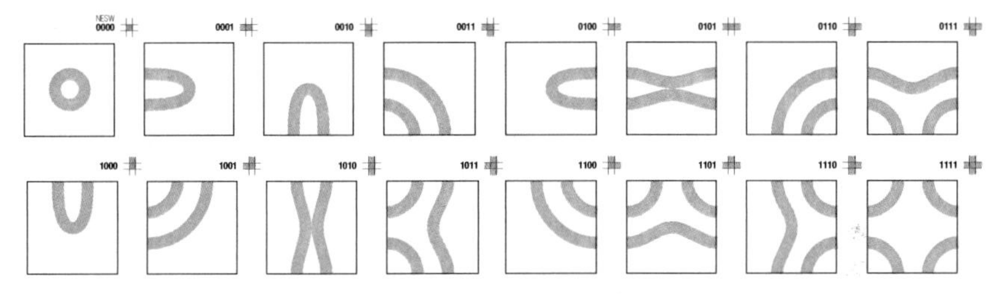

Fig. 13.5 Cellular tiling set for *Eden*'s biomass. Each cell considers the four immediate neighboring cells (north, south, east, and west). The neighboring relations determine the image used for each cell. A function returns the bit pattern representing the neighborhood state for the cell and the tile is selected based on the supplied index. Four bits are required, each representing the four directions. The bits are encoded NESW (from most to least significant bit). The symbols above each cell pattern shown here illustrate the bit pattern and corresponding neighborhood relationships.

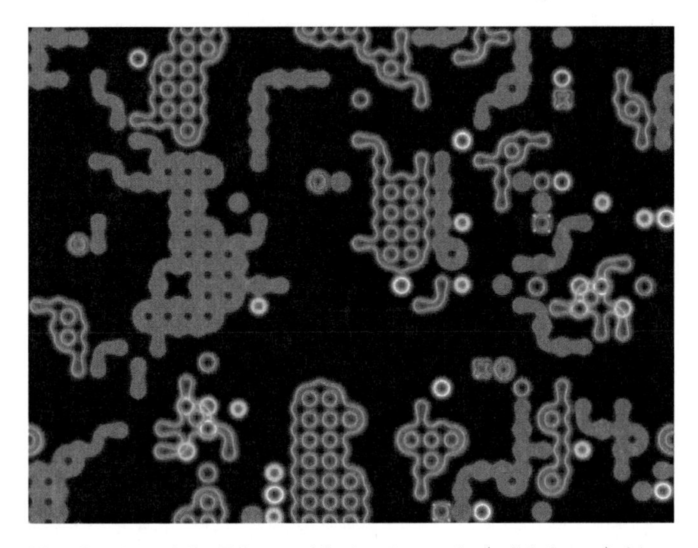

Fig. 13.6 Visualisation of the Eden world, showing rocks (solid shapes), biomass (outline shapes), and agents (circular elements).

13.3.4 Sound

One of the key elements of *Eden* is the ability of agents to create and listen to sound. A large proportion of sensor bandwidth is devoted to sound, allowing orthogonal sensing of both frequency and sound pressure (volume). Some basic physical modeling is performed on sound pressure levels, however many physical sound propagation aspects are simplified in the interests of efficiency.

13.3.4.1 Sound Generation

Actuator messages requesting sound generation need to be converted into a generated sound. As described in Section 13.3.2, actuator messages are bit strings. A portion of the string encodes the sound generation command ("sing"); the remainder encodes the sound generation data (spectral levels over a range of frequency bands). The current implementation has three distinct frequency bands, each occupying one-third of the total of the sound generation data for the "sing" actuator message (see Fig. 13.7).

When an agent "sings," the spectral signature determined by the sing data in the actuator message is registered for the current time step. In addition, the same signature is used to drive a sonification process, so that people in the exhibition space can hear sounds that correspond to the "singing" activities of the agents. To drive this sonification process, the three frequency bands are assigned labels L, M, and H corresponding to low-, medium-, and high-pitched sounds (e.g., the majority of spectral energy in the 100, 1000,

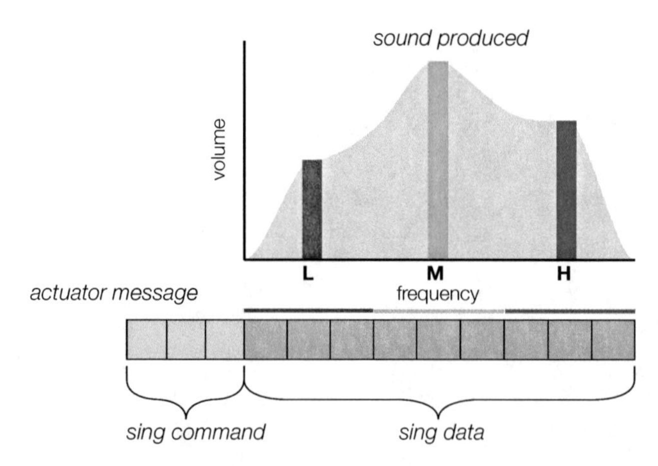

Fig. 13.7 The "sing" actuator message contains two parts. The first is the command requesting the agent to perform a sing operation; the remainder contains the sing data: volume levels for three distinct frequency bands. Using three bits per frequency band results in 2^9, or 512, distinct sounds.

and 10,000 Hz regions respectively). When an agent makes a sound, the corresponding selection from a precomputed library of sounds is triggered and sent to the audio subsystem. The audio subsystem does basic sound spatialisation using the four-channel audio system that is part of the artwork. Sounds are spatialised according to the position of the agent making the sound on the screen. Thus, as an agent making sound moves across the screen, that sound will appear to move with the agent to human observers. The audio subsystem allows many agents to be making sound simultaneously.

13.3.4.2 Sound Reception

Agents have a significant amount of sensor bandwidth devoted to sound reception. An agent's sound reception area is a forward-facing conical pattern that, like the sound generation, is sensitive across three separate frequency bands (see Figure 13.8). Each band has the same propagation and reception pattern, that is, there are no frequency-dependent differences in the modelling.

At each time step, the conical reception area for each agent is checked for any other agent that is singing within that area. A simple physical model [28] controls the propagation of sound through the environment.[5] Sounds arriving at the agent's cell are summed on a per-frequency basis and the resultant sensor message is instantiated.

13.4 Interaction

The *Eden* system has a unique relationship between the physical and virtual components of the system. As shown in Figure 13.1, an infrared video camera monitors the immediate space of the installation environment, recognising the presence and movement of people in the space. A video digitisation sub-system performs basic image processing and analysis, such as background removal and feature detection. This data is converted into a stream of vectors indicating the location and movement of individuals in the exhibition area, and used to drive environmental parameters in the virtual simulation. Before discussing the details of the mappings between position and movement vectors and the simulated environment, we will present a background discussion on the rationale for the mappings used.

---

[5] When sound propagates in a medium such as air at standard temperature and pressure, the perceptual mechanism for loudness behaves in an exponential way, as it does for humans. The relationship between distance and perceived levels is $L = 20\log_{10}(P/P_o)$, where L is the sound pressure level in decibels (dB), and P_o is a reference pressure corresponding roughly to the threshold of hearing in humans [27].

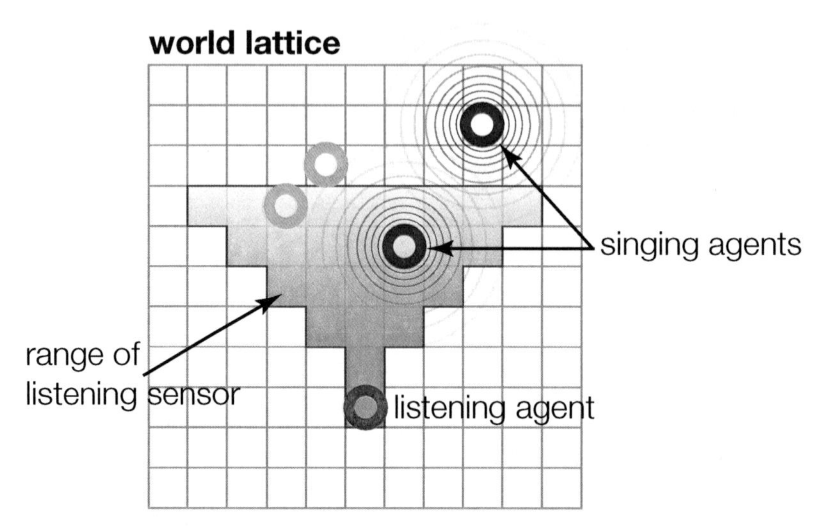

Fig. 13.8 The reception area of an agent. The listening agent will hear agents who are making sound within the blue area only. A simple physical model controls the perceptual volume levels for the agent.

13.4.1 The Problem of Aesthetic Evolution

Typically, genetic algorithms evolve toward finding maxima in *fitness*, where fitness is some criterion that can be evaluated for each phenotype of the population. Many systems define an explicit *fitness function* that can be machine evaluated for every phenotype at each generation [22].

The *Interactive Genetic Algorithm* (IGA, also known as *aesthetic evolution* or *aesthetic selection*) is a popular technique that replaces the machine-evaluated fitness function with the subjective criteria of the human operator. Aesthetic evolution was first used by Dawkins [10] in his "Blind Watchmaker" software to evolve two-dimensional, insect-like shapes. Aesthetic selection has been used to successfully evolve images [29, 30], dynamic systems [31], morphogenic forms [34, 17], even musical patterns and structures [5]. Regardless of the system or form being evolved, aesthetic selection relies on the user to explicitly select the highest-fitness phenotypes at each generation. Users typically evolve to some subjective criteria – often described as "beautiful," "strange" or "interesting" – criteria that prove difficult to quantify or express in a machine-representable form (hence the use of the technique in the first place).

However, aesthetic evolution has two significant problems:

- The number of phenotypes that can be evaluated at each generation is limited by both screen area (in the case of visual representation) and the ability of people to perform subjective comparisons on large numbers of

objects (simultaneously comparing 16 different phenotypes is relatively easy; comparing 10,000 would be significantly more difficult).
- The subjective comparison process, even for a small number of phenotypes, is slow and forms a bottleneck in the evolutionary process. Human users may take hours to evaluate many successive generations that in an automated system could be performed in a matter of seconds.

What we would like is a system that combines the ability to subjectively evolve toward phenotypes that people find "interesting" without the bottleneck and selection problems inherent in traditional aesthetic evolution.

13.4.2 Eden *as a Reactive System*

The solution to the problem described in the previous section is to map the presence and motion data of people experiencing the artwork to the environmental parameters of the virtual environment. Thus, the virtual world in which sonic agents live and evolve is dependent not only on the simulated qualities discussed so far, but also on the presence (or absence) of people experiencing the work and their behavior within the exhibition space. This couples the real and virtual worlds into a kind of ecosystem where behaviours in one world influence behaviours in the other, creating *feedback loops* of behaviour.

Eden has no explicit fitness function. Agents continue to be part of the system based on how well they can survive and mate in the current environment. If certain selection pressures are applied, such as food becoming scarce, only those agents who can adapt and find food will prosper. By driving environmental conditions from the presence and movement of people in the exhibition space, agents must implicitly adapt to an environment that includes aspects of the world outside the simulation.

In the current system, the following mappings are used:

- *Presence in the real environment maps to biomass growth rates.* The presence of people around the screen area affects the rate of biomass growth in that local area of the *Eden* world. Areas with no people correspond to a barren environment: little biomass will grow without the presence of people in the real environment.
- *Movement in the real environment maps to genotype mutation rates.* The greater the movement of people in the space, the higher the mutation rate for rule evolution (see Section 13.3.2.5).

These mappings are based on certain assumptions. First, people will generally spend time experiencing something only if it interests them. In the context of experiencing an artwork, people generally may spend a short time evaluating their interest in an artwork, but, after a short time, if it no longer interests them, they will leave. There may be other reasons for leaving, but in general, the duration of stay will have some relation to how "interesting" the experience is.

Agents require food to survive. If people are in the real environment, then food will grow at a more rapid rate. An agent who is making "interesting" noises, for instance, would have a better chance of keeping a person's attention than one who is not. Moreover, an agent making a progression of sounds, rather than a just a single, repeating sound, is likely to hold a person's attention even longer. Agents who encourage and hold a person's attention in the space implicitly give the environment a more plentiful food supply.

The mapping of people's movement in the space to mutation rates is based on the assumption that people will move over an area looking for something that interests them and, when they find it, will stay relatively still and observe it. Hence, the movement of people within the real space serves to inject "noise" into the genome of agents who are close to the source of movement. Higher mutation rates result in more variation of rules.[6] If an agent or group of agents are holding the viewer's attention, then less rule discovery is needed in the current environment, whereas if people are continually moving, looking for something "interesting," this will aid in the generation of new rules.

Further details on the dynamics of this component of the system can be found in [19].

13.5 Results

At the time of this writing, a number of exhibitions of the work have been completed. An image from a recent exhibition of the work is shown in Fig. 13.9. A typical exhibition may last several weeks, giving plenty of opportunity for the agent evolutionary system to take into account the behavior of people experiencing the work. Certain factors have a marked effect on this behavior and require specific compensations in the software. For example, when the gallery is closed, there will be no people in the space, which diminishes the food supply for the agents. Thus without compensation for gallery opening hours, the entire population dies out each night!

Analysis of the rules agents use shows that sound is often used to assist in mating, as would be expected [37], and with the influence of people, sound is used in other ways as well. Once the environmental pressures from audience behavior are incorporated into the system, the generation of sound shows a marked increase and analysis of the rules discovered shows that making sound is not only used for mating purposes.

---

[6] Most child rules that mutate will not be "better" than the parent rule, but, in general, the use of mutation does provide the possibility for the system to discover rules that would not be possible by crossover alone.

Fig. 13.9 An image of *Eden* in operation.

The resulting sounds produced by the system *do* appear to interest and engage the audience, and combined with the visual experience, *Eden* does incite, from visitors, an interest and curiosity in what it is doing. In many cases, people are not aware of the learning system, camera sensing, even the fact that what they are experiencing is a complex artificial life system.

Since the system is reactive to people (rather than interactive), there is no correct or incorrect way to behave except to appreciate the experience. Anecdotal accounts from people who have experienced the work describe it as "having a sense that it is somehow alive," or "like being in a strange forest at night." In a number of exhibitions, people returned to the work over a period of several days, to see how the qualitative behaviour of the virtual environment had changed. In one recent exhibition, a local businessman visited the work during his lunch-hour every day for 3 weeks, describing the experience as "fascinating... one that made me more sensitive to my own environment." While these are, of course, subjective evaluations, it does appear that *Eden* is able to adapt and evolve to create an ongoing interest for its audience.

13.6 Conclusion

This chapter has described a novel evolutionary ecosystem, where agents make use of sound to assist in survival. While the main impetus and methodologies are based around the development of an artistic system, it is hoped that some of the ideas presented here may be of interest to those working in artificial life from other perspectives or with different agendas and applications.

In summary, a system has been produced that attempts to integrate the open-ended nature of synthetic evolutionary systems into a reactive virtual space. The approach used has been to measure components of the real environment, incorporating them into that of the virtual one, thus enabling a symbiotic relationship between virtual agents and the artwork's audience, without need for explicit selection of phenotypes that engage in "interesting" behavior.

Further information, including sample sound recordings and video documentation of the work, is available on-line [20].

References

1. Ashby WR (1956) An Introduction to Cybernetics. Chapman & Hall, London.
2. Bentley PJ (1999) Evolutionary Design by Computers. Morgan Kaufmann Publishers, San Francisco, CA.
3. Bentley PJ and Corne DW (eds.) (2002) Creative Evolutionary Systems. Academic Press, London.
4. Bowler PJ (1992) Lamarckism, In: Keller EF and Lloyd EA (eds.). Keywords in Evolutionary Biology, Harvard University Press, Cambridge, MA, pp. 188–193.
5. Bulhak A (1999) Evolving Automata for Dynamic Rearrangement of Sampled Rhythm Loops. In: Dorin A and J McCormack (eds.) First Iteration: A Conference on Generative Systems in the Electronic Arts, CEMA, Melbourne, Australia, pp. 46–54.
6. Cariani P (1991) Emergence and Artificial Life. In: Langton CG et al. (eds.) Artificial Life II, SFI Studies in the Sciences of Complexity, Vol 10, Addison-Wesley, Reading, MA, pp. 775–797.
7. Cariani P (1993) To Evolve an Ear: Epistemological Implications of Gordon Pask's Electrochemical Devices. Systems Research 10:19–33.
8. Conrad M and Pattee HH (1970) Evolution Experiments with an Artificial Ecosystem. Journal of Theoretical Biology 28:393.
9. Dahlstedt P and Nordahl MG (1999) Living Melodies: Coevolution of Sonic Communication. In: Dorin A and McCormack J. (eds.) First Iteration: A Conference on Generative Systems in the Electronic Arts, Centre for Electronic Media Art, Melbourne, Australia, pp. 56–66.
10. Dawkins R (1986) The Blind Watchmaker. Longman Scientific & Technicial, Essex, UK.
11. Emmeche C, Køppe S and Stjernfelt F (1997) Explaining Emergence: Towards an Ontology of Levels. Journal for General Philosophy of Science 28:83–119.
12. Goldberg DE (1989) Genetic Algorithms in Search, Optimization, and Machine Learning. Addison-Wesley, Reading, MA.

13. Grünbaum B and Shephard GC (1993) Interlace Patterns in Islamic and Moorish Art. In: Emmer M (ed.) The Visual Mind: Art and Mathematics, MIT Press, Cambridge, MA.
14. Holland, JH (1992) Adaptation in Natural and Artificial Systems: An Introductory Analysis with Applications to Biology, Control, and Artificial Intelligence. MIT Press, Cambridge, MA.
15. Holland JH (1995) Hidden Order: How Adaptation Builds Complexity, Addison-Wesley, Reading, MA.
16. Langton CG (1989) Artificial Life. In: Langton CG (ed.) Artificial Life, SFI Studies in the Sciences of Complexity, Vol. 6, Addison-Wesley, Reading, MA, pp. 1–47.
17. McCormack J (1993) Interactive Evolution of L-System Grammars for Computer Graphics Modelling. In: Green D and Bossomaier T. (eds.) Complex Systems: From Biology to Computation, ISO Press, Amsterdam, pp. 118–130.
18. McCormack J (2001) Eden: An Evolutionary Sonic Ecosystem. In: Kelemen J and Sosík P. (eds.) Advances in Artificial Life, 6th European Conference, ECAL 2001, Springer, Berlin, pp. 133–142.
19. McCormack J (2002) Evolving for the Audience. International Journal of Design Computing 4 (Special Issue on Designing Virtual Worlds). Available online: `http://www.arch.usyd.edu.au/kcdc/journal/vol4/index.html`.
20. McCormack J (2006) Eden: An Evolutionary Sonic Ecosystem, web site, `http://www.csse.monash.edu.au/~jonmc/projects/eden/eden.html`.
21. McCormack J and Dorin A (2001) Art, Emergence and the Computational Sublime. In: Dorin A (ed.) Second Iteration: A Conference on Generative Systems in the Electronic Arts, CEMA, Melbourne, Australia, pp. 67–81.
22. Mitchell M (1996) Introduction to Genetic Algorithms. MIT Press, Cambridge, MA.
23. Nagel E (1961) The Structure of Science: Problems in the Logic of Scientific Explanation. Routledge, London.
24. Pattee HH (1988) Simulations, Realizations, and Theories of Life. In: Langton CG (ed.) Artificial Life, Vol. VI, Addison-Wesley, Reading, MA, pp. 63–77.
25. Reichardt J (1971) Cybernetics, Art and Ideas. New York Graphic Society, Greenwich, CT.
26. Reichardt J (1971) Cybernetics, Art and Ideas. In: Reichardt J (ed.) Cybernetics, Art and Ideas, Studio Vista, London, pp. 11–17.
27. Roads C (1996) The Computer Music Tutorial. MIT Press, Cambridge, MA.
28. Roederer J (1975) Introduction to the Physics and Psychophysics of Music. Springer-Verlag, New York.
29. Rooke S (2002) Eons of Genetically Evolved Algorithmic Images. In: Bentley PJ and Corne DW (eds.) Creative Evolutionary Systems, Academic Press, London, pp. 339–365.
30. Sims K (1991) Artificial Evolution for Computer Graphics. Computer Graphics 25:319–328.
31. Sims K (1991) Interactive Evolution of Dynamical Systems. In: First European Conference on Artificial Life. MIT Press, Cambridge, MA, pp. 171–178.
32. Sommerer C and Mignonneau L (eds.) (1998) Art@Science. Springer-Verlag, Wien, Austria.
33. Todd PM and Werner GM (1998) Frankensteinian Methods for Evolutionary Music Composition. In: Griffith N and Todd PM (eds.) Musical Networks: Parallel Distributed Perception and Performance. MIT Press/Bradford Books, Cambridge, MA.
34. Todd S and Latham W (1992) Evolutionary Art and Computers. Academic Press, London.
35. Ulam S (1952) Random Processes and Transformations. Proceedings of the International Congress on Mathematics, Vol. 2, pp. 264–275.
36. Watson AJ and JE Lovelock (1983) Biological Homeostasis of the Global Environment: The Parable of Daisyworld. Tellus 35B:284–289.

37. Werner GM and Dyer MG (1991) Evolution of Communication in Artificial Systems. In: Langton CG (ed.) Artificial Life II, Addison-Wesley, Redwood City, CA, pp. 659–682.
38. Wiggins G et al (1999) Evolutionary Methods for Musical Composition. Proceedings of the CASYS98 Workshop on Anticipation, Music & Cognition.
39. Wilson S (2002) Information Arts: A Survey of Art and Research at the Intersection of Art, Science, and Technology. MIT Press, Cambridge, MA.
40. Witten TA and Sander LM (1981) Diffusion-Limited Aggregation, a Kinetic Critical Phenomenon. Physical Review Letters 47:1400–1403.

Chapter 14
Enriching Aesthetics with Artificial Life

Alan Dorin

Perhaps one day our machines will be able to take us beyond environments that look more and more familiar as we traverse them, into spaces that continue to surprise us as we explore. Can our engineered artifacts ever rival the intricacy of nature?

This chapter explores the application of artificial life software for the purposes of generating dynamic works of audio-visual art that act to some degree independently of the artist. Autonomous artworks slot neatly into artificial life's fascination with the construction of digitally instantiated living systems. This interest itself reflects the long-standing human desire to construct "living" images, clay statues and idols, hydraulically powered figures, mechanical toys, and electrical robots. The chapter focuses on the elements of dynamic software art that relate to the concept of the *sublime* in nature and computation. It is suggested that artificial life and virtual ecosystems provide an avenue for the production of rich, dynamic works of software art suited to this exploration.

14.1 Introduction

Artificial life is studied largely as a means of furthering our understanding of biology and of complex adaptive systems in general. While it has demonstrated potential in a number of fields, in particular as a means of solving engineering problems, artificial life techniques have also been applied by some in the art community. The field promises to continue to enrich artistic practice and our approach to contemporary aesthetics, even as its initial flash of popularity wanes. It is this continued application to aesthetics which the present chapter begins to address.

The techniques of art based in artificial life form a subset of *generative art*. This is an artistic practice that adopts an aesthetic of *process*. This means that although the final outcome of a work may depend for its appeal on the aesthetics of an image, sound, sculpture, or other form, the process that generates it is also significant. The generative artist is responsible for setting up initial conditions and a process to act upon them. The work that unfolds is the result of this series of changes. This is analogous to a biological phenotype (usually an organism and its behaviour) being the result of the physical and chemical interactions that govern its development from a genotype (its genetic material as stored in DNA). Hence, there are conceptual connections between generative art and artificial life as well as practical ones.

Generative/process-based art is no longer treated as a fresh field for aesthetic exploration. Fascination with its possibilities seemed to fade sometime after its heyday in the late 1960s. This of course was the time of the *Cybernetic Serendipity* exhibition in London, Jack Burnham's text *Beyond Modern Sculpture*, and much other activity linking art and computer technology in innovative ways. Recently, artists involved in artificial life have retrodden some of the ground cleared by their predecessors in cybernetic art and have cleared some new space for themselves. Now that artificial life is also unfashionable, perhaps a serious assessment of its past and future contributions can be made while side-stepping the hype that initially accompanied the field.

The following section begins by discussing the sense of wonder people often feel while contemplating nature or simple physical systems. This is then related to the tradition of the *sublime* in aesthetics. The chapter then discusses means of employing artificial life techniques to explore the *computational sublime* such that a computational system emulating the physical world's capability to generate complexity and novelty might be devised. In particular, the chapter describes the application of metaphors adopted from Ecology to create virtual ecosystems. This avenue has been explored by a few artists to whom it appears to offer the potential to generate and control complexity in creative software systems.

14.2 Wonder and the Sublime in Art and Nature

People stare contemplating the ocean as it swells and crashes on a rocky shore. They gaze fixedly into a fire until the sting of smoke raises their awareness. People may lie on their backs and follow passing clouds or marvel at the glittering of stars. There are many things that fascinate us, that mesmerise us, that cause us to forget ourselves and our situation as we become lost in a timeless appreciation of nature.

There are also circumstances under which we may have an experience that reminds or forces us to consider our insignificance: the feeling that accompanies the vastness of spaces such as the ocean, a desert plain, or an endless mountain range, or the feeling of insignificance when contemplating the age of the universe or the years over which a trickle of water has eroded a canyon, for example. These expanses can nevertheless be made subject to reason. We are mentally equipped to discuss the concept of infinity, even if we are unable to quite fathom it intuitively. This paradoxical experience Kant has labeled the *mathematical sublime* [15].

Likewise, Kant introduced the *dynamically sublime* as relating to encounters with the ferocity of nature and the sense of vulnerability this entails, coupled with the triumph of reason over fear. In his own writings on the sublime, Edmund Burke took the view that the sensation of incomprehensibility, the fear of hopelessness or of danger, coupled with the knowledge that one *is* able to reason about something beyond one's senses or one is not inadequate or in danger, causes a kind of delight through internal conflict – a sublime experience [19]. That is to say, there is an element of the sublime, perhaps not an artistic element, in resisting the urge to flee as a tiger roars behind bars. One's body responds with fear to the situation, but reason easily overcomes this and forces the body to stand its ground – generating a sense of delight tinged with terror. In the case of a painting, the viewer's separation from a wild scene of stormy seas or a vast desert is not enforced by iron bars, but by the picture plane. This is coupled with the knowledge that "it is only a picture."

How can a work of art, specifically a work of computer-based, generative software, approach the sublime? While there is a vast amount of literature on the sublime in art dating back as far as Longinus [23], the focus of this chapter is narrower. An earlier publication [25] outlined the *computational sublime*. This arises from viewing the computer in terms of its capability to perform logical operations at a rate and on a scale vastly outside our own abilities in this area. Due to its speed, the computer is able to mediate between our human perceptual apparatus and (practically) infinite computationally defined spaces. Yet, since we are the makers and programmers of these machines, our power of reason is not only able to overcome but also to *define* these very spaces that our senses are unable to grasp fully.

Before discussing computer-based art in more detail, some much simpler artifacts, possibly just for a moment, rival natural wonders in the fascination they hold. How does this relate to the sublime? It seems that there are a number of methods by which the sensation may be encouraged and that some of these have been recreated in artefacts specifically for this purpose. They are categorised below for convenience.

Marking Time

Maybe the simplest examples of our mesmerizing creations include a stream of liquid running from a water-clock or fountain, the shifting sands of the hourglass, and the oscillating pendulum. Each of these marks time in its own manner: one a continuous stream; the next a finely particulate flow;

and the last in clear, discrete stages. The eternal flow, the innumerable sand particles, the never-ceasing oscillations confront us with the infinite through this visible marking of time. Although these natural processes were utilised in desk ornaments to amuse the bored office worker of the 1980s, even in "serious" art, simple processes like this may give a sense of the sublime.

Exposing Spaces

The wire-suspended *mobiles* of Calder successfully employ mechanical processes in a manner accepted within the art world [29]. Calder's playful pieces are captivating and elegant for all their simplicity. Their workings are laid plainly before the viewer, all that they are is apparent at a glance – and yet this is not so, for their movement brings a vitality and opens a space the static sculpture does not possess. The universe a mobile sweeps out is contained within its wires, rods, and solid forms, so in one sense they may be held in (and created from) the palm of a hand. Yet, as they are touched by invisible air currents, their inner complexity is exposed.

Intricacy

Many a visitor has been fascinated by the button-driven clockwork and gearing of exhibits in traditional science museums. Here, a sense of wonder at the machine in its entirety arises, but also a fascination with the intricacy of the mechanism. Each gear meshes with another, each component is configured "just so," and together the pistons and wheels turn in harmony to produce a composite that might drive a clock, crush a quartz boulder, pump water, or power a vehicle. This is not the beauty of a crashing ocean or a sunset, but the charm of peering into an ant's nest or through a microscope at a drop of pond water. There is something in these systems that causes one to marvel at a complexity that is just beyond grasp.

Defying the Natural Order

There is still another wonder to be described here, that of somehow defying the natural order of things. It appears wonderful that a huge boulder might come to be balanced on a slender rock column, that a gyroscopic top remains on its tip despite interference, that a bird or a massive aeroplane can remain suspended in the air or a colossal steel ship can float, and that a magnet can push another away without touching it. These things, of course, are dealt with to some extent by simple science. Sadly, only children may wonder at these things. Yet, implicitly, these interactions remain in need of continual re-explanation since each instils apprehension through not being quite "right." That boulder or spinning top might topple at any moment. That bird should fall from the sky and the ship ought to take on water. The magnets are behaving unnaturally. All of these systems cause one to ponder "How can this be?" even if one can reason about the answer through science.

Curiosity

Related to all of the above phenomena, in particular the previous one, attention can also be held by riddles and intellectual pursuits. Included in this are mind games such as chess, paradoxes, and mathematical puzzles, but also scientific enquiry. These all captivate us through our drive to learn why something behaves as it does, especially if it seems to defy the natural order of things. Of the algorithmic art exhibition at Xerox PARC in 1994, Bern writes: "The appearance of algorithmic visual art should raise the temporal questions of 'How is this made?' and 'Can this be extended?' " [3]. While the use of the term *should* is of questionable accuracy, nevertheless there is something of relevance in the idea that algorithmic art may invite a question and arouse curiosity as much as it satisfies visual aesthetic criteria associated with traditional media.

Any of the preceding devices (and no doubt many others) may be used to convey a sense of the sublime. All of them expose the limitations in our comprehension, experience, or endurance. However, simultaneously, we have the power to reason about these phenomena, to encapsulate them in a word or other representation, or we may just have the opportunity to walk away unharmed. Of course, the context in which the preceding occurrences and objects are encountered will influence the extent to which the sensation derived relates to art. However, the potential to employ any of these basic devices for artistic purposes in search of the sublime experience is present.

14.3 Sublime Software

Now we return to the issue at hand: writing artificial life software for artistic purposes. There are, of course, infinitely many reasons why an artist might wish to do this. Let us suppose in this instance that the goal was software that surprised not only a viewer but also the artist who fashioned the work. This is in keeping with a proposal made elsewhere by the author [10]. The surprise required might not be a trivial one-off shock, but a genuine and continuing fascination with the novelty of the outcome generated, combined with a sense of being lost in a complex world beyond one's grasp. The work requires an aesthetic quality that exhibits the character listed above as *intricacy* and also at the level listed as *curiosity*.

It so happens, of course, that the techniques of artificial life are well suited to this application, Conway's *Game of Life* cellular automata (CA) being a case in point [17]. CAs are fascinating for their ability to (practically, if not theoretically) produce ever-fresh patterns. Like clockwork automata or Calder's mobiles, the patterns produced may fall well within a limited domain, often even a cycle. However, especially in a large-scale configuration, there is always some aspect of the system that remains to be discovered in the vast universe these interactions define.

The innovative design pair, the Eames, understood this aspect of a complex visual field. They used their understanding effectively in the 1964 New York World Fair exhibit for IBM to build a multiscreen cinema. The screens displayed related material simultaneously in such a way as to make it impossible to view all of the footage in detail. Viewers could watch the spectacle as a whole and see it as a multiscreen montage or they could pick and choose elements to observe in detail – much as one examines a complex clockwork machine, a microscopic world, or a beehive – piecemeal.

On a related topic, Tufte indicated the benefits of presenting information in parts grouped carefully together, "Panorama, vista and prospect deliver to viewers the freedom of choice that derives from an overview, a capacity to compare and sort through detail" [33]. Such considerations also assist painters in creating worlds that reward careful study of their detail. For example, Hieronymous Bosch's *Garden of Delights* (ca. 1500) portrays a world over which the eye may wander at will as it takes in relationships between couples, groups of people, sections of garden, and so on. Of course, where the object under study is static, a viewer may take the time to examine and re-examine the various parts of a display and avoid missing anything of importance (as Tufte would prefer). When the object is constantly changing, as with multiscreen cinema or a *Game of Life* run, an investigation of one element results in irrevocably missed information about another. Under these circumstances, no matter where a viewers' attention is focused, they are left with the impression that an abundance of chances were missed. Their faculties are fundamentally unable to absorb all that is before them.

While one of the attractions of the *Game of Life* occurs at the visual level, the software is also an attractive subject for philosophical discussions concerning emergence and complex systems [5]. It contains a vast number of possibilities within the bounds of its simple virtual physics. While its transition rules may be considered in terms of biological analogy (overcrowding and mutual support of living cells, for instance), the result is still an intrinsically digital system. Cells are in one of exactly two states, and their behaviour is completely deterministic. Yet, from this emerges a milieu of flickering forms that somehow *suggest* life, without mimicking it.

The question "How does it work?" is implied by any given run of the *Game of Life* in which it is recognised that the output is not at all random, but highly organised and structured according to its own internal rules. Even once the viewer know these rules, that they are capable of creating such a bewildering outcome remains a source of fascination to the artificial life researcher and to the newly informed student alike.

14.4 The Betrayal of Points and Lines

Having briefly discussed the concepts of wonder and the sublime, this section now addresses the representational schemes that may be used to bring the digital realm to the realm of sublime experience. Of particular interest is the way in which points of light may be used to give the bounded computer screen the appearance of *intricacy* and an ability to expose *unbounded* space (see the earlier categorization of methods for approaching the sublime). This has already been touched upon while discussing the *Game of Life* in the previous section.

Taking the lead of Kandinsky, the discussion begins with the geometric point, an invisible thing too perfect to exist [21]. An artist may approximate it by instigating a collision between a sharp tool and a flat plane. The artist's point has character. It is not one of Plato's Ideal Forms, but it is beautiful for all its imperfection.

The point may also be displayed on a CRT or LCD monitor. The size and shape of the smallest displayable shape are dictated by the display technology. Beyond a single pixel, the character of the point may also be altered, albeit with less potential for variation than in the analog world.

Anyone familiar with Islamic or Roman mosaics, medieval tapestry, Pointillist painting, the television screen, or a computer monitor is aware that, in vast numbers, the point may vanish in an ocean of its kind. Here it becomes simply one atom in a larger texture – mobs of points are not as honest as individuals.

If the draftsman drags a pen across the page, another form results: a line. Like the point, the line seems unpretentious, especially if it is straight and completely surrounded by white space. However, place a few straight lines together and important transformations may take place. For example, the *triangle* is born of three lines and is eventful for its ability to enclose space on the plane and thereby define a boundary and, with it, an object. The second transformation of relevance here is the apparent shift into three dimensions brought about by the *Cartesian axes*. This is also born of three lines and is eventful for its ability to *suggest* space that extends beyond the plane in which the lines are drawn.

Under some circumstances the three lines that represent Cartesian space may appear simple and flat; under others, however, the third dimension "pops" out of the plane at the viewer. It is outside the scope of this chapter to discuss how this second transformation occurs; the interested reader is referred to [20]. For now, it is sufficient if it is acknowledged that at least in some cases, three Cartesian axes suggest a volume where there is none, this being (of course) the principle that underlies Renaissance painting and a host of styles up to and, of relevance to this discussion, including modern computer graphics. Using only lines, one may build a two-dimensional representation of a solid object in unbounded three-dimensional space.

To return at this stage to the *Game of Life*, the system is typically displayed as, strangely enough, a grid of points. The choice of the point as the basic element of the simulation excludes the misleading suggestions of line drawings. Still, even within this limited visual space, the CA is successful at its evocation of life and with it *intricacy*. How can similar systems expose or suggest spaces that are infinitely large *and* intricate?

14.5 Moving Beyond Two Dimensions

The *Game of Life* and CA-based generative art such as *IMA Traveler* [12] work in the domain of points and suggestion, as outlined. With a computer monitor it is easy to represent lines and surfaces using a multitude of similarly lit pixels.

IMA Traveler, while working with points in a plane, recursively sub-divides these, giving the viewer the sensation of a bottomless plummet. This is an interesting perceptual phenomenon when it is considered that by continuously zooming in on a two-dimensional image (the points are not modeled in three-dimensional space), one can instil the sensation of motion through a continuous and infinitely detailed universe. The fractal zooms of the 1980s are perhaps the most familiar and overused form of this same trick. In their work *Powers of Ten* (1968), the Eames also utilised rapid scale shifts to produce a film of great effect.

The CA and its artistic derivatives draw attention for evoking natural phenomena and for prompting the viewer to determine the underlying principles by which they operate. Not all artificial life software is of this type, even if it is important to the research field for other reasons. One may propose a full range of works of which the CA-based software falls near the center. This spectrum runs from abstract systems such as Ray's *Tierra* [28] to clearly representational systems such as Sims' *Virtual Creatures* [31].

Ray's work emphasised the principles of selection and evolution through rapid reproduction and has itself proved to be a useful model and a starting point for other researchers such as Pargellis [26] (discussed later) to further the study of evolutionary systems. However, Ray's work does not have the visceral appeal that visual cellular automata may have. *Tierra*'s space of machine code instructions is abstract and not typically represented in such a way as to be comprehended by the eye. A fascination with *Tierra* arises through careful study and is predicated upon an understanding of the core workings of the system and the way in which the instructions interact and occupy memory. This contrasts with the CA, in which the rules need not be known to understand the system at one level. It is this first level of visceral comprehension that prompts theorizing about what underlies the behaviour of the CA cells.

At the opposite end of the spectrum, Sims' *Virtual Creatures* fascinate researcher and layperson alike. His jumping, swimming, running and limping creations are so full of character that one easily takes a leap of faith by referring to them as "creatures," even though their bodies are clearly rendered cuboids. This leap erases any chance of a viewer posing the question "How does it work?" because it is implied by the visualization that these *are* creatures! While a graphics researcher or software engineer may see Sims' video and wonder at the means of producing such marvelous results, this question is not implied by the creatures' visualization as it is by the CA. The virtual creatures do not operate according to unknown rules; we are tricked into believing that they operate according to the rules all creatures obey. There is seemingly nothing here to discover.

This aspect of Sims' creatures is largely due to the way in which they are visualised – the representations of three-dimensional space, of solids and their surfaces, of friction and other forces are all customary. In this case, the visualization is prescriptive, rather than evocative. McCormack's *Eden*, which was constructed as an artwork (see Chapter 13 in this volume and [24]), similarly represents organisms and their real-world behaviours. However, it displays them using iconic forms on a two-dimensional grid (mapped to an "X" in three-dimensional space) and in a world nevertheless governed by rules of survival, energy, mating, and space, based on those of the real world. In contrast with Sims' work, this underlying link to the physical world is not prescribed and takes some experience to decipher – if a viewer is able to decipher it at all. Although the system is still considerably more representational than the *Game of Life*, what remains is an aesthetic visual experience akin to viewing an intricate CA, coupled with the implicit question "How does it work?"

This discussion of Ray's, Sims', and McCormack's work is in no way intended as an evaluation of their worth. These examples only serve to illustrate the various ways in which a viewer may contemplate a generative process and its outcome and, therefore, the various ways in which the devices may be employed for artistic purposes. The next section takes current generative artworks and artificial life software as a starting point. It explores areas into which artists might move to further expand the sense of wonder their works inspire.

14.6 Spaces That Build Themselves

An artist aiming to produce a system capable of sustaining a continuous increase in complexity shares a goal with many artificial life researchers [2]. Why would an artist wish to do such a thing? To return to the ideas discussed earlier, the artist searching for the computational sublime would find it, perhaps in its ultimate form, by generating an experience of open-ended complexity governed by processes instantiated on a computer. The loss of control would be complete – the system would build its own structures and define its own universe. It would do this according to human-engineered rules, yet in a way that defies humans to anticipate its outcome.

This system would be following the code a programmer laid down and run on a machine an engineer designed. Would this device behave in a way about which we could easily reason? In this conflict between control and riot lies the sublime. Mary Shelley knew this well – her friends and contemporaries were much interested in the sublime – when she vividly penned *Frankenstein* and his monster run amok [30]. For a historical overview of theories of the sublime in this period, see [19].

How *does* one write code that will produce the hierarchically organised composite structures associated with life and ever-increasing complexity? If we consider the cells of a CA grid as analogous to molecules and, consider higher-level emergent structures such as gliders as analogous to organelles (a far stretch when one considers the complexity of interactions a physical molecule or an organelle may undergo, and the feeble interactions between neighbouring cellular automata), can we code the system so that still larger-scale groupings of structure occur? Can it produce structures at the level of single cells, a multicellular organism, or an ecosystem?

In theory, even gliders, spinners, and other structures of the *Game of Life* may be carefully arranged into larger-scale units (such as a self-reproducing machine incorporating a universal Turing machine, if one has the patience to arrange its 10^{13} cells [5]). The question remains though "Is it possible that such a higher-level structure will appear of its own accord?" If software could be arranged to facilitate this, the structures that arose would do so on their own terms and *might* therefore behave in ways not envisaged by the creator. This *might* provide an effective source of complexity to assist an artist in his search for the computational sublime.

If theories about the process of natural increase in complexity (such as those extensively discussed by Kauffman, for example [22]) hold true in virtual systems, then the elements in the virtual space might self-assemble into simple stable structures. Perhaps, if the virtual physics and chemistry of the world allowed it, simple reactions might occur, possibly some in auto-catalytic and cross-catalytic sets such as those described by Dorin [6]. These might form the basis of a recognizable topology with the bare bones of a metabolism. What next?

As natural evolution demonstrates, one way to achieve an increase in complexity beyond this is to have the structures engage in a reproductive battle against one another for resources (see [32]). Pargellis' system *Amoeba* manages to initiate reproduction randomly. It establishes conditions in which practical CPU and memory resources are sufficient for a replicator to appear spontaneously from the prebiotic soup [26]. In a *Tierra* run, such an event is much less likely than in *Amoeba*. Hence, Ray initially seeded *Tierra*'s population with a replicator to allow evolution to commence. The problem of coding a simulation that can make the leap from self-organization to spontaneous evolution of structure seems (as far as this author is aware) unmade by any researcher.

Setting aside the leap from self-organization to evolution, even though systems solely employing artificial evolution are readily implemented, getting these to mimic natural evolution's progression from molecules to organelles to cells and on to multicellular creatures and ecosystems has proved a stumbling block. Although worlds such as *SOCA* give rise to auto-catalytic and cross-catalytic sets, *Amoeba* may randomly give rise to replicators and *Polyworld* [35] gives rise to simple communities, there has been limited success (arguably no success) in writing software that encompasses more than one of these important level shifts without resorting to abstractions so high that the simulations they are contained within become trivial.

The reasons for this difficulty are not yet clear. In part, current computational resources may be to blame. However, this is quite likely only a part of the story, and maybe only a small part at that. Rasmussen et al., for instance, has suggested that the bottom level of our simulations are not complex enough to give rise to the kind of multilayered outcomes we desire. They proposed that only by adding complexity to the bottom-level elements of a simulation can we expect to gain extra levels of organised structure on top of any earlier ones [27]. This claim seems to contradict the "complex systems dogma" that explicitly treats complex phenomena as emergent from simple interactions. Since in Rasmussen et al.'s work, notions of "complexity" and "adding complexity" are only loosely defined, it is not clear exactly how and to what extent this might be the case. This is discussed in detail elsewhere [8].

Rasmussen et al. supported their view with a claim about a model they had constructed. Gross and McMullin [18] argued that this model does *not*, in fact, demonstrate the emergence of multiple levels and that a similar outcome can be obtained without adding complexity at the base layer.

It is outside the scope of this chapter to become too deeply embroiled in this battle. Rasmussen et al.'s suggestion may in some sense prove true. Either way, further questions need to be addressed simultaneously. Might there be a limit to all this "complexity adding"? How much added complexity is necessary to move from one level to the next? Questions like these remain open and continue to be debated.

Besides issues raised above, there may be fundamental problems with current approaches to solving the problem. It is possible that simulations on current computer architectures and employing computer programs as they are currently understood will turn out to be practically limited in their ability to produce the kind of truly open-ended complexity increase required. Issues of available resource consumption are an obvious reason why infinite increase is impossible; however, are there reasons why *any* interesting string of increases in complexity may be impossible with current programming and computer technology?

One potential means for overcoming these limitations has been identified – several artists, the author included, have employed *virtual ecosystems* in order to harness the complexity of interactions between organisms and their environments over evolutionary time periods. The goal of one day creating a virtual space as multifaceted as nature may perhaps be realised by adopting ideas from Ecology. This idea is discussed next.

14.7 An Ecosystemic Approach to Writing Artistic Software

A virtual ecosystem is a software simulation of organism and environmental interactions within a real ecosystem. There is a tradition in artificial life, ecology, and even other fields, of constructing such models. For instance, some early systems include *Polyworld* [35], *Tierra* [28], *Echo* [16], *Sugarscape* [14] and *Avida* [1]. These all model relationships amongst organisms within virtual environments inspired by biological counterparts. Purely creative and artistic applications of these systems have also been devised. The author has been directly involved with a few as follows:

Listening Sky is an interactive sonic virtual reality environment in which evolving inhabitants, represented visually as patches of particles, move across the surface of a globe singing to one another and passing their songs onto their offspring (Fig. 14.1). Songs are employed to attract mates and therefore govern the fitness of individuals [4].

Meniscus is an interactive work in which virtual invertebrates breed and swim. The invertebrates are visualised as a series of connected discs with tufts of cilia-like hair (Fig. 14.2). Humans can adjust the depth and agitation of the water to differentially favour the creatures that evolve. The creatures have preferred depths and certain levels of agitation they find favourable for breeding; hence, humans indirectly control the virtual population and therefore the appearance of the work at any time [9].

Eden is an immersive sonic and visual composition written by Jon McCormack in which an evolving population of creatures roam a space of virtual plants and minerals that is projected into a gallery onto central screens arranged in a large X. The creatures communicate their whereabouts and the location of resources to one another audibly. As they produce offspring adapted to their environment, musical patterns emerge for mating calls, food

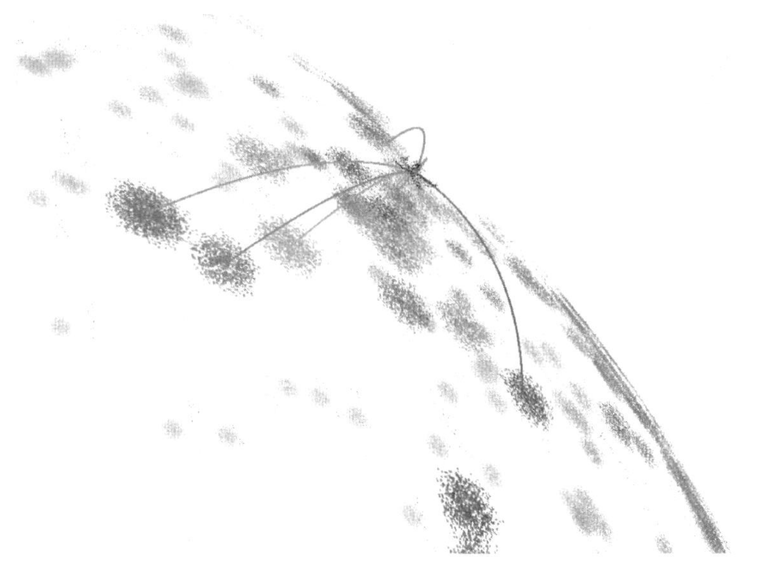

Fig. 14.1 *Listening Sky*, generative, interactive virtual-reality installation (detail) [4].

Fig. 14.2 *Meniscus*, generative, interactive installation (detail) [9].

indicators, and so forth. Humans indirectly alter the conditions in the virtual environment by moving in physical space around the screens monitored by infrared sensors [24].

Autumn Squares is a textural, tapestry-like generative animation in which populations of coloured rectangles roam a two-dimensional grid. The grid is representative of the paths through any human construction (a city, an office building); the rectangles (people who populate the construction) wander down its paths meeting and avoiding one another depending on the kind of “boxes” they are (or “fit into”). Rectangle communities form fanning, intermingling clusters of colour in an evolving visual field [10].

Plague (Diseased Squares) is an epidemiological visualisation and sonification exploring the impact of disease on the population dynamics of simple, evolving coloured agents. A replicating disease is spread by inter-agent contact. It infects agents of particular colours more or less virulently depending on the disease’s co-evolved preferences for hosts. Agents attempt to evolve immunity to the co-evolving diseases whilst being constrained by their need to attract mates [11].

The interactive aesthetic selection and artificial evolutionary technique introduced by Richard Dawkins and popularised for computer graphics by Karl Sims has been widely used in artificial life art. One merit of the technique is its avoidance of prespecified, explicit fitness functions [7]. Whereas the need for these functions may not present many difficulties in engineering applications where the final goal is well understood (e.g., a bridge must span a river without collapsing under load, a sorting algorithm must order numbers, etc.), in artistic and creative endeavours these functions are as troublesome to specify as the question “What is beauty?” is to answer. A drawback of the interactive technique for artistic exploration is the bottleneck introduced by the user’s manual selection of fit individuals for reproduction. Only a single evolutionary chain may be explored in this manner at any one time. Selections of one or a few fit individuals are usually made from a tiny population of only around sixteen to twenty-five individuals. Many potentially interesting avenues will always remain unexplored and even unimagined.

In contrast, the “ecosystemic” approach permits simultaneous, multidirectional and automatic exploration of a space of virtual agent traits without any need for a prespecified fitness function. Instead, the fitness function is implicit in the design of the agents, their virtual environment, and its physics and chemistry. If the system is designed to mimic real ecosystems in its formation of niches and habitats, it gives rise to parallel and simultaneous evolutionary trees that interact with one another over evolutionary time periods. Although even this system cannot explore vast areas of the potential phenotypic space, it is at least able to cope with large, interacting populations of phenotypes. There are several ways in which the design of such a system may be used as the basis for generative software art. A few include the following:

- Agents’ individual appearance or sonic outbursts may be artistic works (*Meniscus* and *Listening Sky* above).
- Agents’ individual behaviours or constructions may be artistic works (swarm-built architecture, ant pheromone trails).
- Agents may generate (draw or sound) a collective structure that becomes the observable artistic output (*Eden*).
- A property of the ecosystem as a whole (e.g., its population dynamics) may constitute or generate a dynamic artistic output (*Plague*).

Combinations and alternatives to these ideas are also potentially interesting avenues for artistic expression. The character of the virtual ecosystem that the author has found most appealing is also fourfold:

- It demonstrates complex dynamics over fine and coarse timescales.
- It may explore large search spaces independently of human input.
- It has the potential for user-events to influence its behaviour.
- It has the potential to allow artist-laid constraints on the search spaces.

These traits are discussed in detail elsewhere [10] and the third, in particular, is discussed in [13]. These are by no means intended to represent universally appreciated characteristics of generative art, merely aspects of computer-based works of interest to the author. They are also traits he has observed in the works of others. It should come as no surprise then that an artist in search of computationally generated creativity described by these traits would settle on ecosystems as an avenue worthy of exploration. Ecosystems are, after all, the dominant source of the Earth's (maybe even the Universe's) novelty and complexity. It is this complexity that has given rise to us and also our concept of the sublime! Since the evolutionary process can be simulated in isolation and turned to aesthetic purposes, it seems only a small step to attempt to turn the processes of Ecology to creative endeavour.

14.8 Conclusion

Although the limitations of our abilities to code multiple-level hierarchies are apparent, clearly this does not imply that our art is similarly constrained. The element of the sublime in a Caspar David Friedrich canvas does not arise from the intricacy of its mechanism, but from contemplating nature and our place in it from behind the safety of a picture plane. Even more apparent is the irrelevance of intricacy and nature (taken literally) to postmodern interpretations of the sublime such as those discussed by Jean-Francois Lyotard [34]. Hence, works such as the dark canvasses produced in Mark Rothko's later years may be discussed in terms of their contribution to the *postmodern sublime*. The sublime does not lie *in* a work, rather the work may act to trigger a sublime experience in a viewer. In the case of Lyotard's ideas, this relates to a sense of formlessness and therefore of things that may be better left unpresented.

Since our machines are faster at mathematics than we are, they will always maintain the ability to play the role of mediator between us and the vast computational spaces outside our direct experience. Perhaps one day these same machines will be able to take us beyond spaces that look more and more familiar as we travel through them, into spaces that increase in complexity and continue to surprise us. Here the sublime experience of nature's vastness and ferocity may be rivaled through a sense of the computational sublime. We will be sensing a space rendered maybe with points on a plane and computed on-the-fly by our fastest machines, and it will seem to us as terrible and delightful as standing on an icy summit surveying all the world.

References

1. Adami C, Brown CT (1994) Evolutionary Learning in the 2D Artificial Life System Avida. In: Artificial Life IV. Brooks RA and Maes P (eds.), MIT Press, pp 377–381.
2. Bedau MA, McCaskill JS, Packard NH, Rasmussen S, Adami C, Green DG, Harvey I, Ikegami T, Kaneko K, Ray TS (2000) Open Problems in Artificial Life. Artificial Life 6:363–376.
3. Bern M (1999) Art Shows at PARC. In: Art and Innovation, Harris C (ed). MIT Press, pp 259–277.
4. Berry R, Rungsarityotin W, Dorin A, Dahlstedt P, Haw C (2001) Unfinished Symphonies – Songs of 3.5 worlds. In: Workshop on Artificial Life Models for Musical Applications, Sixth European Conference on Artificial Life, Bilotta E, et al. (eds.). Editoriale Bios, pp 51–64.
5. Dennett DC (1991) Real Patterns. Journal of Philosophy 88:27–51.
6. Dorin A (2000) Creating a Physically-Based, Virtual-Metabolism with Solid Cellular Automata. In: Proceedings Artificial Life 7, Bedau M, et al. (eds.). MIT Press, pp 13–20.
7. Dorin A (2001) Aesthetic Fitness and Artificial Evolution for the Selection of Imagery from the Mythical Infinite Library. In: Advances In Artificial Life, 6th European Conference on Artificial Life, Kelemen J and Sosik P (eds.). Springer-Verlag, pp 10–14.
8. Dorin A, McCormack J (2002) Self-Assembling Dynamical Hierarchies. In: Proceedings of Artificial Life 8, Standish R, et al. (eds.). MIT Press, pp 423–428.
9. Dorin A (2003) Meniscus. In: Experimenta, House of Tomorrow Catalogue, Taylor A (ed.). Experimenta Media Arts, Australia, p 32.
10. Dorin A (2004) The Virtual Ecosystem as Generative Electronic Art. In: Proceedings of 2nd European Workshop on Evolutionary Music and Art, Applications of Evolutionary Computing: EvoWorkshops 2004, Coimbra, Portugal, April 5–7, Günther RR, et al. (eds.). Springer-Verlag, pp 467–470.
11. Dorin A (2006) The Sonic Artificial Ecosystem. In: Proceedings of the Australasian Computer Music Conference (ACMC 2006), Haines C (ed.). ACMA, Adelaide, Australia, pp 32–37.
12. Driessens E, Verstappen M (2001) Keynote presentation in Proceedings of Second Iteration, Second International Conference on Generative Systems in the Electronic Arts, Dorin A (ed.). CEMA, Melbourne, Australia, pp 12–13.
13. Eldridge A, Dorin A, McCormack J (2008) Manipulating Artificial Ecosystems. In: 6th European Workshop on Evolutionary and Biologically Inspired Music, Sound, Art and Design: EvoWorkshops 2008, Napoli, Italy, Giacobini M, et al. (eds.). Springer-Verlag, pp 392–401.
14. Epstein JM, Axtell R (1996) Growing Artificial Societies, Social Science from the Bottom Up. Brookings Institution, MIT Press.
15. Feagin SL (1995) Sublime. In: The Cambridge Dictionary of Philosophy, Audi R (ed.). Cambridge University Press, p 774.
16. Forrest S, Jones T (1994) Modelling Adaptive Systems with Echo In: Complex Systems: Mechanisms of Adaptation. IOS Press, pp 3–21.
17. Gardner M (1970) The Fantastic Combinations of John Conway's New Solitaire Game, "Life". Scientific American 223:120–123.
18. Gross D, McMullin B (2001) Is It the Right Ansatz? Artificial Life 7:355–365.

19. Hipple WJ (1957) The Beautiful, the Sublime, and the Picturesque in Eighteenth-Century British Aesthetic Theory. Southern Illinois University Press.
20. Hoffman DD (2000) Visual Intelligence: How We Create What We See. Norton.
21. Kandinsky W (1979) Point and Line to Plane. Dover Publications, (first published 1926).
22. Kauffman SA (1993) The Origins of Order. Oxford University Press.
23. Longinus (1963) On the Sublime. Translated by Havell. In: Aristotle's Poetics, Demetritus on Style, Longinus on the Sublime. Everyman's Library, Dutton, pp 133–202 (first written ca. 250 AD).
24. McCormack J (2001) Eden: An Evolutionary Sonic Ecosystem. In: Advances in Artificial Life, 6th European Conference, Kelemen J and Sosik P (eds.). Springer, pp 133–142.
25. McCormack J, Dorin A (2001) Art, Emergence, and the Computational Sublime. In: Proceedings of Second Iteration, Second International Conference on Generative Systems in the Electronic Arts, Dorin A (ed.), CEMA, Melbourne, Australia, pp 67–81.
26. Pargellis AN (2001) Digital Life Behaviour in the Amoeba World. Artificial Life 7:63–75.
27. Rasmussen S, Baas NA, Mayer B, Nilsson M, Olesen MW (2001) Ansatz for Dynamical Hierarchies, Artificial Life 7:329–353.
28. Ray TS (1991) An Approach to the Synthesis of Life. In: Artificial Life II, Langton C, et al. (eds.). Addison-Wesley, pp 371–408.
29. Rower ASC (1998) Calder Sculpture, National Gallery of Art, Washington, Universe.
30. Shelley M (1989) Frankenstein, the Modern Prometheus. Joseph MK (ed.). Oxford Univeristy Press (first published 1818).
31. Sims K (1994) Evolving Virtual Creatures. In: Proceedings of SIGRRAPH 1994, ACM Press, pp 15–34.
32. Taylor T (2002) Creativity in Evolution: Individuals, Interactions and Environments. In: Creative Evolutionary Systems, Bentley PJ and Corne DW (eds.). Academic Press, pp 79–108.
33. Tufte ER (1990) Envisioning Information. Graphics Press.
34. van de Vall R (1995) Silent Visions, Lyotard on the Sublime. In: The Contemporary Sublime, Sensibilities of Transcendence and Shock, Hodges (ed.). Art and Design. VCH Publishers, pp 69–75.
35. Yaeger L (1994) Computational Genetics, Physiology, Metabolism, Neural Systems, Learning, Vision and Behavior or Polyworld: Life in a New Context. In: Proceedings, Artificial Life III, SFI Studies in the Sciences of Complexity, Langton C (ed.). Addison-Wesley, pp 263–298.

Appendix: Artificial Life Software

Related chapter, software name and license type	Description	Web site (`http://`), availability, requirements
1. Avida. GPL.	A digital world in which simple computer programs mutate and evolve. Avida can be used to study questions and perform experiments in evolutionary dynamics and theoretical biology.	`avida.devosoft.org` Linux, Unix, MacOS X, MS Windows
2. Recursive Porous Agent Simulation Toolkit (Repast). Free and open source (BSD).	An advanced agent-based modeling toolkit. Repast supports the development of extremely flexible models of living social agents, but is not limited to modeling living social entities alone.	`repast.sourceforge.net` Repast Simphony, Repast 3 for Java and Repast for Python Scripting: all platforms (requires JRE to run models and JDK to create models). Repast.Net 3: Windows 2000/XP (requires a .Net framework to run models and .Net compiler to create models)
3. Sodarace. Free to use and develop.	The online olympics pitting human creativity against machine learning in a competition to design robots that race over 2D terrains using the Sodaconstructor virtual construction kit.	`www.sodarace.net` All platforms (Java plug-in required)
4. breve Simulation Environment. Free, open source (GPL).	A software package for building 3D simulations of multiagent systems and artificial life. breve supports realistic 3D physical simulation, an OpenGL display engine for visualization and scripting using Python or a simple language called steve.	`www.spiderland.org` MacOS X, Linux and MS Windows. Requires OpenGL.
5. Framsticks GUI. Free for educational and research use, otherwise shareware.	A powerful graphical user interface (GUI) for the Framsticks simulator of 3D life forms.	`www.framsticks.com` MS Windows

5.	Framsticks Theater. Free for educational and research use, otherwise shareware.	Easy-to-use application that illustrates basic phenomena like genes and genetics, mutation, evolution, user-driven evolution and artificial selection, etc. Recommended for presentation, illustration, entertainment, screen-saving mode, visualization of genotypes, etc.	`www.framsticks.com` Linux, Unix, MacOS, MS Windows and more
5.	Framsticks simulator CLI, server, clients, editor (FRED). Free. Clients and FRED are open source.	CLI is a command-line interface for the Framsticks simulator and includes a network server. Clients are remote user interfaces to the server. FRED is a user-friendly graphical editor of creatures.	`www.framsticks.com` FRED and network clients: all platforms (Java required) Other programs: Linux, MacOS, MS Windows, and more
6.	StarLogo TNG. Free for educational use.	A 3D agent-based simulation environment suitable for novices, designed to make modeling and simulation (as well as game creation) accessible to students and teachers.	`education.mit.edu/starlogo-tng` Windows XP/Vista, Mac OS X 10.4 and up, Linux (with some limitations)
7.	NetLogo. Freeware. For commercial distribution, please contact CCL for license.	A programmable integrated agent-based modeling environment designed both to enable novices to easily build models and for more advanced users to build research quality models. It is in widespread use both in education and in research.	`ccl.northwestern.edu/netlogo` All platforms (Java required)
8.	Discrete Dynamics Lab (DDLab). Free for personal use. License required for commercial or institutional use. Source code may be available on request.	Interactive graphics software for studying multivalue discrete dynamical networks including cellular automata and random Boolean networks. Unique in its ability to generate basins of attraction. Space-time patterns can be generated in 1D, 2D (square or hex), or 3D. Many tools and functions are available for creating the network (its rules and wiring), setting the initial state, analyzing the dynamics, and amending parameters on-the-fly.	`www.ddlab.org` Linux, Unix, MacOS X, DOS, IRIX or MS Windows command line

Related chapter, software name and license type	Description	Web site (http://), availability, requirements
9. EINSTein. Freely available; source code will be released.	Multiagent-based simulation of land combat; includes agent personalities, weapons behaviors, intelligent targeting, and terrain effects.	www.cna.org/isaac/ MS Windows
10. OrgTools. GPL (source code available upon request).	Tools for the computation of chemical organizations in reaction networks integrating into the Systems Biology Workbench.	orgtheo.artificialchemistries.org/ All platforms (Java plugin required). Binaries available for MS Windows and Linux
11. ESPresSo. GPL.	Highly versatile software package for the scientific simulation and analysis of coarse-grained atomistic or bead-spring models as they are used in soft matter research.	www.espresso.mpg.de Linux, Unix, MacOS X, MS Windows
11. Spartacus. GPL.	Experimental simulator for simulation and analysis of spatially resolved artificial chemistry with graphical user interface and Python extensibility.	flint.sdu.dk/spartacus Linux, Unix, MacOS X
11. Smoldyn. LGPL.	Interactive computer program for simulating spatially resolved diffusion and reaction networks at the mesoscopic scale with special emphasis on molecular biology.	www.smoldyn.org Linux, Unix, MacOS X, MS Windows
11. LAMMPS. GPL.	Large-scale Atomic/Molecular Massively Parallel Simulator with support for coarse-grained simulations.	lammps.sandia.gov Linux, Unix, MacOS X, MS Windows

12. SBART. Free.	A design support tool to create an abstract 2D computer graphics image based on artificial selection, which was originally proposed as Artificial Evolution by Karl Sims.	www.intlab.soka.ac.jp/~unemi/sbart/ Ver. 1: Unix (Linux, FreeBSD, Solaris, HP-UX, IRIX), Motif 1.2 (or LessTif) and jpeg6 library Ver. 2: MacOS 8, 9, QuickTime 3 or later Ver. 3: MacOS X 10.4 or later, QuickTime 5 or later, G4 or better recommended. For Intel-based machine, 10.4.8 or later.
12. SBEAT. Free.	A composition support tool to create short musical phrases and rhythms based on artificial selection. Includes utility mechanisms to build up a complete tune.	www.intlab.soka.ac.jp/~unemi/sbeat/ MacOS 8, 9, X. Requires CarbonLib 1.3 or later for MacOS 8 and 9, and QuickTime 5 or later
13. Eden. Copyrighted, personal edition available on CD-ROM.	An interactive evolutionary system where creatures evolve different strategies using sound. Individual creatures use a classifier-based learning system. Over time, they learn to adapt and survive in their environment. In many instances this involves using sound, so as the system evolves, users hear the musical compositions of the creatures.	www.csse.monash.edu.au/~jonmc/projects/eden/eden.html MacOS X. OpenGL and stereo speakers required

Index